Student Solutions CALCULUS and Analytic Geometry 4th Edition

DAVID WEND
Montana State University

Wadsworth Publishing Company
Belmont, California

A division of Wadsworth, Inc.

Printed in the United States of America

1 2 3 4 5 6 7 8 9 10---88 87 86 85 84

ISBN 0-534-04584-7

Contents

Chapter 1
Plane Analytic Geometry

CHAPTER 1, SECTION 1 (pp. 8-10)

1. $d = \sqrt{(2 - 1)^2 + (5 + 3)^2} = \sqrt{65}$.

5. $x = \dfrac{x_1 + x_2}{2} = \dfrac{4 + 3}{2} = \dfrac{7}{2}$, $y = \dfrac{y_1 + y_2}{2} = \dfrac{-1 + 3}{2} = 1$; $(\frac{7}{2}, 1)$.

9. $\sqrt{(x - 1)^2 + (x - 4)^2} = \sqrt{5}$,

   $x^2 - 2x + 1 + x^2 - 8x + 16 = 5$, $x = 2, 3$.

13. $A = (1, -1)$, $B = (3, 4)$, $C = (-1, -6)$;

    $\overline{AB} = \sqrt{2^2 + 5^2} = \sqrt{29}$, $\overline{BC} = \sqrt{(-4)^2 + (-10)^2} = 2\sqrt{29}$,

    $\overline{AC} = \sqrt{(-2)^2 + (-5)^2} = \sqrt{29}$; $\overline{AB} + \overline{AC} = \overline{BC}$, collinear.

17. $A = (0, 2)$, $B = (-2, 4)$, $C = (1, 3)$;

    $\overline{AB} = \sqrt{(-2)^2 + 2^2} = \sqrt{8}$, $\overline{BC} = \sqrt{3^2 + (-1)^2} = \sqrt{10}$,

    $\overline{AC} = \sqrt{1^2 + 1^2} = \sqrt{2}$; $\overline{AB}^2 + \overline{AC}^2 = \overline{BC}^2$, right triangle.

21. $P = (5, 2)$ is on the perpendicular bisector of the segment AB if and only if $\overline{AP} = \overline{BP}$.

    $\overline{AP} = \sqrt{4^2 + (-1)^2} = \sqrt{17}$, $\overline{BP} = \sqrt{1^2 + 4^2} = \sqrt{17}$.

25. $A = (1, 1)$, $B = (4, 1)$, $C = (3, -2)$, $D = (0, -2)$.

    $\overline{AB} = \overline{DC} = \sqrt{3^2 + 0^2} = 3$, $\overline{BC} = \overline{AD} = \sqrt{(-1)^2 + (-3)^2} = \sqrt{10}$.

29. Midpoint $= (\frac{3 + x}{2}, 1)$. $5 = \sqrt{(\frac{1 + x}{2})^2 + 9}$,

    $25 = \dfrac{x^2 + 2x + 1}{4} + 9$, $x = 7, -9$.

33.

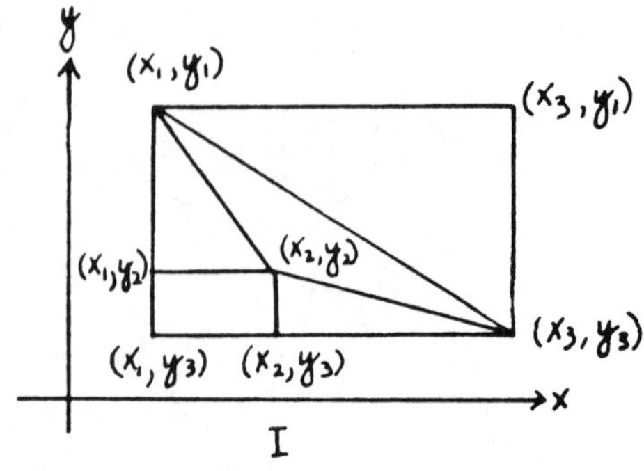

Case I: $A = (x_3 - x_1)(y_1 - y_3) - \frac{1}{2}(x_3 - x_1)(y_1 - y_3)$

$- \frac{1}{2}(x_3 - x_2)(y_2 - y_3) - (x_2 - x_1)(y_2 - y_3)$

$- \frac{1}{2}(x_2 - x_1)(y_1 - y_2)$

$= \frac{1}{2}(x_3y_1 + x_1y_3 - x_1y_1 - x_3y_3 - x_3y_2 - x_2y_3$

$+ x_2y_2 + x_3y_3 - 2x_2y_2 - 2x_1y_3 + 2x_1y_2 + 2x_2y_3$

$- x_2y_1 + x_1y_1 + x_2y_2 - x_1y_2)$

$= \frac{1}{2}(x_1y_2 + x_2y_3 + x_3y_1 - x_1y_3 - x_2y_1 - x_3y_2).$

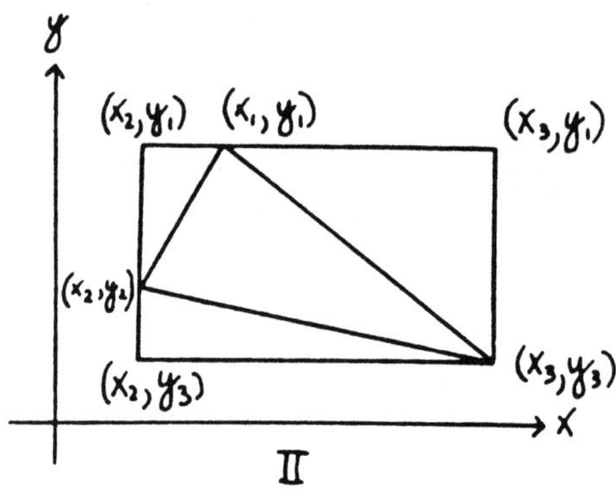

Other positions of the vertices and Case II give the same result with the subscripts permuted. The terms can then be rearranged to give the result above or its negative. Thus,

$$A = \frac{1}{2}|x_1y_2 + x_2y_3 + x_3y_1 - x_1y_3 - x_2y_1 - x_3y_2|.$$

CHAPTER 1, SECTION 2 (pp. 14-15)

1. $m = \frac{8 - 3}{5 - 2} = \frac{5}{3}$, $\tan\theta = 1.6667$, $\theta = 59°$.

5. $m = \frac{5 - 2}{-4 + 4}$, no slope, $\theta = 90°$.

9. $m_1 = \frac{-11 + 2}{-2 - 1} = 3$, $m_2 = \frac{8 - 2}{2 - 0} = 3$, (1, -2), (2, 8),
$m_3 = \frac{8 + 2}{2 - 1} = 10$, parallel.

13. $m_1 = \frac{-1 - 1}{4 - 1} = -\frac{2}{3}$, $m_2 = \frac{-3 - 3}{7 + 2} = -\frac{2}{3}$, $m_3 = \frac{3 - 1}{-2 - 1} = -\frac{2}{3}$,
coincident.

17. $3 = \frac{5 - 3}{x - 4}$, $3x - 12 = 2$, $x = \frac{14}{3}$.

21. $\frac{4 - 7}{x - 3} = \frac{-1 - 1}{x - 5}$, $\frac{-3}{x - 3} = \frac{-2}{x - 5}$, $2x - 6 = 3x - 15$, $x = 9$.

25. $b = \sqrt{x^2 + a^2}$

$b^2 = x^2 + a^2$

$x^2 = b^2 - a^2$

$x = -\sqrt{b^2 - a^2}$

$m_{AC} = \frac{a}{b + \sqrt{b^2 - a^2}}$

$m_{BD} = \frac{-a}{b - \sqrt{b^2 - a^2}}$

$= \frac{-a(b + \sqrt{b^2 - a^2})}{b^2 - (b^2 - a^2)} = -\frac{b + \sqrt{b^2 - a^2}}{a}$.

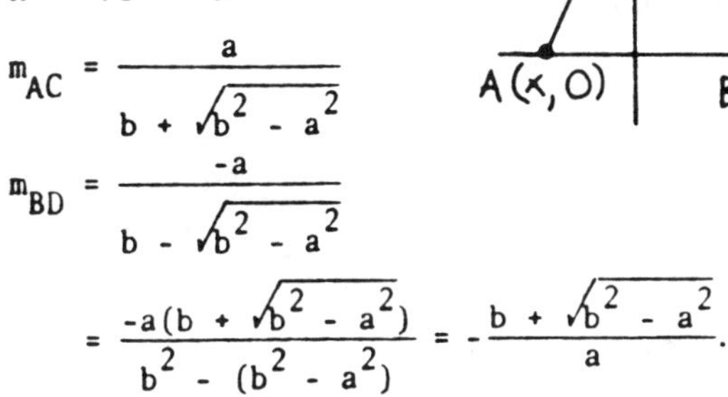

29. $\theta = \alpha_2 - \alpha_1$

$$\tan\theta = \tan(\alpha_2 - \alpha_1)$$

$$= \frac{\tan\alpha_2 - \tan\alpha_1}{1 + \tan\alpha_1 \tan\alpha_2}$$

$$= \frac{m_2 - m_1}{1 + m_1 m_2} = \frac{|m_1 - m_2|}{1 + m_1 m_2}$$

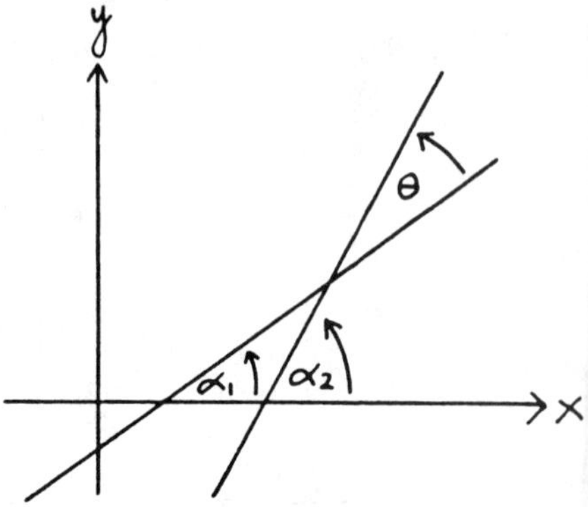

$$\theta = \pi - (\alpha_2 - \alpha_1)$$

$$\tan\theta = \tan[\pi - (\alpha_2 - \alpha_1)]$$

$$= -\tan(\alpha_2 - \alpha_1)$$

$$= \frac{m_1 - m_2}{1 + m_1 m_2}.$$

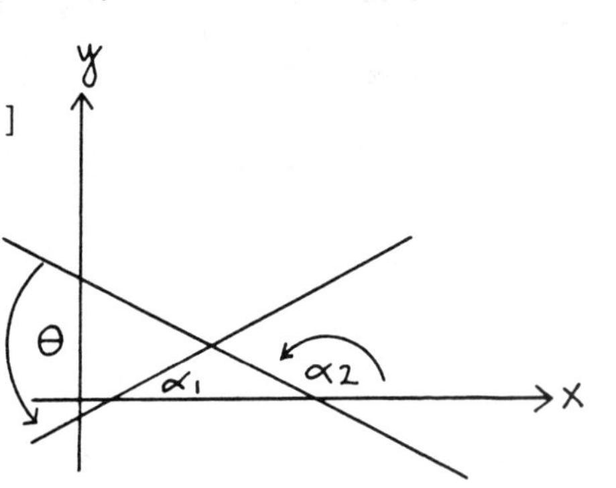

CHAPTER 1, SECTION 3 (pp. 23-24)

1.

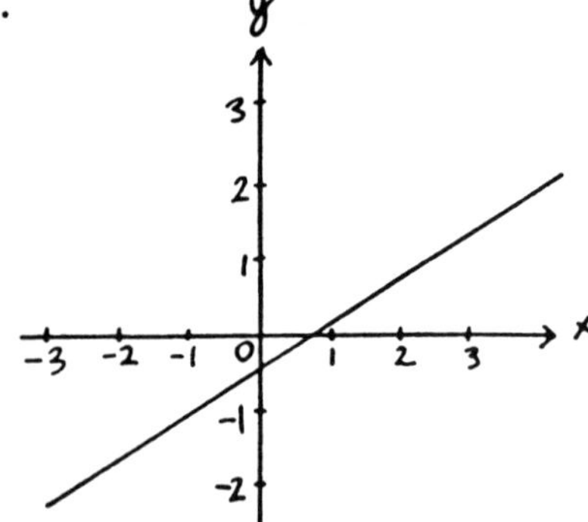

5.

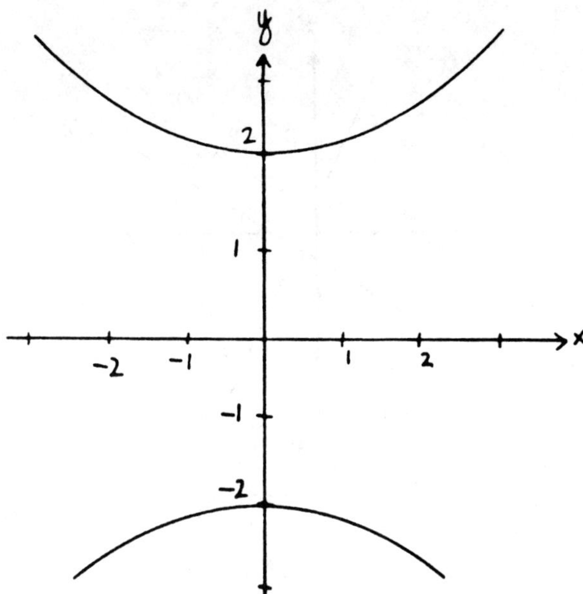

9.

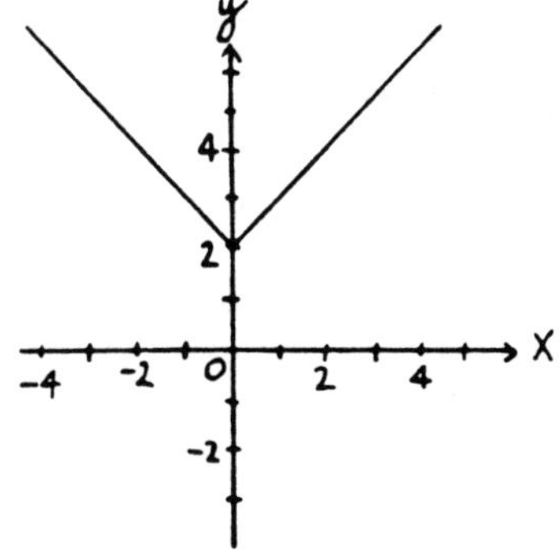

13. $\begin{cases} 3x - 5y = 2 \\ 4x + 2y = 1 \end{cases}$

$6x - 10y = 4$

$20x + 10y = 5$

$26x = 9$

$x = \frac{9}{26};$

$\frac{27}{26} - 5y = \frac{52}{26}$

$y = -\frac{5}{26}.$

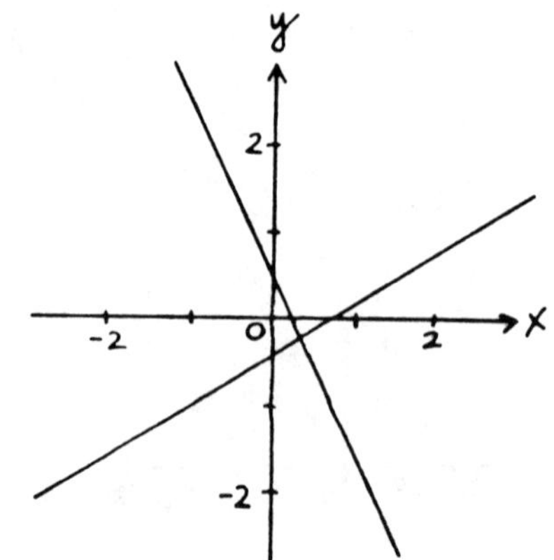

17.

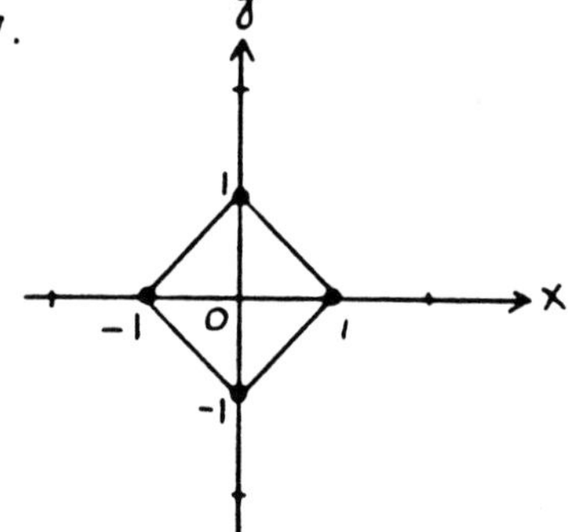

21.

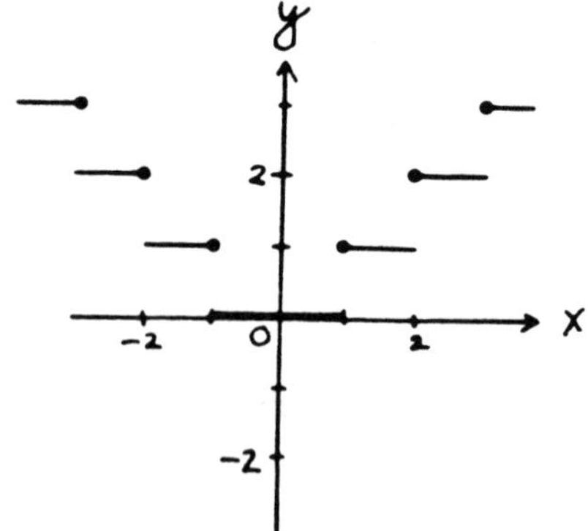

25.

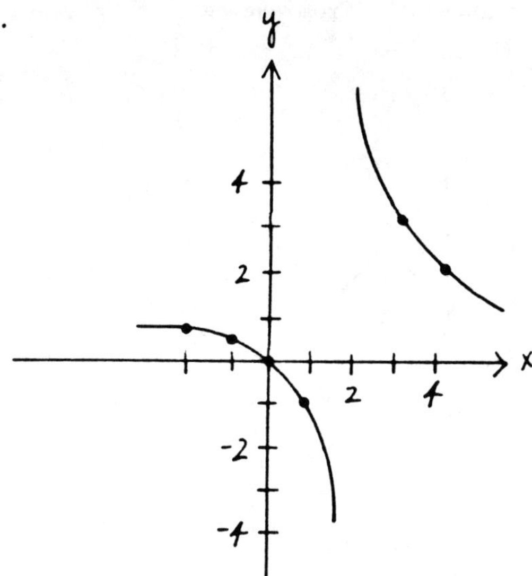

29.

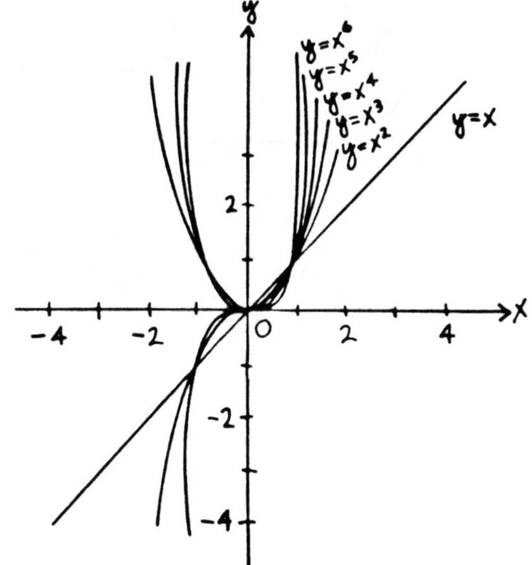

33. $y = \{x\}$ is the distance of x from the nearest integer.

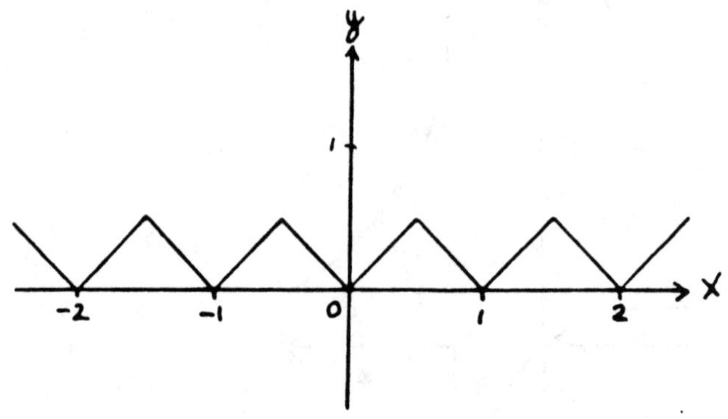

CHAPTER 1, SECTION 4 (pp. 32-34)

1. $y - (-4) = -2(x - 2)$

   $2x + y = 0.$

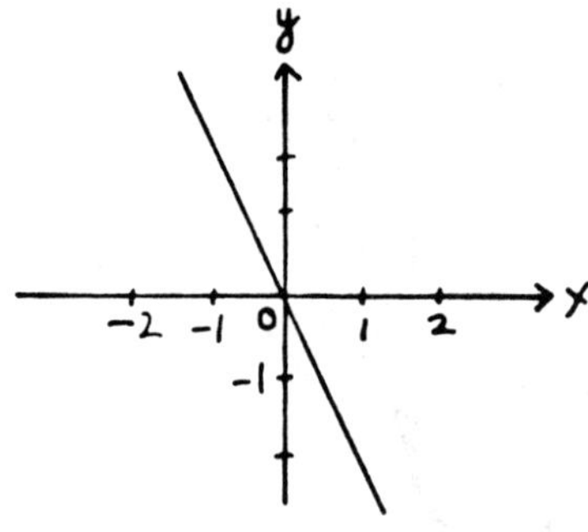

5. $y - 4 = \frac{5 - 4}{3 - 1}(x - 1)$

$2y - 8 = x - 1$

$x - 2y + 7 = 0.$

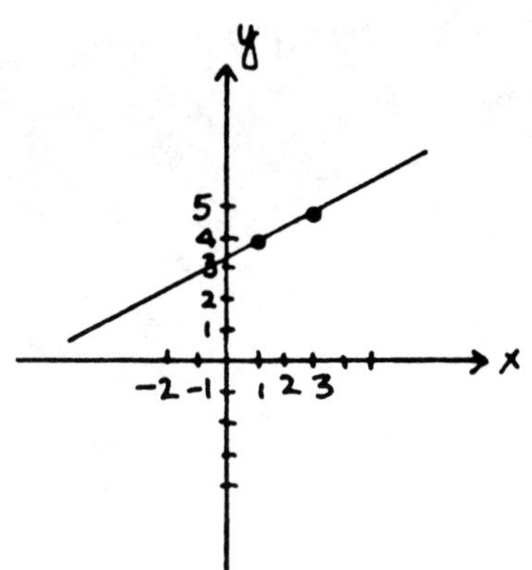

9. $y = 4x + 2$

$4x - y + 2 = 0.$

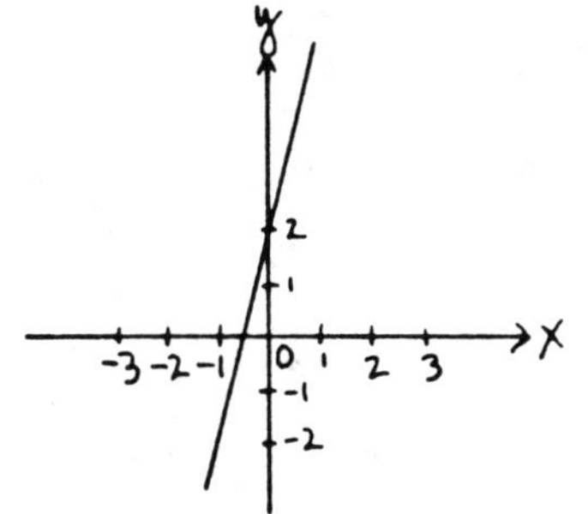

13. $\frac{x}{4} + \frac{y}{2} = 1$

$x + 2y - 4 = 0.$

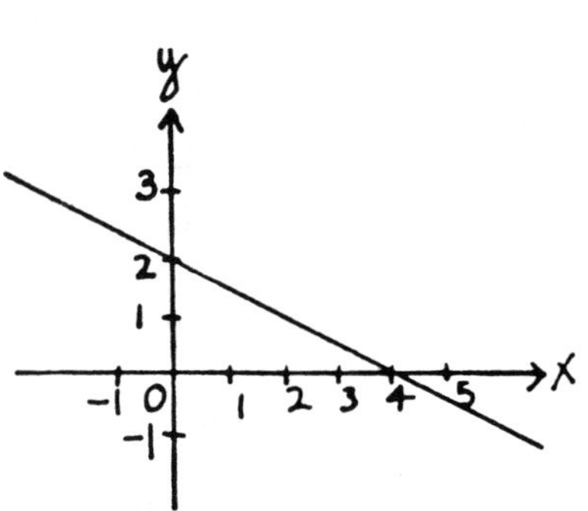

17. $x = a$ when $y = 0$.

$y = 3(x - 4)$

$3x - y - 12 = 0$.

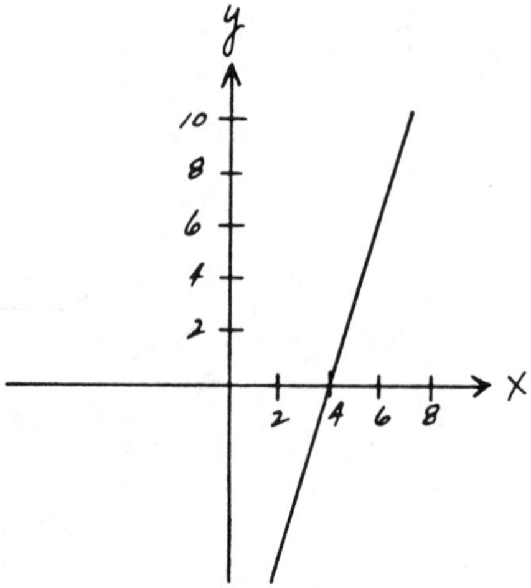

21. $m = \frac{3}{2}$, $y - 1 = \frac{3}{2}(x - 5)$, $3x - 2y - 13 = 0$.

25. (1, 4), (3, 0)

$m = \frac{4 - 0}{1 - 3} = -2$

$y - 4 = -2(x - 1)$

$2x + y - 6 = 0$

(3, 0), (-1, -2)

$m = \frac{0 + 2}{3 + 1} = \frac{1}{2}$

$y = \frac{1}{2}(x - 3)$

$x - 2y - 3 = 0$

(-1, -2), (1, 4)

$m = \frac{-2 - 4}{-1 - 1} = 3$

$y + 2 = 3(x + 1)$

$3x - y + 1 = 0$.

29. (1, 3), (4, -2)

$m_1 = \frac{3 + 2}{1 - 4} = -\frac{5}{3}$, $m_2 = \frac{3}{5}$

Midpoint: $(\frac{5}{2}, \frac{1}{2})$

$y - \frac{1}{2} = \frac{3}{5}(x - \frac{5}{2})$

$6x - 10y - 10 = 0$

(4, -2), (-2, 1)

$m_1 = \frac{-2 - 1}{4 + 2} = -\frac{1}{2}$, $m_2 = 2$

Midpoint: $(1, -\frac{1}{2})$

$y + \frac{1}{2} = 2(x - 1)$

$4x - 2y - 5 = 0$

$6x - 10y - 10 = 0$

$20x - 10y - 25 = 0$

$14x - 15 = 0$

$x = \frac{15}{14}$, $y = -\frac{5}{14}$

Center $(\frac{15}{14}, -\frac{5}{14})$.

33. $m_1 = \frac{0 - 4}{-3 - 5} = \frac{1}{2}$, $m_2 = -2$, $(1, 2)$

$y - 2 = -2(x - 1)$, $2x + y - 4 = 0$.

37. $\sqrt{(x - 2)^2 + (y - 5)^2} = \sqrt{(x - 4)^2 + (y + 1)^2}$

$x^2 - 4x + 4 + y^2 - 10y + 25 = x^2 - 8x + 16 + y^2 + 2y + 1$

$4x - 12y + 12 = 0$, $x - 3y + 3 = 0$.

41. $xy = 0$

$x = 0$, $y = 0$.

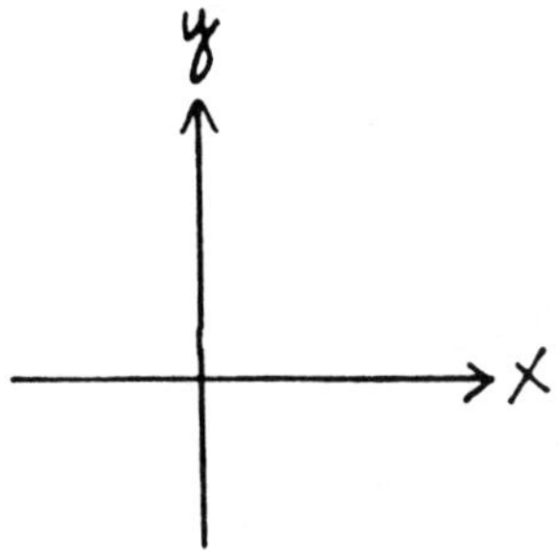

45. $\frac{|2 \cdot 4 + 1 \cdot 2 - 5|}{\sqrt{4 + 1}} = \frac{5}{\sqrt{5}} = \sqrt{5}$.

49. $\log_{10} y = \log_{10} k + \frac{1}{n} \log_{10} C$

$0.431 = \frac{1}{n}$ $\log_{10} k = -0.796 = 9.204 - 10$

$n = \frac{1}{0.431} = 2.32$ $k = 0.160$.

CHAPTER 1, SECTION 5 (pp. 39-40)

1. $(x - 1)^2 + (y - 3)^2 = 25$

$x^2 + y^2 - 2x - 6y - 15 = 0.$

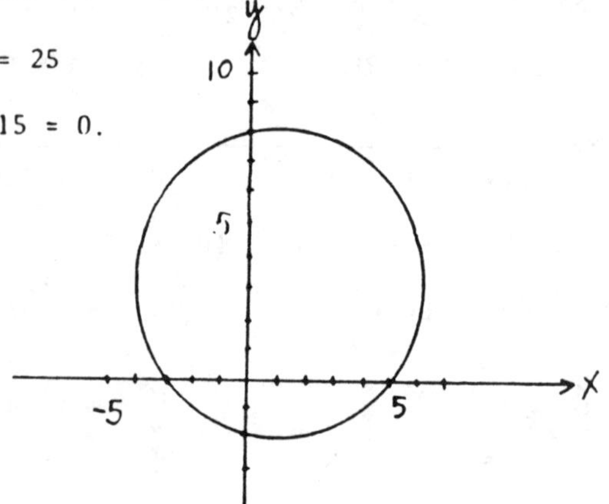

5. $(x - \frac{1}{2})^2 + (y + \frac{3}{2})^2 = 4$

$x^2 + y^2 - x + 3y - \frac{3}{2} = 0$

$2x^2 + 2y^2 - 2x + 6y - 3 = 0.$

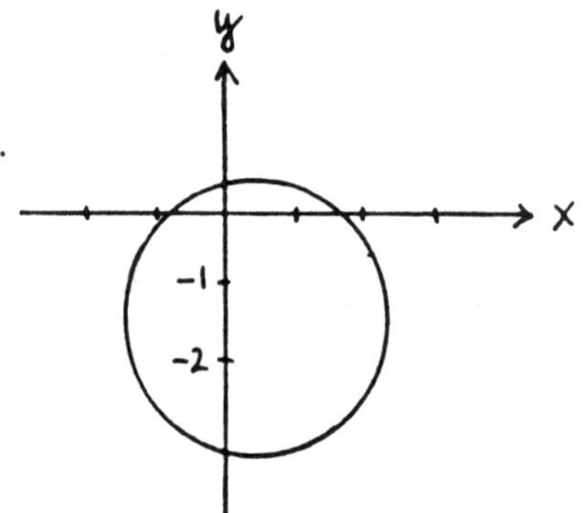

9. $C = (0, -\frac{3}{2}),\ r^2 = 2^2 + \frac{9}{4} = \frac{25}{4}$

$x^2 + (y + \frac{3}{2})^2 = \frac{25}{4}$

$x^2 + y^2 + 3y - 4 = 0.$

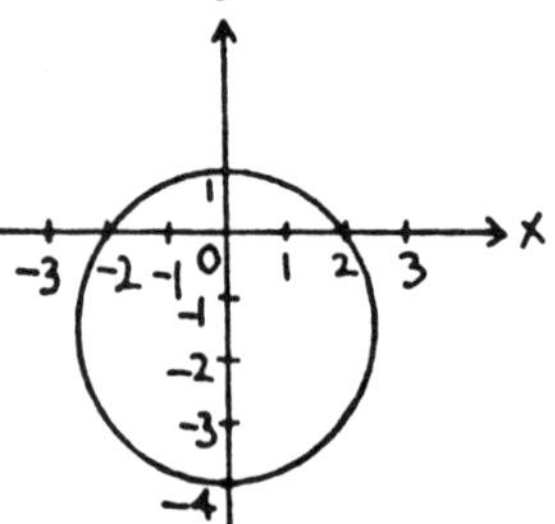

13. $x^2 + y^2 - 2x - 4y + 1 = 0$

$x^2 - 2x + 1 + y^2 - 4y + 4$

$= -1 + 1 + 4$

$(x - 1)^2 + (y - 2)^2 = 4.$

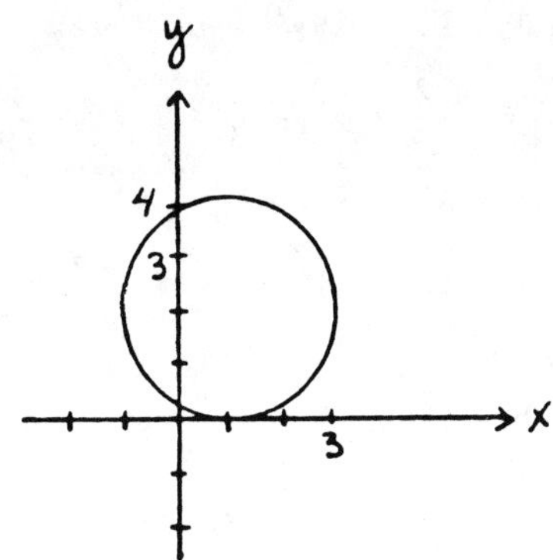

17. $(x - 4)^2 + (y - 1)^2 = 4$

$x^2 + y^2 - 8x - 2y + 13 = 0.$

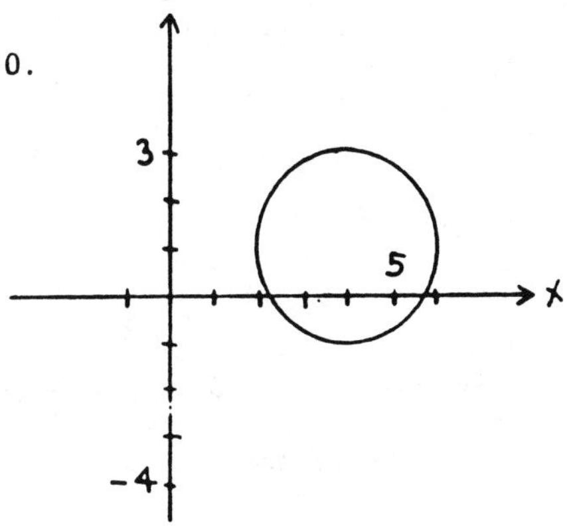

21. $4x^2 + 4y^2 - 4x - 12y + 1 = 0$

$x^2 + y^2 - x - 3y + \frac{1}{4} = 0$

$x^2 - x + \frac{1}{4} + y^2 - 3y + \frac{9}{4}$

$= -\frac{1}{4} + \frac{1}{4} + \frac{9}{4}$

$(x - \frac{1}{2})^2 + (y - \frac{3}{2})^2 = \frac{9}{4}.$

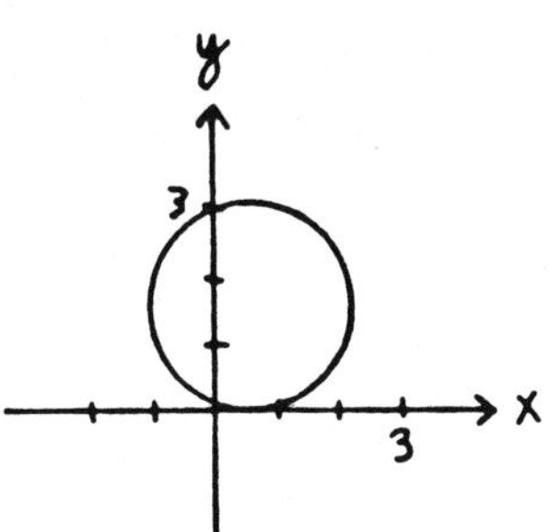

25. $9x^2 + 9y^2 - 6x + 18y + 11 = 0$

$x^2 + y^2 - \frac{2}{3}x + 2y + \frac{11}{9} = 0$

$x^2 - \frac{2}{3}x + \frac{1}{9} + y^2 + 2y + 1 = -\frac{11}{9} + \frac{1}{9} + \frac{9}{9}$

$(x - \frac{1}{3})^2 + (y + 1)^2 = -\frac{1}{9}.$

29. $x^2 + y^2 - x - 3y - 6 = 0,$

$4x - y - 9 = 0$

$y = 4x - 9, \quad x^2 + 16x^2 - 72x + 81 - x - 12x + 27 - 6 = 0$

$17x^2 - 85x + 102 = 0, \quad x^2 - 5x + 6 = 0$

$(x - 3)(x - 2) = 0$

$x = 3 \qquad x = 2$

$y = 3 \qquad y = -1.$

33. $x = 2y - 2, \quad 4y^2 - 8y + 4 + y^2 - 4y + 4 + 4y + 1 = 0$

$5y^2 - 8y + 9 = 0, \quad y = \frac{8 \pm \sqrt{64 - 180}}{2} = 4 \pm \sqrt{-29}$

The circle and line do not intersect.

37. $x^2 + y^2 = a^2$

$m_1 = \frac{y}{x + a}$

$m_2 = \frac{y}{x - a}$

$m_1 m_2 = \frac{y^2}{x^2 - a^2} = \frac{y^2}{-y^2} = -1.$

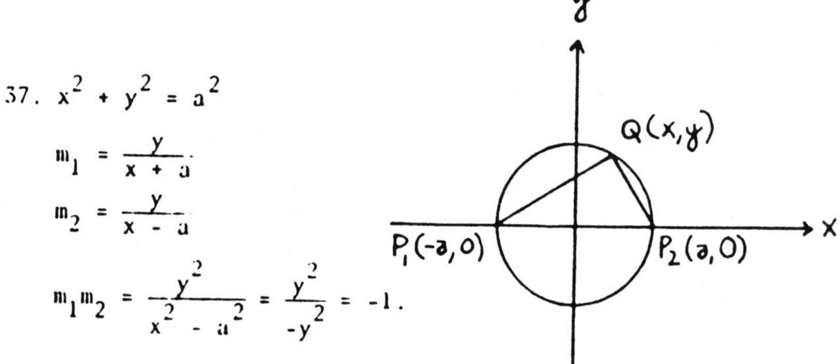

41. $Ax^2 + Ay^2 + Dx + Ey + F = 0$

$$x^2 + y^2 + \frac{D}{A}x + \frac{E}{A}y + \frac{F}{A} = 0$$

$$x^2 + \frac{D}{A}x + \frac{D^2}{4A^2} + y^2 + \frac{E}{A}y + \frac{E^2}{4A^2} = -\frac{4AF}{4A^2} + \frac{D^2}{4A^2} + \frac{E^2}{4A^2}$$

$$(x + \frac{D}{2A})^2 + (y + \frac{E}{2A})^2 = \frac{D^2 + E^2 - 4AF}{4A^2}$$

a) $D^2 + E^2 - 4AF > 0$

b) $D^2 + E^2 - 4AF = 0$

c) $D^2 + E^2 - 4AF < 0$.

$$h = -\frac{D}{2A}, \quad k = -\frac{E}{2A}, \quad r = \sqrt{\frac{D^2 + E^2 - 4AF}{4A^2}}.$$

CHAPTER 1, REVIEW (pp. 41-43)

1. $A = (1, 5)$, $B = (-2, -1)$, $C = (4, 10)$.

$\overline{AB} = \sqrt{9 + 36} = \sqrt{45} = 3\sqrt{5}$, $\overline{BC} = \sqrt{36 + 121} = \sqrt{157}$,

$\overline{AC} = \sqrt{9 + 25} = \sqrt{34}$. Since $\overline{BC} \neq \overline{AB} + \overline{AC}$, they are non-collinear. $m_{AB} = \frac{5 + 1}{1 + 2} = \frac{6}{3} = 2$, $m_{BC} = \frac{10 + 1}{4 + 2} = \frac{11}{6}$.

Since $m_{AB} \neq m_{BC}$, they are noncollinear.

5. $\begin{cases} 2x + y = 5 \\ x - 3y = 7 \end{cases}$

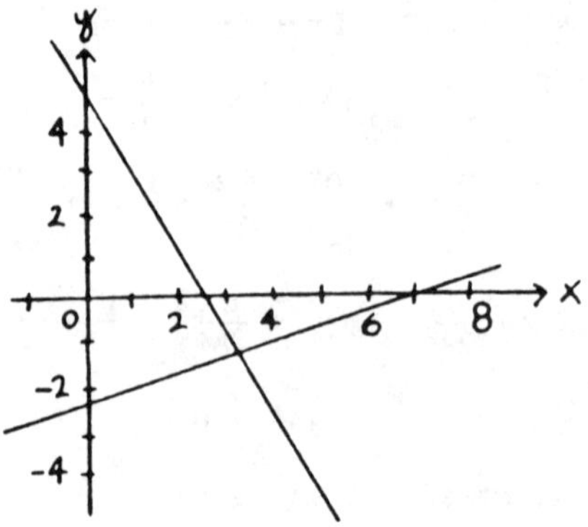

$6x + 3y = 15$

$x - 3y = 7$

$7x = 22$

$x = \frac{22}{7}$

$y = 5 - 2x$

$= \frac{35}{7} - \frac{44}{7}$

$= -\frac{9}{7}.$

y
4
2
0
2
4
6
8
x
-2
-4

9. (a) $3x - y = k$

$k = 3 \cdot 4 - 2 = 10$

$3x - y - 10 = 0.$

(b) $\frac{x}{1/2} + \frac{y}{-5/4} = 1$

$10x - 4y - 5 = 0.$

(c) $y = -2$

$y + 2 = 0.$

(d) $m = -\frac{1}{2}, \quad b = \frac{2}{3}$

$y = -\frac{1}{2}x + \frac{2}{3}$

$3x + 6y - 4 = 0.$

13. (a) $m = \frac{1}{4}, \quad a = -1, \quad b = \frac{1}{4}.$

(b) $m = -\frac{2}{3}, \quad a = -\frac{5}{2}, \quad b = -\frac{5}{3}.$

(c) $m = -\frac{5}{2}, \quad a = 0, \quad b = 0.$

(d) No slope, $a = -\frac{1}{3}$, no y intercept.

17. $\begin{cases} x^2 + y^2 - 4x + 6y - 12 = 0 \\ x - 7y + 2 = 0 \end{cases}$

$x = 7y - 2$

$(7y - 2)^2 + y^2 - 4(7y - 2) + 6y - 12 = 0$

$50y^2 - 50y = 0$

$50y(y - 1) = 0$

$y = 0, 1$

$x = -2, 5.$

$(x - 2)^2 + (y + 3)^2 = 25.$

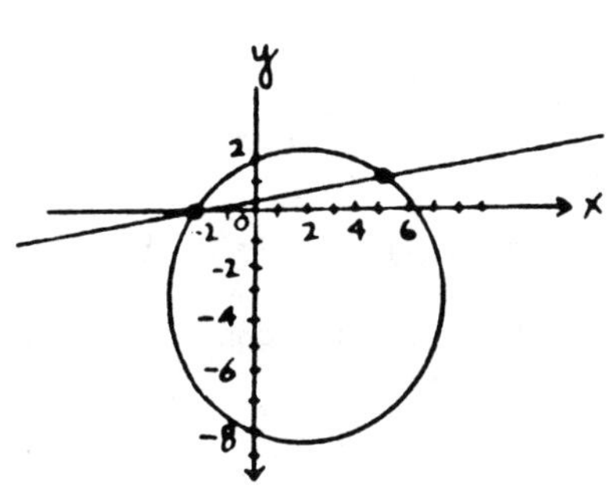

21. The midpoint M of (5, y) and (-1, 1) is $(2, \frac{y + 1}{2})$.

$P = (5, -2)$

$\overline{MP} = \sqrt{(2 - 5)^2 + (\frac{y + 1}{2} + 2)^2} = \sqrt{13}$

$9 + (\frac{y + 5}{2})^2 = 13, \ (y + 5)^2 = 4(13 - 9) = 16$

$y + 5 = \pm 4; \quad y = -9, \ y = -1.$

25. Subtracting the first equation from the second,

$6x - 2y + 8 = 0, \quad 3x - y + 4 = 0.$

29. Let the circle have equation $x^2 + y^2 = r^2$. Let $(a, \sqrt{r^2 - a^2})$, $(b, \sqrt{r^2 - b^2})$ be the endpoints of a chord of the circle. The perpendicular bisector of the chord is the set of points equidistant from the endpoints of the chord:

$$[(x - a)^2 + (y - \sqrt{r^2 - a^2})^2]^{1/2}$$

$$= [(x - b)^2 + (y - \sqrt{r^2 - b^2})^2]^{1/2}$$

$$x^2 - 2ax + a^2 + y^2 - 2y\sqrt{r^2 - a^2} + r^2 - a^2$$

$$= x^2 - 2bx + b^2 + y^2 - 2y\sqrt{r^2 - b^2} + r^2 - b^2$$

$$(b - a)x + (\sqrt{r^2 - b^2} - \sqrt{r^2 - a^2})y = 0$$

(0, 0), the coordinates of the center, satisfy this equation.

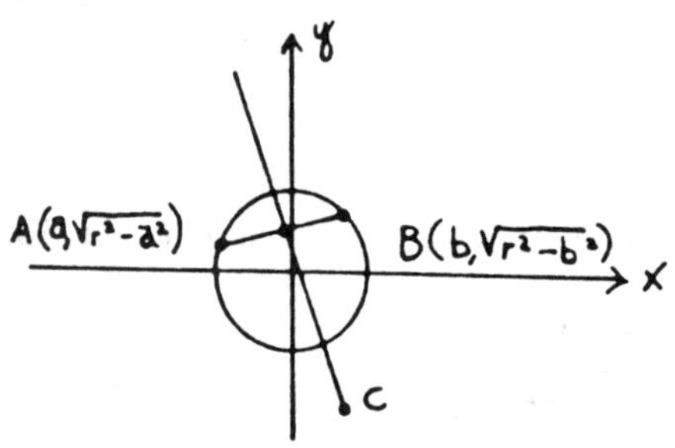

33. $m_{AD} = m_{BC}$; $\frac{y - 4}{x - 3} = \frac{3 - 0}{6 - 1} = \frac{3}{5}$

$m_{BD} = m_{AC}$; $\frac{y - 3}{x - 6} = \frac{4 - 0}{3 - 1} = 2$

$3x - 5y + 11 = 0$, $2x - y - 9 = 0$

Solving for x and y,
x = 8 and y = 7.
Thus, D = (8, 7).

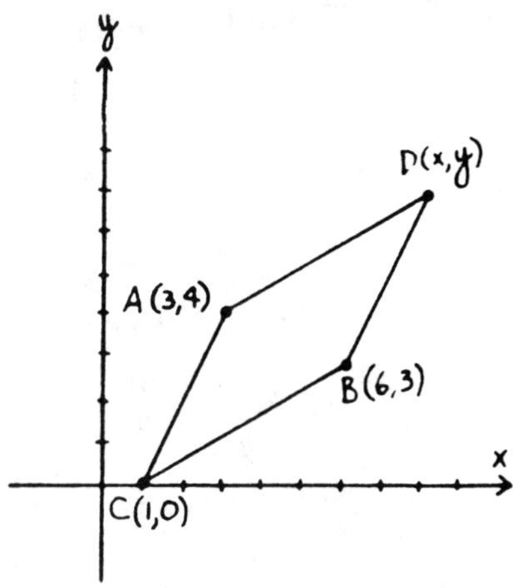

Chapter 2
Functions, Limits, and Continuity

CHAPTER 2, SECTION 1 (pp. 51-53)

1. Function.

5. No function.

9. No function.

13. $D = \{x|\ x \neq -1,\ x \neq 1\}$

17. $D = \{x|\ x \geq 1 \text{ or } x < 0\}$.

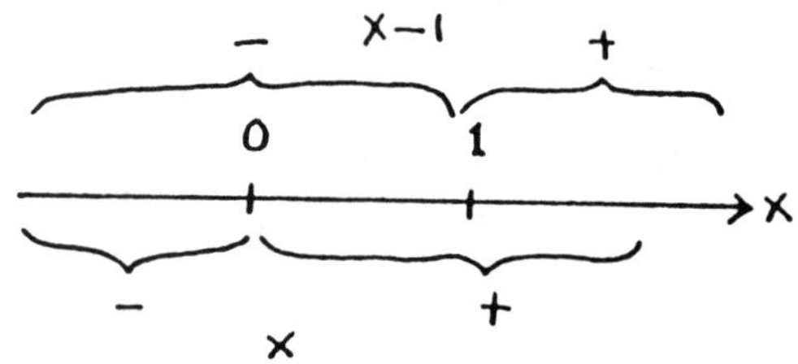

21. $D = \{x|\ x \geq 2\}$, $R = \{y|\ y \geq 0\}$.

25. $D = \{x|\ x > 0\}$, $R = \{y|\ y > 0\}$.

29. $D = \{x|\ x \neq 0\}$, $R = \{y|\ y \geq 1 \text{ or } y < 0\}$.

33. $f(x) = x - 3$, $D = \{x|\ 0 < x \leq 6\}$, $R = \{y|\ -3 < y \leq 3\}$.

37. $f(x) = 2x$, $D = \{x|\ x \text{ a positive integer}\}$,

$R = \{y|\ y \text{ an even positive integer}\}$.

41. $f(-1) = -1$, $f(0)$ does not exist, $f(2) = 1/2$,

$f(x + 1) = 1/(x + 1)$.

45. $f(y) = y^2 + 1$, $f(x + h) = (x + h)^2 + 1$.

49. $A = \pi r^2$; $C = 2\pi r$.

53. $h = 2r$; $V = \frac{1}{3}\pi r^2 h$

$$= \frac{1}{3}\pi r^2 \cdot 2r$$

$$= \frac{2}{3}\pi r^3.$$

57. $f(2t + 5) = f(t^2 - 3)$, $a(2t + 5) + b = a(t^2 - 3) + b$

$2t + 5 = t^2 - 3$, $t^2 - 2t - 8 = 0$,

$(t - 4)(t + 2) = 0$, $t = 4$, $t = -2$.

CHAPTER 2, SECTION 2 (pp. 57-58)

1. $$\frac{f(x + h) - f(x)}{h} = \frac{(x + h)^2 + 1 - (x^2 + 1)}{h}$$

$$= \frac{x^2 + 2xh + h^2 + 1 - x^2 - 1}{h} = \frac{2xh + h^2}{h} = 2x + h.$$

5. $$\frac{f(2 + h) - f(2)}{h} = \frac{[2(2 + h) + 3]^2 - (2\cdot 2 + 3)^2}{h}$$

$$= \frac{(7 + 2h)^2 - 49}{h} = \frac{28h + 4h^2}{h} = 28 + 4h.$$

9. $$\frac{f(x) - f(1)}{x - 1} = \frac{1/x - 1}{x - 1} = \frac{1 - x}{x(x - 1)} = -\frac{1}{x}.$$

13. $$\frac{f(h) - f(0)}{h} = \frac{h \sin 1/h - 0}{h} = \sin 1/h.$$

17. $$\frac{1}{f(x) + f(y)} = \frac{1}{1/x + 1/y} = \frac{xy}{y + x}.$$

21. $(f + g)(x) = \sqrt{x} + \sqrt{1 - x}\ (0 \le x \le 1)$,

$(f - g)(x) = \sqrt{x} - \sqrt{1 - x}\ (0 \le x \le 1)$,

$(fg)(x) = \sqrt{x}\sqrt{1 - x} = \sqrt{x - x^2}\ (0 \le x \le 1)$,

$(f/g)(x) = \dfrac{\sqrt{x}}{\sqrt{1 - x}} = \sqrt{\dfrac{x}{1 - x}}\ (0 \le x < 1)$,

$(g \circ f)(x) = g(f(x)) = g(\sqrt{x}) = \sqrt{1 - \sqrt{x}}\ (0 \le x \le 1)$.

25. $(f + g)(x) = x^3 + x + 4 \quad (0 \le x < 2)$,

$(f - g)(x) = x^3 - x - 4 \quad (0 \le x < 2)$,

$(fg)(x) = x^3(x + 4) = x^4 + 4x^3 \quad (0 \le x < 2)$,

$(f/g)(x) = \dfrac{x^3}{x + 4} \quad (0 \le x < 2)$,

$(g \circ f)(x) = g(x^3) = x^3 + 4 \quad (0 \le x < 2)$.

29. $f(x) = 1/x, \quad f(x)g(x) = x, \quad g(x) = \dfrac{x}{1/x} = x^2$.

33. (a) $(fg)(-x) = f(-x)g(-x) = (-f(x))(-g(x)) = f(x)g(x)$

$= (fg)(x)$, even.

(b) $(fG)(-x) = f(-x)G(-x) = -f(x)G(x) = -(fG)(x)$, odd.

(c) $(FG)(-x) = F(-x)G(-x) = F(x)G(x) = (FG)(x)$, even.

(d) $(f + g)(-x) = f(-x) + g(-x) = -f(x) - g(x)$

$= -(f + g)(x)$, odd.

(e) $(f + G)(-x) = f(-x) + G(-x) = -f(x) + G(x)$, neither.

(f) $(F + G)(-x) = F(-x) + G(-x) = F(x) + G(x) = (F + G)(x)$, even.

(g) $(f \circ g)(-x) = f(g(-x)) = f(-g(x)) = -f(g(x))$

$= -(f \circ g)(x)$, odd.

(h) $(f \circ G)(-x) = f(G(-x)) = f(G(x)) = (f \circ G)(x)$, **even.**

(i) $(F \circ G)(-x) = F(G(-x)) = F(G(x)) = (F \circ G)(x)$, **even.**

CHAPTER 2, SECTION 3 (pp. 64-66)

1. $\lim_{x\to 1} x^2 = 1.$

5. $\lim_{x\to 2} \frac{x^2 - 1}{x - 1} = \lim_{x\to 2} (x + 1) = 3.$

9. $\lim_{x\to 2} |x| = 2.$

13. $\lim_{x\to 1} f(x) = \lim_{x\to 1} (x + 1) = 2.$

17. $$\lim_{h\to 0} \frac{(x + h + 1)^2 - (x + 1)^2}{h}$$
$$= \lim_{h\to 0} \frac{(x + 1)^2 + 2h(x + 1) + h^2 - (x + 1)^2}{h}$$
$$= \lim_{h\to 0} [2(x + 1) + h] = 2(x + 1).$$

21. $$\lim_{h\to 0} \frac{\sqrt{x + h} - \sqrt{x}}{h} = \lim_{h\to 0} \frac{x + h - x}{h(\sqrt{x + h} + \sqrt{x})} = \lim_{h\to 0} \frac{h}{h(\sqrt{x + h} + \sqrt{x})}$$
$$= \lim_{h\to 0} \frac{1}{\sqrt{x + h} + \sqrt{x}} = \frac{1}{2\sqrt{x}}.$$

25. $$\lim_{h\to 0} \frac{|-1 + h| - 1}{h} = \lim_{h\to 0} \frac{1 - h - 1}{h} \; (h < 1) = \lim_{h\to 0} (-1) = -1.$$

29. $$\lim_{x\to a} \frac{\sqrt{x} - \sqrt{a}}{x - a} = \lim_{x\to a} \frac{x - a}{(x - a)(\sqrt{x} + \sqrt{a})} = \lim_{x\to a} \frac{1}{\sqrt{x} + \sqrt{a}} = \frac{1}{2\sqrt{a}}.$$

33. $$\lim_{x\to a} \frac{P(x)}{Q(x)} = \frac{\lim_{x\to a} P(x)}{\lim_{x\to a} Q(x)} = \frac{P(a)}{Q(a)},$$

$Q(a) \neq 0.$

CHAPTER 2, SECTION 4 (pp. 72-73)

1. $\lim_{x\to 0} x^3 = 0$, $0^3 = 0$, continuous.

5. $f(0) = 0$, $\lim_{x\to 0} f(x)$ does not exist since a different number is approached from the two sides, discontinuous.

9. $\lim_{x\to 0} f(x) = 0$, $f(0) = 0$, continuous.

13. $\lim_{x\to 0} x^3 = 0^3 = 0$.

17. $\lim_{x\to 2} \dfrac{x^2 + x - 6}{x - 2} = \lim_{x\to 2} (x + 3) = 2 + 3 = 5$.

21. $f(0) = 0$. $\lim_{x\to 0} f(x)$ does not exist since there are both rational numbers arbitrarily close to 0 (e.g., $1/10$, $1/10^2$, $1/10^3$, ...) and irrational numbers arbitrarily close to 0 (e.g., $\sqrt{2}/10$, $\sqrt{2}/10^2$, $\sqrt{2}/10^3$, ...). Discontinuous.

25. $\dfrac{\frac{1}{2+h} - \frac{1}{2}}{h} = \dfrac{2 - 2 - h}{2(2 + h)h} = \dfrac{-1}{2(2 + h)}$ if $h \neq 0$, so

$$\lim_{h\to 0} \frac{\frac{1}{2+h} - \frac{1}{2}}{h} = \lim_{h\to 0} \frac{-1}{2(2 + h)} = -\frac{1}{4}.$$

29. Since $g(x) = x$ is continuous, $x^2 = (gg)(x)$ is continuous. Inductively, if x^n is continuous, then $x^{n+1} = x \cdot x^n$ is continuous, $n = 1,2,3,\ldots$. Thus, kx^n is a continuous function for constant k and positive integer n. Again, by mathematical induction, the sum of any finite number of continuous functions is continuous on their common domain. If $f(x) = a_0x^n + a_1x^{n-1} + \ldots + a_{n-1}x + a_n$, then a_ix^{n-i} is continuous for $i = 0,1,\ldots,n$, so their sum $f(x)$ is continuou

CHAPTER 2, SECTION 5 (pp. 84-85)

1. $\lim_{x\to 1} \frac{x}{(x-1)^2} = +\infty.$

5. $\frac{x}{x-1} \to +\infty$ as x approaches 1 from the right;

$\frac{x}{x-1} \to -\infty$ as x approaches 1 from the left.

$\lim_{x\to 1} \frac{x}{x-1}$ is neither $+\infty$ nor $-\infty$.

9. $\lim_{x\to-\infty} \frac{x-100}{x+100} = \lim_{x\to-\infty} \frac{1-100/x}{1+100/x} = \frac{1-0}{1+0} = 1.$

13. $x = 3$, $y = 0$.

17. $x = 1$, $y = 3$.

21. $y = 2$.

25. True.

29. False.

33. $\lim_{t\to+\infty} 40{,}000 \frac{2t+1}{t+1} = \lim_{t\to+\infty} 40{,}000 \frac{2+1/t}{1+1/t} = 80{,}000$

$(0.95)(80{,}000) = 40{,}000 \frac{2t+1}{t+1} \qquad 1.9 = \frac{2t+1}{t+1}$

$1.9t + 1.9 = 2t + 1, \qquad 0.1t = 0.9, \qquad t = 9$ hr.

37. $y = \frac{4x^2 - (4x^2+3)}{2x + \sqrt{4x^2+3}} = \frac{-3}{2x + \sqrt{4x^2+3}}.$

$y = 0$. (The curve approaches it at the positive end only.)

CHAPTER 2, REVIEW (pp. 86-87)

1. (a) $D = \{x \mid x \text{ real}\}$, $R = \{y \mid y \text{ real}, y \geq 2\}$.

   (b) $D = \{x \mid x \text{ real}, x \geq 4\}$, $R = \{y \mid y \text{ real}, y \geq 0\}$.

   (c) $D = \{x \mid x \text{ real}, x \neq -5\}$, $R = \{y \mid y \text{ real}, y \neq 0\}$.

   (d) $f(x) = \dfrac{x^2}{x^2 + 1} = 1 - \dfrac{1}{x^2 + 1}$

   $D = \{x \mid x \text{ real}\}$, $R = \{y \mid y \text{ real}, 0 \leq y < 1\}$.

   (e) $D = \{x \mid x \text{ real}\}$, $R = \{y \mid y \text{ real}\}$.

   (f) $D = \{x \mid x \text{ real}, x \leq 0 \text{ or } x \geq 2\}$, $R = \{y \mid y \text{ real}, y \geq 0\}$.

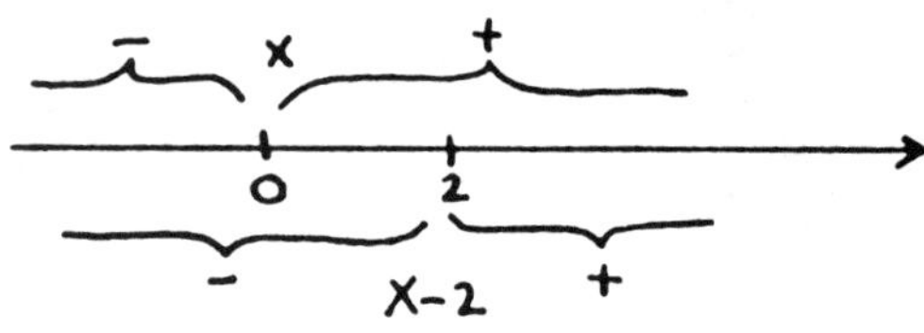

5. (a) $f(0) = 0$. (b) $f(4) = 2$.

   (c) $f(-1)$ does not exist. (d) $f(x^2) = \sqrt{x^2} = |x|$.

   (e) $\dfrac{f(1 + h) - f(1)}{h} = \dfrac{\sqrt{1 + h} - 1}{h} = \dfrac{1 + h - 1}{h(\sqrt{1 + h} + 1)}$

   $= \dfrac{1}{\sqrt{1 + h} + 1}$.

9. $(f \circ g)(x) = f(g(x)) = f(-x) = -2x + 3 \quad (x \leq 0)$,

   $(g \circ f)(x) = g(f(x)) = g(2x + 3) = -2x - 3$

   $2x + 3 \leq 0, \qquad x \leq -3/2$

   But since the domain of f is $x \geq 0$, the domain of $g \circ f$ is empty; $g \circ f$ does not exist.

13. (a) No. $f(x)$ approaches a different limit from the two sides of $x = 1$.

    (b) Yes. $f(1) = 0$.

17. (a) The limit does not exist.

(b) The limit does not exist.

(c) $\lim_{x\to 2}\left(\frac{x^2}{x-2} - \frac{2x}{x-2}\right) = \lim_{x\to 2}\frac{x(x-2)}{x-2} = \lim_{x\to 2} x = 2.$

No. The first two limits do not exist.

Chapter 3
The Derivative

CHAPTER 3, SECTION 1 (pp. 94-95)

1. $A = (-1, 1)$, $P = (-1 + h, (-1 + h)^2)$,

$$m_{AP} = \frac{(-1 + h)^2 - 1}{-1 + h + 1} = \frac{-2h + h^2}{h} = -2 + h \to -2.$$

5. $A = (2, 4)$, $P = (2 + h, 3(2 + h)^2 - 5(2 + h) + 2$,

$$m_{AP} = \frac{12 + 12h + 3h^2 - 10 - 5h + 2 - 4}{2 + h - 2} = \frac{7h + 3h^2}{h}$$

$$= 7 + 3h \to 7.$$

9. $A = (2, 4)$, $P = (2 + h, (2 + h)^2)$,

$$m_{AP} = \frac{(2 + h)^2 - 4}{2 + h - 2} = \frac{4h + h^2}{h} = 4 + h \to 4$$

Tangent:	Normal:
$y - 4 = 4(x - 2)$	$y - 4 = -\frac{1}{4}(x - 2)$
$y - 4 = 4x - 8$	
$4x - y - 4 = 0.$	$x + 4y - 18 = 0.$

13. $A = (0, -1)$, $P = (h, \frac{h + 1}{h - 1})$,

$$m_{AP} = \frac{\frac{h + 1}{h - 1} + 1}{h} = \frac{h + 1 + h - 1}{h(h - 1)} = \frac{2}{h - 1} \to -2.$$

17. $A = (2, 0)$, $P = (2 + h, h/(3 + h))$,

$$m_{AP} = \frac{\frac{h}{3 + h} - 0}{2 + h - 2} = \frac{1}{3 + h} \to \frac{1}{3}$$

Tangent:	Normal:
$y - 0 = \frac{1}{3}(x - 2)$	$y - 0 = -3(x - 2)$
$x - 3y - 2 = 0.$	$3x + y - 6 = 0.$

21. $A = (x, x^2 - 4x)$, $P = (x + h, (x + h)^2 - 4(x + h))$,

$$m_{AP} = \frac{x^2 + 2hx + h^2 - 4x - 4h - x^2 + 4x}{x + h - x}$$

$$= \frac{2hx + h^2 - 4h}{h} = 2x + h - 4 \to 2x - 4 = 0.$$

$x = 2$, $y = -4$.

25. $f(x) = |x|$, $A = (0, 0)$.

As P approaches A from the right, $m_{AP} = 1 \to 1$.
As P approaches A from the left, $m_{AP} = -1 \to -1$. (See Problem 19.)

---

19. $A = (1, 1)$, $P = (1 + h, |1 + h|)$,

$$m_{AP} = \frac{|1 + h| - 1}{1 + h - 1} = \frac{1 + h - 1}{1 + h - 1} = 1 \to 1 \ (h > -1).$$

$A = (3, 3)$, $P = (3 + h, |3 + h|)$,

$$m_{AP} = \frac{|3 + h| - 3}{3 + h - 3} = 1 \to 1 \ (h > -3).$$

$A = (8, 8)$, $P = (8 + h, |8 + h|)$,

$$m_{AP} = \frac{|8 + h| - 8}{8 + h - 8} = 1 \to 1 \ (h > -8).$$

$A = (-2, 2)$, $P = (-2 + h, |-2 + h|)$,

$$m_{AP} = \frac{|-2 + h| - 2}{-2 + h + 2} = \frac{2 - h - 2}{-2 + h + 2} = -1 \to -1 \ (h < 2).$$

$A = (-5, 5)$, $P = (-5 + h, |-5 + h|)$,

$$m_{AP} = \frac{|-5 + h| - 5}{-5 + h + 5} = \frac{5 - h - 5}{-5 + h + 5} = -1 \to -1 \ (h < 5).$$

$A = (-7, 7)$, $P = (-7 + h, |-7 + h|)$,

$$m_{AP} = \frac{|-7 + h| - 7}{-7 + h + 7} = \frac{7 - h - 7}{h} = -1 \to -1 \quad (h < 7).$$

$A = (0, 0)$, $P = (h, |h|)$,

$$m_{AP} = \frac{|h|}{h} = \begin{cases} \frac{h}{h} = 1 & (h > 0) \\ \frac{-h}{h} = -1 & (h < 0) \end{cases} \to \text{no single number.}$$

y

x

CHAPTER 3, SECTION 2 (pp. 101-103)

1. $$f'(x) = \lim_{h\to 0} \frac{f(x + h) - f(x)}{h}$$

$$= \lim_{h\to 0} \frac{[2(x + h)^2 - 4(x + h) + 1] - [2x^2 - 4x + 1]}{h}$$

$$= \lim_{h\to 0} \frac{2x^2 + 4hx + 2h^2 - 4x - 4h + 1 - 2x^2 + 4x - 1}{h}$$

$$= \lim_{h\to 0} (4x + 2h - 4) = 4x - 4.$$

5. $f(s) = s^4 + 2s^2 + 1$

$$f'(s) = \lim_{h\to 0} \frac{f(s + h) - f(s)}{h}$$

$$= \lim_{h\to 0} \frac{[(s + h)^4 + 2(s + h)^2 + 1] - [s^4 + 2s^2 + 1]}{h}$$

$$= \lim_{h\to 0} \frac{s^4+4hs^3+6h^2s^2+4h^3s+h^4+2s^2+4hs+2h^2+1-s^4-2s^2-1}{h}$$

$$= \lim_{h\to 0} (4s^3 + 6hs^2 + 4h^2s + h^3 + 4s + 2h) = 4s^3 + 4s.$$

9. $f'(v) = \lim_{h \to 0} \frac{f(v + h) - f(v)}{h}$

$= \lim_{h \to 0} \frac{\frac{v + h - 4}{v + h + 1} - \frac{v - 4}{v + 1}}{h}$

$= \lim_{h \to 0} \frac{(v + h - 4)(v + 1) - (v - 4)(v + h + 1)}{h(v + h + 1)(v + 1)}$

$= \lim_{h \to 0} \frac{v^2 + hv - 4v + v + h - 4 - v^2 - hv - v + 4v + 4h + 4}{h(v + h + 1)(v + 1)}$

$= \lim_{h \to 0} \frac{5}{(v + h + 1)(v + 1)} = \frac{5}{(v + 1)^2}.$

$f'(0) = 5, \quad f'(2) = 5/9.$

13. $s = t^2 + t - 1 = 19, \quad v = 2t + 1 = 9.$

17. $s = t^4 - 1 = 15, \quad v = 4t^3 = 32.$

21. $s = t^2(t + 1)^2 = t^4 + 2t^3 + t^2 = 144,$

$v = 4t^3 + 6t^2 + 2t = 168.$

25. $P = R - C, \quad P' = R' - C'$

$P(300) = \$18,000$, $P'(300) = \$60$; increase production.

$P(400) = \$19,000$, $P'(400) = -\$50$; decrease production.

29. $y' = 4x^3 - 4x = 0, \quad x(x^2 - 1) = 0,$

$x = 0, \pm 1$, $(0, 0)$, $(1, -1)$, $(-1, -1)$.

33. $y' = 3x^2 + 6x - 5 = 4, \quad 3(x + 3)(x - 1) = 0$

$x = -3$	$x = 1$
$y = 21$	$y = 5$
$y - 21 = 4(x + 3)$	$y - 5 = 4(x - 1)$
$4x - y + 33 = 0.$	$4x - y + 1 = 0.$

37. $y = -4.9t^2 + v_0t = 0,\ t = v_0/4.9$

$y' = -9.8\,t + v_0 = -2v_0 + v_0 = -v_0.$

41.

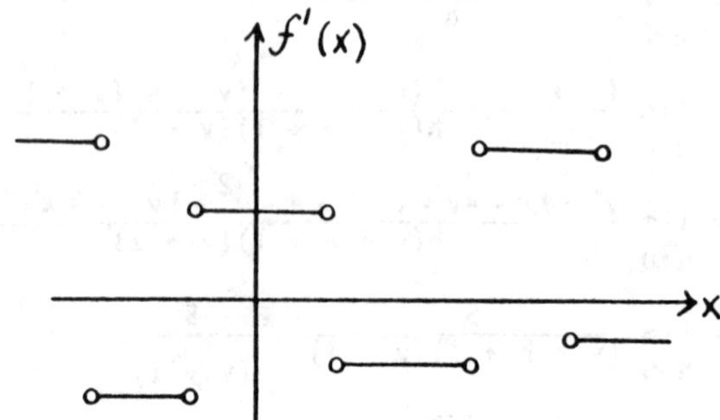

CHAPTER 3, SECTION 3 (pp. 108-109)

1. $y' = 6x + 5.$

5. $y' = 35x^4 + 15x^2 + 1.$

9. $y' = 6x - 5.$ At $x = 1,\ y' = 6 - 5 = 1.$

13. $y = [x^2 + (2x - 1)][x^2 - (2x - 1)] = x^4 - (2x - 1)^2$

$= x^4 - 4x^2 + 4x - 1,\quad y' = 4x^3 - 8x + 4.$

17. $y' = (x^4 + 1)2x + (x^2 - 1)4x^3 = 4.$

21. $y' = 3x^2 - 24x + 45 = 0,\quad x^2 - 8x + 15 = 0$

$(x - 3)(x - 5) = 0,\quad x = 3, 5,\ (3, -1),\ (5, -5).$

25. $s = 5 - 3t + 4t^2 - t^3,\quad v = -3 + 8t - 3t^2$

At $t = 1$: $s = 5 - 3 + 4 - 1 = 5$

$v = -3 + 8 - 3 = 2.$

At $t = 4$: $s = 5 - 12 + 64 - 64 = -7$

$v = -3 + 32 - 48 = -19$, speed $= 19.$

29. $y' = (x+1)(x+2)\cdot 1 + (x+1)\cdot 1(x+3) + 1\cdot(x+2)(x+3)$

$= x^2 + 3x + 2 + x^2 + 4x + 3 + x^2 + 5x + 6 = 3x^2 + 12x + 11.$

33. $f(x) = u(x) - v(x)$

$$f'(x) = \lim_{h\to 0} \frac{f(x+h) - f(x)}{h}$$

$$= \lim_{h\to 0} \frac{[u(x+h) - v(x+h)] - [u(x) - v(x)]}{h}$$

$$= \lim_{h\to 0} \left(\frac{u(x+h) - u(x)}{h} - \frac{v(x+h) - v(x)}{h}\right)$$

$$= u'(x) - v'(x).$$

37. Let $x = a + h$, $h = x - a$. Then $x - a \to 0$ as $h \to 0$ and so $x \to a$ as $h \to 0$.

$\lim_{h\to 0} f(a+h) = \lim_{x\to a} f(x) = f(a)$ so $f(x)$ is continuous at $x = a$.

CHAPTER 3, SECTION 4 (pp. 119-121)

1. $y = x^{-2}$, $y' = -2x^{-3} = -\frac{2}{x^3}.$

5. $y' = \frac{2}{3}x^{-1/3} = \frac{2}{3x^{1/3}}.$

9. $\frac{ds}{dt} = \frac{2}{3}t^{-1/3} + \frac{1}{3}t^{-4/3} = \frac{(2t+1)}{3t^{4/3}}.$

13. $\frac{dy}{dx} = \frac{dy}{du}\cdot\frac{du}{dx} = \frac{1}{2}u^{-1/2}(4x) = \frac{2x}{\sqrt{2x^2 - 3}}.$

17. $y' = 2\cdot\frac{1}{2}x^{-1/2} - 3\cdot\frac{1}{3}x^{-2/3} = \frac{1}{x^{1/2}} - \frac{1}{x^{2/3}} = 0.$

21. $y' = \frac{(x^2-2)3x^2 - (x^3+1)2x}{(x^2-2)^2} = \frac{x^4 - 6x^2 - 2x}{(x^2-2)^2} = \frac{-3}{(-1)^2} = -3.$

25. $y = \frac{(x^2 + 1)(2x - 3)}{x} = \frac{2x^3 - 3x^2 + 2x - 3}{x}$

$= 2x^2 - 3x + 2 - 3x^{-1}$

$y' = 4x - 3 + 3x^{-2} = \frac{4x^3 - 3x^2 + 3}{x^2}.$

29. $y = 2x^3 - 3x^2 + 1, \quad y' = 6x^2 - 6x = 12$

$x^2 - x - 2 = 0, \qquad (x - 2)(x + 1) = 0$

$x = 2, \quad x = -1, \qquad (2, 5), (-1, -4).$

33. $y' = \frac{1}{2\sqrt{x}} = \frac{1}{4}, \quad y - 2 = \frac{1}{4}(x - 4),$

$x - 4y + 4 = 0.$

37. $PV = k, \quad 1 \cdot 10 = k, \quad PV = 10, \quad (1 + t)V = 10$

$V = \frac{10}{t + 1}, \quad \frac{dV}{dt} = \frac{-10}{(t + 1)^2} = \frac{-10}{10^2} = -\frac{1}{10}$ liter/min.

41. $u(x) = f(x) \cdot v(x)$

$u'(x) = f(x) \cdot v'(x) + f'(x) \cdot v(x)$

$f'(x) = \frac{u'(x) - f(x) \cdot v'(x)}{v(x)}$

$= \frac{u'(x) - \frac{u(x)}{v(x)} v'(x)}{v(x)}$

$= \frac{v(x)u'(x) - u(x)v'(x)}{[v(x)]^2}.$

CHAPTER 3, SECTION 5 (pp. 124-125)

1. $y' = 4(x + 1)^3.$

5. $y' = \frac{4}{2\sqrt{4x + 2}} = \frac{2}{\sqrt{4x + 2}}.$

9. $y' = \frac{1}{2}\left(\frac{x + 1}{x - 1}\right)^{-1/2} \frac{(x - 1) - (x + 1)}{(x - 1)^2}$

$= \frac{-1}{(x + 1)^{1/2}(x - 1)^{3/2}}.$

13. $y' = \frac{2}{3}\left(\frac{2x - 1}{2x + 1}\right)^{-1/3} \frac{(2x + 1)2 - (2x - 1)2}{(2x + 1)^2}$

$= \frac{8}{3(2x - 1)^{1/3}(2x + 1)^{5/3}}.$

17. $y' = 4(x^2 + 1)^3(2x) = 4(2)^3(2) = 64.$

21. $y' = -3(x^{-2} + x)^{-4}(-2x^{-3} + 1)$

$= \frac{-3\left(1 - \frac{2}{x^3}\right)}{\left(\frac{1}{x^2} + x\right)^4} = \frac{3x^5(2 - x^3)}{(1 + x^3)^4}.$

25. $y' = \frac{(x^2 - 4)\left(\frac{x}{2\sqrt{x + 1}} + \sqrt{x + 1}\right) - x\sqrt{x + 1} \cdot 2x}{(x^2 - 4)^2}$

$= \frac{(x^2 - 4)(x + 2x + 2) - 4x^2(x + 1)}{2\sqrt{x + 1}(x^2 - 4)^2}$

$= \frac{3x^3 + 2x^2 - 12x - 8 - 4x^3 - 4x^2}{2\sqrt{x + 1}(x^2 - 4)^2}$

$= \frac{-x^3 - 2x^2 - 12x - 8}{2\sqrt{x + 1}(x^2 - 4)^2} = -\frac{x^3 + 2x^2 + 12x + 8}{2\sqrt{x + 1}(x^2 - 4)^2}.$

29. $y' = \frac{(x - 3) - (x + 2)}{(x - 3)^2} = \frac{-5}{(x - 3)^2} = -5$

$(x - 3)^2 = 1, \quad x - 3 = \pm 1, \quad x = 3 \pm 1 = 2, 4.$

33. $y' = 2(x - 4) = 2(-3) = -6$

$y - 9 = -6(x - 1), \quad 6x + y - 15 = 0.$

37. $f(x) = (x - a)^n \cdot P(x)$

$$f'(x) = (x - a)^n \cdot P'(x) + n(x - a)^{n-1} \cdot P(x)$$
$$= (x - a)^{n-1}[(x - a)P'(x) + nP(x)]$$
$$= (x - a)^{n-1}Q(x).$$

The graph of $f(x) = (x - 1)^2(x + 3)$ has a horizontal tangent at (1, 0).

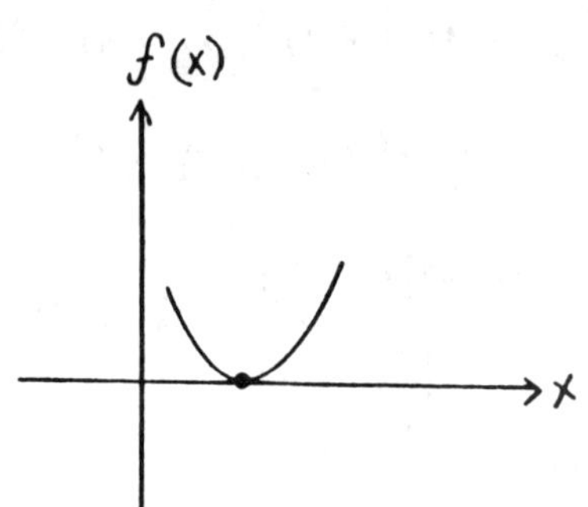

CHAPTER 3, SECTION 6 (pp. 129-130)

1. $y = \dfrac{1 - x^2}{2} \qquad 2x + 2y' = 0$

$y' = -x = -1. \ y' = -x = -1.$

5. $3x^2 + 3y^2y' = 0, \quad y' = -\dfrac{x^2}{y^2}.$ 6. $6x - 12y^2y' = 0, \quad y' = \dfrac{x}{2y^2}.$

9. $\frac{1}{3} x^{-2/3} - \frac{1}{3} y^{-2/3}y' = 0, \quad y' = \dfrac{x^{-2/3}}{y^{-2/3}} = (\frac{y}{x})^{2/3}.$

13. $2xy' + 2y - 2yy' = 0, \quad y'(x - y) = -y, \quad y' = \dfrac{y}{y - x}.$

17. $y^2 + (4x - 2)y + 3x^2 + x + 7 = 0$

$$y = \frac{-4x + 2 \pm \sqrt{16x^2 - 16x + 4 - 12x^2 - 4x - 28}}{2}$$

$$y = -2x + 1 + \sqrt{x^2 - 5x - 6}$$

$$y' = -2 + \frac{2x - 5}{2\sqrt{x^2 - 5x - 6}} = -2 + \frac{-7}{2\sqrt{0}},$$

y' does not exist.

$$6x + 4y + 4xy' + 2yy' + 1 - 2y' = 0$$

$$y' = \frac{6x + 4y + 1}{2 - 2y - 4x} = \frac{-6 + 12 + 1}{2 - 6 + 4} = \frac{7}{0},$$

y' does not exist.

21. $3x^2 + 6xy + 3x^2y' + 3y^2y' = 0$

$$y'(x^2 + y^2) = -x^2 - 2xy$$

$$y' = -\frac{x^2 + 2xy}{x^2 + y^2}.$$

25. $\dfrac{(x - y)(1 + y') - (x + y)(1 - y')}{(x - y)^2} = 2x + 2yy'$

$$-2y + 2xy' = (2x + 2yy')(x - y)^2$$

$$y' = \frac{x(x - y)^2 + y}{x - y(x - y)^2}.$$

29. $2x + 2yy' - 6 - 8y' = 0, \qquad y'(y - 4) = 3 - x,$

$$y' = \frac{3 - x}{y - 4} = \frac{3 - 6}{0 - 4} = \frac{-3}{-4} = \frac{3}{4}$$

$y - 0 = \frac{3}{4}(x - 6), \qquad 3x - 4y - 18 = 0.$

33. $2x + 2yy' - 6 - 4y' = 0$ $\quad y'(y - 2) = 3 - x$

$y' = \frac{3 - x}{y - 2}$ $\quad$ Let (x_0, y_0) be the point of tangency.

$y' = \frac{3 - x_0}{y_0 - 2}$

$y - 3 = \frac{3 - x_0}{y_0 - 2}(x + 4)$ is the equation of tangent line.

(x_0, y_0) is on the tangent line.

$y_0 - 3 = \frac{3 - x_0}{y_0 - 2}(x_0 + 4)$

$y_0^2 - 5y_0 + 6 = -x_0^2 - x_0 + 12$

$x_0^2 + y_0^2 + x_0 - 5y_0 - 6 = 0.$ (x_0, y_0) is on the original curve.

$x_0^2 + y_0^2 - 6x_0 - 4y_0 - 12 = 0.$

$7x_0 - y_0 + 6 = 0,$ $\quad y_0 = 7x_0 + 6$

$x_0^2 + 49x_0^2 + 84x_0 + 36 + x_0 - 35x_0 - 30 - 6 = 0$

$50x_0^2 + 50x_0 = 0$ $\quad 50x_0(x_0 + 1) = 0$

$x_0 = 0, \ y_0 = 6$ $\quad x_0 = -1, \ y_0 = -1$

$y - 3 = \frac{3}{4}(x + 4)$ $\quad y - 3 = -\frac{4}{3}(x + 4)$

$3x - 4y + 24 = 0$ $\quad 4x + 3y + 7 = 0$

(The fact that the two tangent lines are perpendicular is accidental.)

37. $4x + xy' + y - 2yy' - 6 + 3y' = 0$

$y'(x - 2y + 3) = -4x - y + 6$

$y' = \frac{6 - 4x - y}{3 + x - 2y}$

At $(2, 4)$, $y' = \frac{6 - 8 - 4}{3 + 2 - 8} = \frac{-6}{-3} = 2.$

At (2, 1), $y' = \frac{6 - 8 - 1}{3 + 2 - 2} = \frac{-3}{3} = -1.$

At (3, 6), $y' = \frac{6 - 12 - 6}{3 + 3 - 12} = \frac{-12}{-6} = 2.$

At (3, 0), $y' = \frac{6 - 12 - 0}{3 + 3 - 0} = \frac{-6}{6} = -1.$

At (1, 2), $y' = \frac{6 - 4 - 2}{3 + 1 - 4} = \frac{0}{0}$, no derivative.

$$2x^2 + xy - y^2 - 6x + 3y = (2x - y)(x + y) - 3(2x - y)$$
$$= (2x - y)(x + y - 3)$$

$2x - y = 0, \quad x + y = 3.$

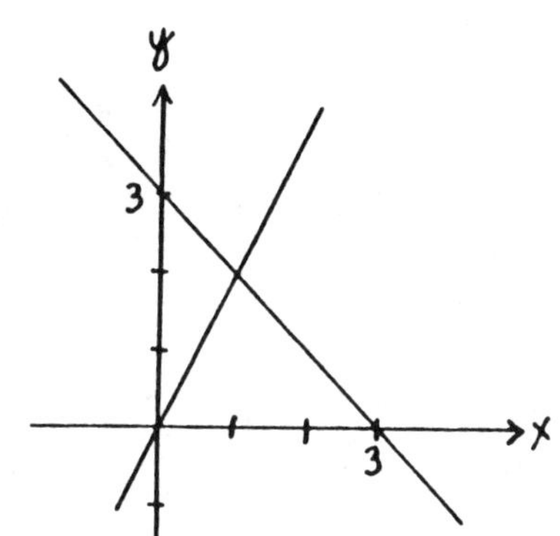

CHAPTER 3, SECTION 7 (pp. 134-136)

1. $y' = 14x + 2, \quad y'' = 14, \quad y''' = 0.$

5. $f'(x) = \frac{-2x}{(x^2 - 1)^2}$

$$f''(x) = \frac{(x^2 - 1)^2(-2) + 2x(x^2 - 1)4x}{(x^2 - 1)^4}$$
$$= \frac{-2x^2 + 2 + 8x^2}{(x^2 - 1)^3} = \frac{6x^2 + 2}{(x^2 - 1)^3}.$$

9. $\frac{2}{3} x^{-1/3} + \frac{2}{3} y^{-1/3} y' = 0$

$y' = -\frac{y^{1/3}}{x^{1/3}}.$

$$y'' = -\frac{x^{1/3}(1/3)y^{-2/3}y' - y^{1/3}(1/3)x^{-2/3}}{x^{2/3}}$$

$$= -\frac{-x^{1/3}y^{-2/3}y^{1/3}x^{-1/3} - y^{1/3}x^{-2/3}}{3x^{2/3}}$$

$$= \frac{y^{-1/3} + y^{1/3}x^{-2/3}}{3x^{2/3}} = \frac{x^{2/3} + y^{2/3}}{3x^{4/3}y^{1/3}} = \frac{1}{3x^{4/3}y^{1/3}}.$$

13. $f(x) = 4x^2 - 2x + 1$ $\qquad f(1) = 3$

$f'(x) = 8x - 2$ $\qquad f'(1) = 6$

$f''(x) = 8$ $\qquad f''(1) = 8.$

17. $2x + 2yy' = 0, \quad y' = -\frac{x}{y} = -\frac{3}{-4} = \frac{3}{4},$

$$y'' = -\frac{y \cdot 1 - xy'}{y^2} = -\frac{-4 - 3(3/4)}{16} = \frac{4 + 9/4}{16} = \frac{25}{64}.$$

21. $x = t^3 - 2t^2 - 4t - 8$

$v = 3t^2 - 4t - 4 = (3t + 2)(t - 2)$

$a = 6t - 4$

	t	x	v	a
P_1	0	-8	-4	-4
P_2	2/3	-304/27	-16/3	0
P_3	2	-16	0	8
P_4	>2		+	+

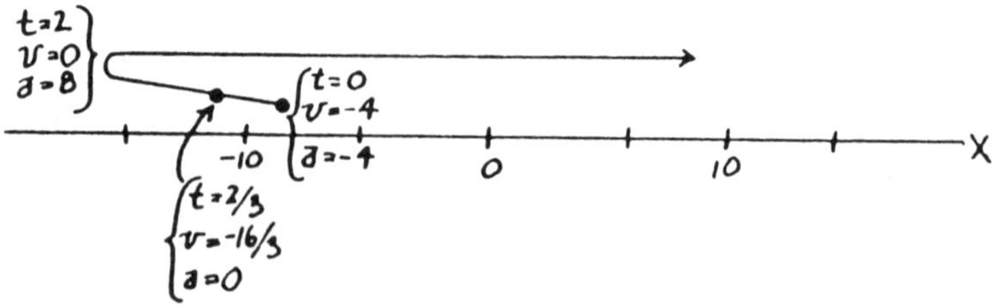

25. $x = 2t^3 - 15t^2 + 24t$

$v = 6t^2 - 30t + 24$

$= 6(t - 1)(t - 4)$

$a = 12t - 30 = 6(2t - 5)$

	t	x	v	a
P_1	0	0	24	-30
P_2	1	11	0	-18
P_3	5/2	-5/2	-27/2	0
P_4	4	-16	0	18
P_5	>4		+	+

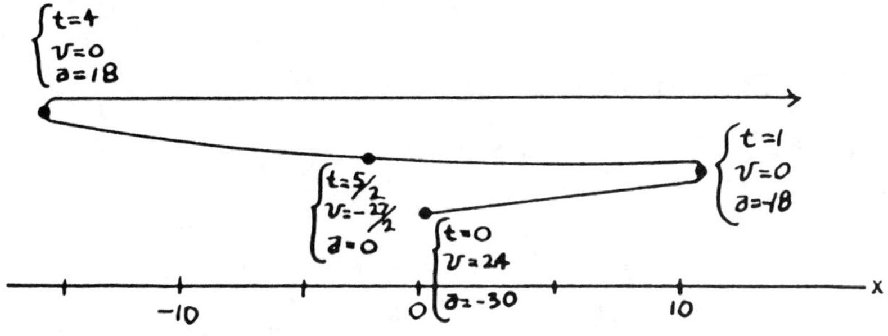

29.

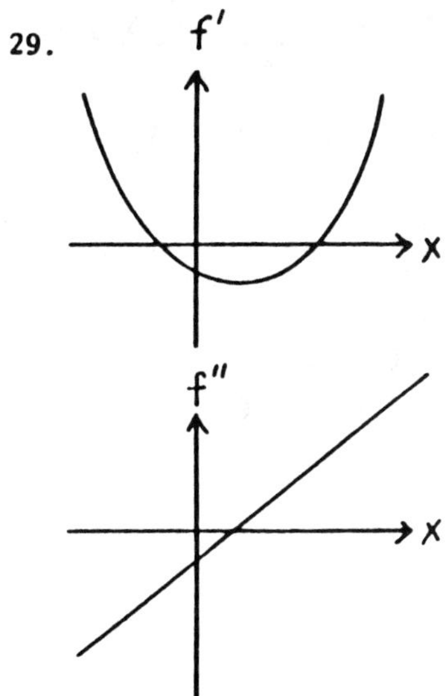

33. $x = 2.5t^2 - 0.08t^3$

$v = 5t - 0.24t^2$

(a) $v(10) = 50 - 24 = 26$m/s.

(b) $v > 0$ for $0 < t \leq 10$, $x(10) = 170$m.

$$1000 = 170 + 26(t - 10)$$

$$t = \frac{1090}{26} = 41.923\text{s}.$$

(c) $a(0) = 5$, $a(5) = 2.6$, $a(10) = 0.2$, $a(15) = -2.2$.

CHAPTER 3, REVIEW (pp. 136-137)

1. $y = 4x^{-3}$, $y' = -12x^{-4} = -\frac{12}{x^4}$.

5. $$y' = \frac{x\dfrac{2x}{2\sqrt{x^2+1}} - \sqrt{x^2+1}}{x^2} = \frac{x^2 - (x^2+1)}{x^2\sqrt{x^2+1}} = \frac{-1}{x^2\sqrt{x^2+1}}.$$

9. $$y = \frac{2x^2 - 5x - 3}{x + 4}, \quad y' = \frac{(x+4)(4x-5) - (2x^2 - 5x - 3)}{(x+4)^2}$$

$$= \frac{4x^2 + 11x - 20 - 2x^2 + 5x + 3}{(x+4)^2} = \frac{2x^2 + 16x - 17}{(x+4)^2}.$$

13. $$y' = \frac{(2x-1)2(x+1) - (x+1)^2 2}{(2x-1)^2}$$

$$= \frac{2(x+1)(2x - 1 - x - 1)}{(2x-1)^2} = \frac{2(x+1)(x-2)}{(2x-1)^2}.$$

17. $\frac{dy}{du} = 2u$

$$\frac{dy}{dx} = \frac{dy}{du}\cdot\frac{du}{dx} = 2u\left(1 + \frac{1}{2\sqrt{x+1}}\right) = 2u\,\frac{1 + 2\sqrt{x+1}}{2\sqrt{x+1}}$$

$$= \frac{u(1 + 2\sqrt{x+1})}{\sqrt{x+1}}.$$

21. $y' = \lim_{h\to 0} \frac{f(x + h) - f(x)}{h} = \lim_{h\to 0} \frac{\frac{x + h + 1}{2x + 2h - 3} - \frac{x + 1}{2x - 3}}{h}$

$= \lim_{h\to 0} \frac{(2x - 3)(x + h + 1) - (x + 1)(2x + 2h - 3)}{h(2x + 2h - 3)(2x - 3)}$

$= \lim_{h\to 0} \frac{2x^2 + 2hx + 2x - 3x - 3h - 3 - 2x^2 - 2hx + 3x - 2x - 2h + 3}{h(2x + 2h - 3)(2x - 3)}$

$= \lim_{h\to 0} \frac{-5h}{h(2x + 2h - 3)(2x - 3)} = \lim_{h\to 0} \frac{-5}{(2x + 2h - 3)(2x - 3)}$

$= \frac{-5}{(2x - 3)^2}.$

25. $y' = 3x^2 - 1 = 2$, $x^2 = 1$, $x = \pm 1$, (1, 0) and (-1, 0).

29 Let $u(x) = (x - a)^n$, $v(x) = g(x)$.

$u(a) = (a - a)^n = 0$, $u'(x) = n(x - a)^{n-1}$, $u'(a) = 0$,

$u''(x) = n(n - 1)(x - a)^{n-2}$, $u''(a) = 0, \ldots,$

$u^{(n-1)}(x) = n(n - 1) \ldots 2(x - a)$, $u^{(n-1)}(a) = 0$,

$u^{(n)}(x) = n(n - 1) \ldots 2 \cdot 1 = n!$

From Problem 36, Exercise 3.7,

$f^{(n)}(x) = u(x)v^{(n)}(x) + nu'(x)v^{(n-1)}(x) + \ldots$

$+ nu^{(n-1)}(x)v'(x) + u^{(n)}(x)v(x)$

Since $u^{(i)}(a) = 0$, $i = 0,1,\ldots,n - 1$, and $u^{(n)}(a) = n!$,

$f^{(n)}(a) = u^{(n)}(a)g(a) = n!g(a) \neq 0.$

If $0 \leq k < n$, then $f^{(k)}(a) = u(a)g^{(k)}(a) +$ $ku'(a)g^{(k-1)}(a) + \ldots + u^{(k)}(a)g(a) = 0$. Thus, $f^{(n)}(a) =$ $n!g(a)$ is the first nonzero derivative at $x = a$.

Chapter 4
Curve Sketching

CHAPTER 4, SECTION 1 (pp. 150-153)

1.

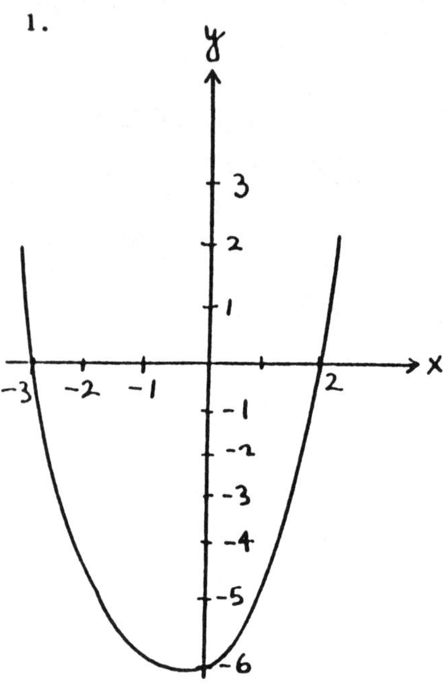

5. $y = x^3(x - 2)(x + 2)$.

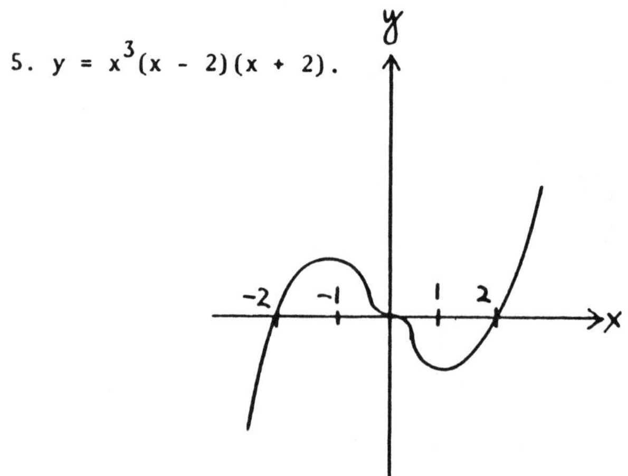

9.

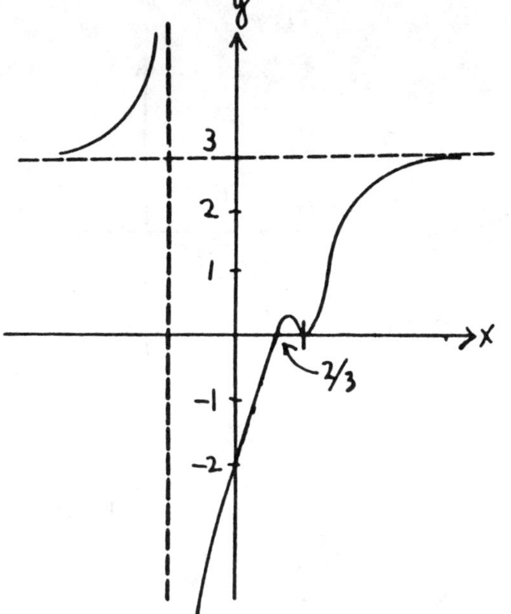

13.

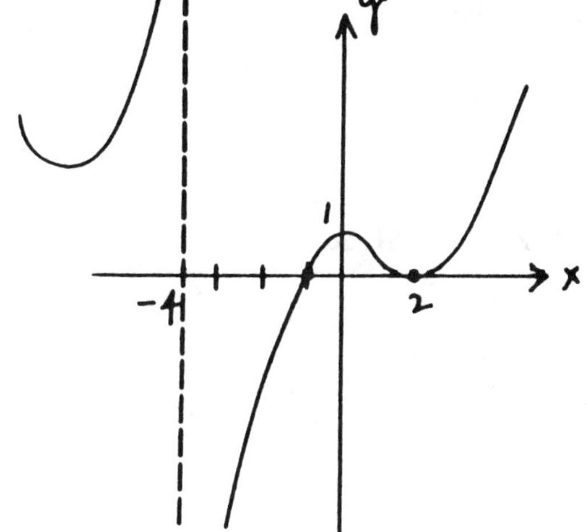

17. $y = \dfrac{3x + 2}{x}$.

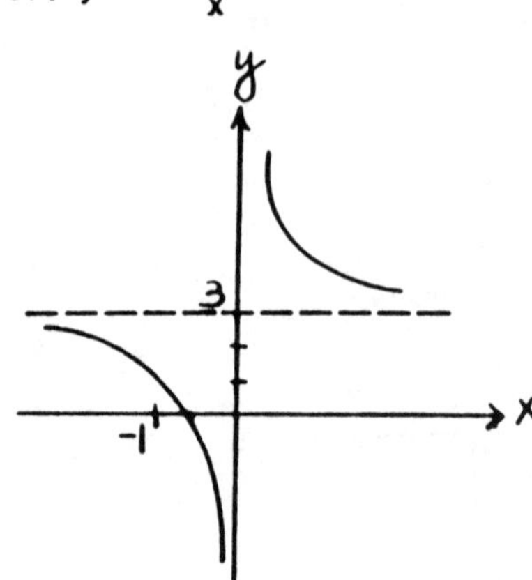

21.

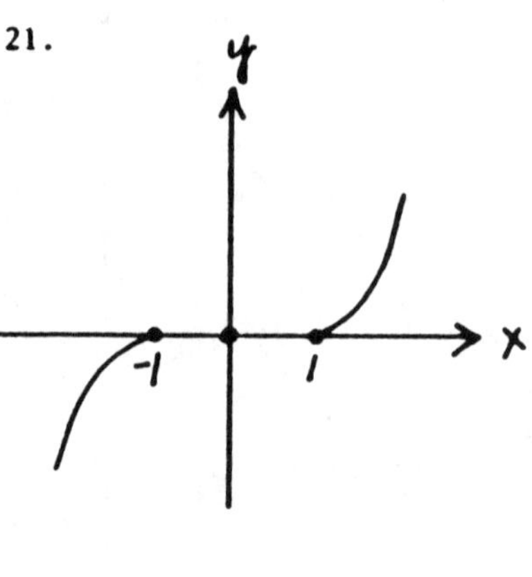

25.

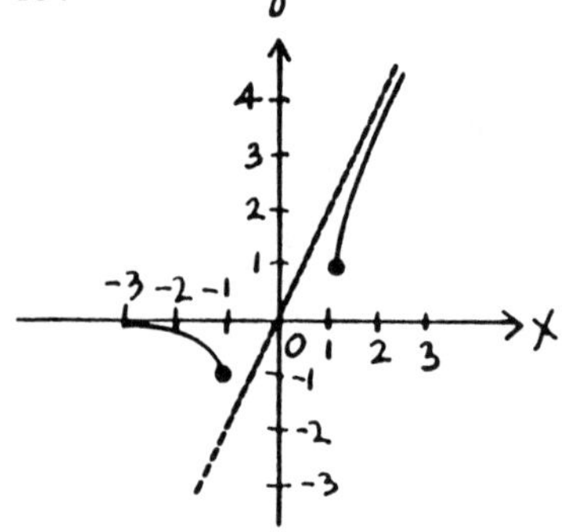

29.

Equation	Figure
(a)	(b)
(b)	(c)
(c)	(a)
(d)	(d)
(e)	(f)
(f)	(e)

CHAPTER 4, SECTION 2 (pp. 161-162)

1. $a = 0$, $b = 3$,

f is continuous on $[0,3]$; $f'(x) = 2x - 3$ exists for all x in $(0,3)$; $f(0) = f(3) = 0$. Thus there is a number x_0 between 0 and 3 such that $f'(x_0) = 0$.

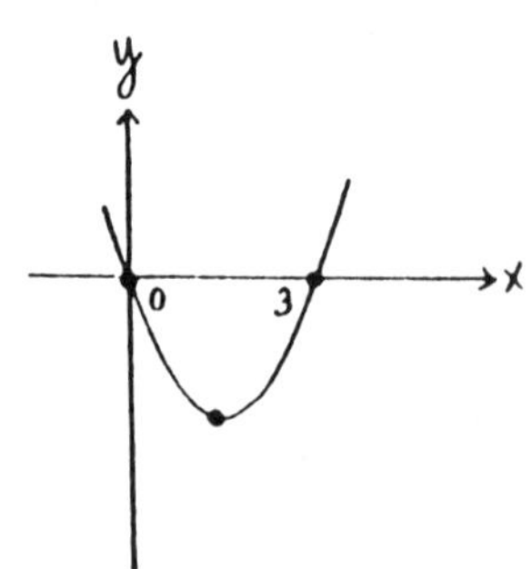

$f'(x_0) = 2x_0 - 3 = 0$ when

$x_0 = \frac{3}{2}$.

5. $f(-1) = 0$, $f(1) = 0$, $f'(x) = 3x^2 - 1 = 0$,

$x_0 = \pm \frac{1}{\sqrt{3}}$

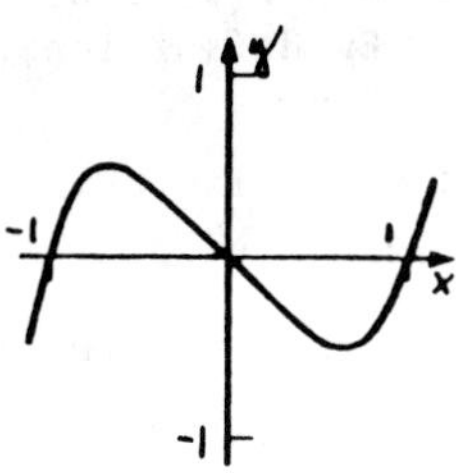

9. Rolle's theorem tells us nothing because $f'(0)$ does not exist ($f(x) = 0$ only at $x = -1$ and $x = 1$).

13. f is continuous on $[1,3]$; $f'(x) = -\frac{1}{x^2}$ exists on $(1,3)$.

Thus there is a number between 1 and 3 such that

$f'(x_0) = \frac{f(3) - f(1)}{3 - 1}$. $-\frac{1}{x_0^2} = \frac{1/3 - 1}{2} = -\frac{1}{3}$ when $x_0 = \sqrt{3}$.

17. The mean-value theorem tells us nothing because f is not defined (and therefore not continuous--nor does f' exist) at $x = 0$.

21. Rolle's theorem tells us nothing because $f'(1)$ does not exist ($f(x) = 0$ only at $x = 0$ and $x = 2$).

25. Case II $f(x) < 0$ for some x between a and b: As in the proof of Case I, there is a number x_0 in $[a,b]$ such that $f(x_0) \leqq f(x)$ for every x in $[a,b]$. Since $f(x) < 0$ for some x between a and b and $f(x_0) \leqq f(x)$, it follows that $f(x_0) < 0$. Thus x_0 is neither a nor b, since $f(a) = f(b) = 0$, and x_0 is between a and b. Now define a function F by

$$F(x) = \frac{f(x_0) - f(x)}{x_0 - x}$$

for all x except x_0 in $[a,b]$. Since $f(x_0) \leqq f(x)$ for all x in $[a,b]$,

$$f(x_0) - f(x) \leqq 0$$

for all such values of x. If $x < x_o$, then $x_o - x > 0$ and $F(x) \leq 0$. If $x > x_o$, then $x_o - x < 0$ and $F(x) \geq 0$. By definition,

$$\lim_{x \to x_o} F(x) = \lim_{x \to x_o} \frac{f(x_o) - f(x)}{x_o - x} = f'(x_o) .$$

Since we are given that $f'(x)$ exists for all x in (a,b), we know that $\lim_{x \to x_o} F(x)$ is some number L. Let us assume L to be positive. If $\varepsilon = L/2$, then, by the definition of limit, there is a positive number δ such that

$$|F(x) - L| < \frac{L}{2}$$

whenever

$$0 < |x - x_o| < \delta.$$

Suppose $x > a$ and $x_o - \delta < x < x_o$. For this choice of x, $|x - x_o| < \delta$. But, since $x < x_o$, $F(x) \leq 0$ and

$$F(x) - L \leq -L .$$

Thus

$$|F(x) - L| \geq L > \frac{L}{2} ,$$

which is a contradiction. Thus the assumption that L is positive is wrong.

Assume now that L is negative. If $\varepsilon = -L/2$, then by the definition of limit, there is a positive number δ such that

$$|F(x) - L| < -\frac{L}{2}$$

whenever

$$0 < |x - x_o| < \delta.$$

Suppose $x < b$ and $x_o < x < x_o + \delta$. For this choice of x, $|x - x_o| < \delta$. But, since $x > x_o$, $F(x) \geq 0$ and

$$F(x) - L \geq -L > 0 .$$

Thus

$$|F(x) - L| \geq -L > -\frac{L}{2},$$

which is a contradiction. The assumption that L is negative is also wrong. Since L is neither positive nor negative, L must be 0, which means $f'(x_0) = 0$.

29. Since f is continuous on [a,b], f'(x) exists for all x in (a,b), and f(a) = f(b) = 0, there is a number p between a and b such that f'(p) = 0. Similarly there is a number q between b and c such that f'(q) = 0.

    Now we use Rolle's theorem for f'(x) on (p,q). We have shown that f'(p) = f'(q) = 0 and f''(x) exists for all x in (p,q). Since f''(x) exists for all x in (a,c) and a < p < q < c, f' is continuous on [p,q]. Hence there is a number d between p and q (and therefore between a and c) such that f''(d) = 0.

CHAPTER 4, SECTION 3 (pp. 169-170)

1. $y' = 2x + 2$

   (-1, -4) min.

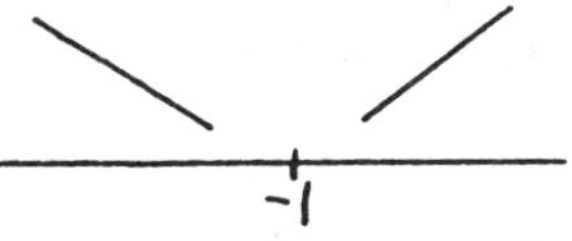

5. $y' = 3x^2 - 6x = 3x(x - 2)$

   (0, 1) max, (2, -3) min.

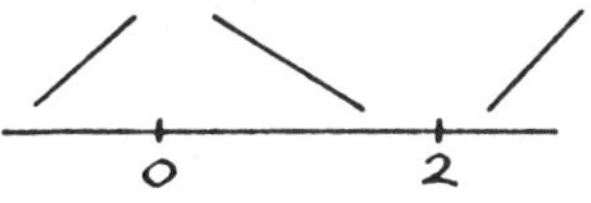

9. $y' = 12x^3 + 12x^2$

   $= 12x^2(x + 1)$

   (-1, -1) min, (0, 0) neither.

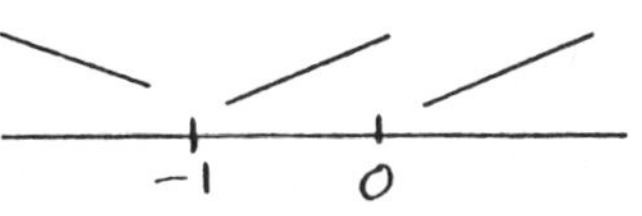

13. $y' = \dfrac{x + 1 - x}{(x + 1)^2} = \dfrac{1}{(x + 1)^2}.$ No critical point.

17. $y' = \dfrac{-2x}{(x^2 - 1)^2}$ (0, -1) max.

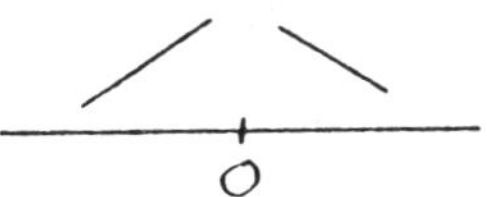

21. $y' = (x + 1)^2 + (x - 2)2(x + 1)$

$= 3(x + 1)(x - 1)$

(-1, 0) max, (1, -4) min.

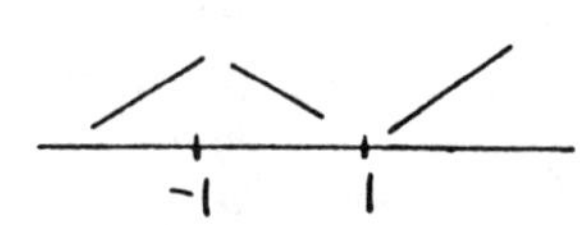

25. $y' = \frac{1}{3x^{2/3}}$ (0, 0) neither.

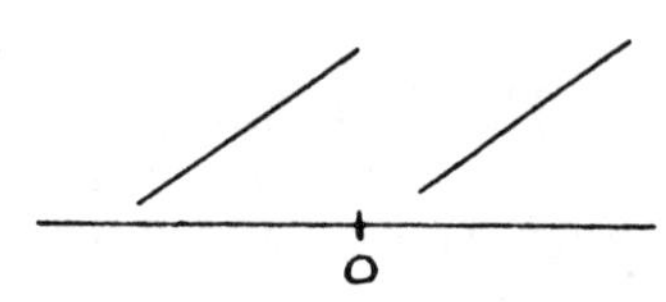

29. $y' = (x - 1)^{1/3} \frac{2}{3}(x + 2)^{-1/3} + (x + 2)^{2/3} \frac{1}{3}(x - 1)^{-2/3}$

$= \frac{2(x - 1) + (x + 2)}{3(x + 2)^{1/3}(x - 1)^{2/3}} = \frac{x}{(x + 2)^{1/3}(x - 1)^{2/3}}$

$(0, -\sqrt[3]{4})$ min, (-2, 0) max,

(1, 0) neither.

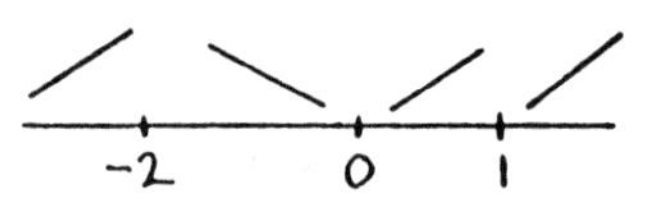

33. $y' = \frac{4}{3} x^{1/3} - \frac{2}{3} x^{-1/3} = \frac{2(2x^{2/3} - 1)}{3x^{1/3}}$

(0, 0) max, $(\frac{1}{2\sqrt{2}}, -\frac{1}{4})$ min,

$(-\frac{1}{2\sqrt{2}}, -\frac{1}{4})$ min.

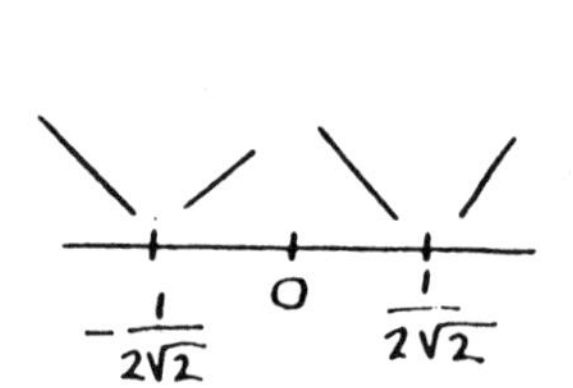

37. $y' = 2Ax + B$, $x = -\frac{B}{2A}$. A < 0 max, A > 0 min.

CHAPTER 4, SECTION 4 (pp. 176-177)

1. $y' = 2x - 1$, $y'' = 2$, $(\frac{1}{2}, -\frac{25}{4})$ absol. min.

5. $y' = 12x^3 - 12x^2 - 24x = 12x(x - 2)(x + 1)$,

$y'' = 36x^2 - 24x - 24$

$(0, 24)$	$(2, -8)$	$(-1, 19)$
$y'' = -24$	$y'' = 72$	$y'' = 36$
rel. max.	absol. min.	rel. min.

9. $y' = 2x - 2$,

$y'' = 2$.

Concave up for all x.

13. $y' = 4x^3 - 18x^2$,

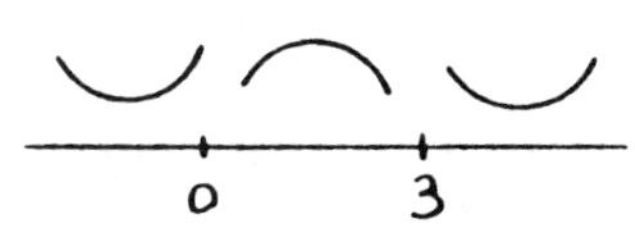

$y'' = 12x^2 - 36x = 12x(x - 3)$

$(0, 0)$, $(3, -81)$. Concave up for $x < 0$ and $x > 3$,

concave down for $0 < x < 3$.

17. $y' = \dfrac{(x + 1)2x - x^2}{(x + 1)^2} = \dfrac{x(x + 2)}{(x + 1)^2}$,

$y'' = \dfrac{(x + 1)^2(2x + 2) - (x^2 + 2x)2(x + 1)}{(x + 1)^4} = \dfrac{2}{(x + 1)^3}$

$(0, 0)$	$(-2, -4)$
$y'' = 2$	$y'' = -2$
rel. min.	rel. max.

21. $y' = \dfrac{x^2 + 1 - 2x^2}{(x^2 + 1)^2} = \dfrac{1 - x^2}{(x^2 + 1)^2}$,

$y'' = \dfrac{(1 + x^2)^2(-2x) - (1 - x^2)2(1 + x^2)2x}{(1 + x^2)^4}$

$= \dfrac{2x(x^2 - 3)}{(1 + x^2)^3}$

$(0, 0)$, $(\sqrt{3}, \frac{\sqrt{3}}{4})$, $(-\sqrt{3}, -\frac{\sqrt{3}}{4})$.

25. $y' = 12x^2 - 30x - 18 = 6(2x + 1)(x - 3)$,

$y'' = 24x - 30 = 6(4x - 5)$

$(-\frac{1}{2}, \frac{59}{4})$	$(3, -71)$	$(\frac{5}{4}, -\frac{225}{8})$
$y'' = -42$	$y'' = 42$	Point of
rel. max.	rel. min.	inflection.

29. $y' = 5x^4 - 20x^3 = 5x^3(x - 4)$,

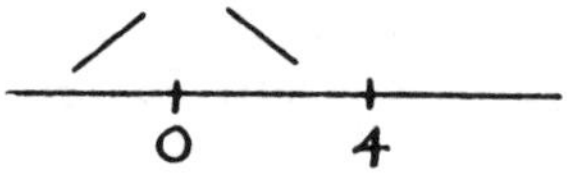

$y'' = 20x^3 - 60x^2 = 20x^2(x - 3)$

$(0, 0)$	$(4, -256)$	$(3, -162)$
$y'' = 0$	$y'' = 320$	Point of
rel. max.	rel. min.	inflection.

33. $y' = \frac{(cx + d)a - (ax + b)c}{(cx + d)^2} = \frac{ad - bc}{(cx + d)^2}$ (no critical point)

$y'' = \frac{(ad - bc)(-2)(c)}{(cx + d)^3} = \frac{-2c(ad - bc)}{(cx + d)^3}$ (no point of inflection)

37. Suppose $f''(x)$ is defined in an open interval containing $x = a$ except possibly at $x = a$, and $f'(a) = 0$. If $f''(a) > 0$, f has a minimum at $x = a$; if $f''(a) < 0$, f has a maximum at $x = a$. Thus $f''(a) = 0$. Also, if f'' does not change sign at $x = a$, then $(a, f(a))$ is a maximum or a minimum, thus $(a, f(a))$ is an inflection point.

CHAPTER 4, SECTION 5 (pp. 181-182)

1. $y = x^2 - 2, \quad -1 \le x \le 2$.

$y' = 2x, \quad y'' = 2$

$(0, -2)$	$(-1, -1)$	$(2, 2)$
absol. min.	rel. max.	absol. max.

5. $y = \frac{1}{x^2}$, $-1 \le x \le 1$.

$y' = \frac{-2}{x^3}$ No critical point. (-1, 1) absol. min. (1, 1) absol. min.

9. $y = \begin{cases} x^2 & \text{if } x < 0, \\ x^3 & \text{if } x \ge 0. \end{cases}$ $y' = \begin{cases} 2x & \text{if } x < 0, \\ 3x^2 & \text{if } x \ge 0. \end{cases}$

(0, 0) absol. min.

13. $y = x^3 - 3x$, $-\sqrt{3} \le x \le 2$.

$y' = 3x^2 - 3$, $y'' = 6x$

(1, -2)	(-1, 2)	$(-\sqrt{3}, 0)$	(2, 2)
$y'' > 0$	$y'' < 0$	rel. min.	absol. max.
absol. min.	absol. max.		

17. $y = \begin{cases} -x - 1 & \text{if } x < -1, \\ 1 - x^2 & \text{if } -1 \le x \le 1, \\ x - 1 & \text{if } x > 1. \end{cases}$

$y' = \begin{cases} -1 & \text{if } x < -1, \\ -2x & \text{if } -1 < x < 1, \\ 1 & \text{if } x > 1. \end{cases}$

(0, 1)	(-1, 0)	(1, 0)
$y'' = -2$	absol. min.	absol. min.
rel. max.		

21. $$y' = \begin{cases} 2 & \text{if } x < -2, \\ -1 & \text{if } -2 < x < 0, \\ 2 - 2x & \text{if } x > 0. \end{cases}$$

(1, 1)	(-2, 2)	(0, 0)
$y'' = -2$	absol. max.	rel. min.
rel. max.		

CHAPTER 4, SECTION 6 (pp. 186-188)

1. $y = x^2(x + 6)$, $y' = 3x^2 + 12x = 3x(x + 4)$,

$y'' = 6x + 12 = 6(x + 2)$

(0, 0)	(-4, 32)	(-2, 16)
$y'' = 12$	$y'' = -12$	Point of
min.	max.	inflection.

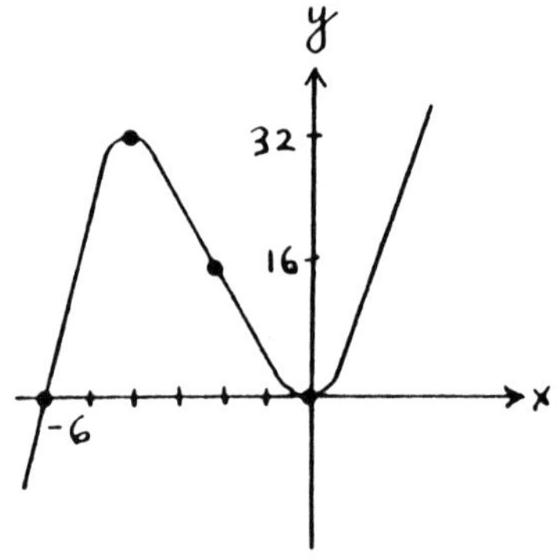

5. $y' = (x + 1)^2 3(x - 3)^2 + (x - 3)^3 2(x + 1)$

$= (x + 1)(x - 3)^2(5x - 3),$

$y'' = (x + 1)(x - 3)^2 5 + (x + 1)2(x - 3)(5x - 3)$

$+ (x - 3)^2(5x - 3)$

$= (x - 3)(20x^2 - 24x - 12) = 4(x - 3)(5x^2 - 6x - 3)$

(-1, 0)	(3, 0)	$(\frac{3}{5}, -35.4)$	(3, 0), (1.6, -18.5) (-0.4, -14.8)
y'' is -	y'' = 0	y'' is +	
max.	neither	min.	Points of inflection.

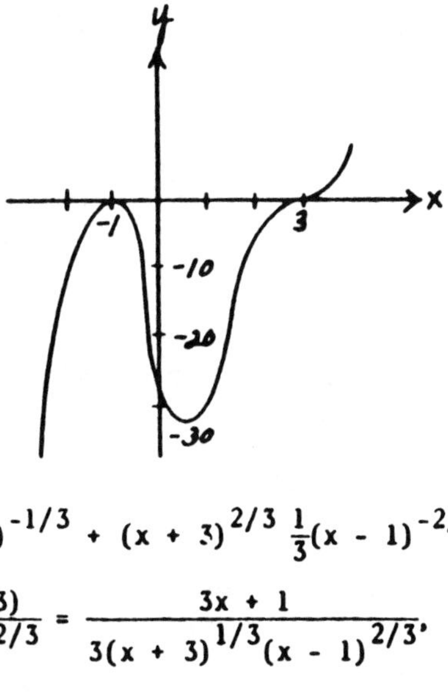

9. $y' = (x - 1)^{1/3} \frac{2}{3}(x + 3)^{-1/3} + (x + 3)^{2/3} \frac{1}{3}(x - 1)^{-2/3}$

$$= \frac{2(x - 1) + (x + 3)}{3(x + 3)^{1/3}(x - 1)^{2/3}} = \frac{3x + 1}{3(x + 3)^{1/3}(x - 1)^{2/3}},$$

$$y'' = \frac{3(x + 3)^{1/3}(x - 1)^{2/3} \cdot 3 - (3x + 1)3 \cdot \left[(x + 3)^{1/3} \frac{2}{3}(x - 1)^{-1/3} + (x - 1)^{2/3} \frac{1}{3}(x + 3)^{-2/3}\right]}{9(x + 3)^{2/3}(x - 1)^{4/3}}$$

$$= \frac{9(x + 3)(x - 1) - (3x + 1)[2(x + 3) + (x - 1)]}{9(x + 3)^{4/3}(x - 1)^{5/3}}$$

$$= \frac{-32}{9(x + 3)^{4/3}(x - 1)^{5/3}}$$

$(1, 0)$	$(-3, 0)$	$(-\frac{1}{3}, -\frac{4\sqrt[3]{4}}{3})$	$(1, 0)$
neither	max.	y'' is +	Point of
		min.	inflection.

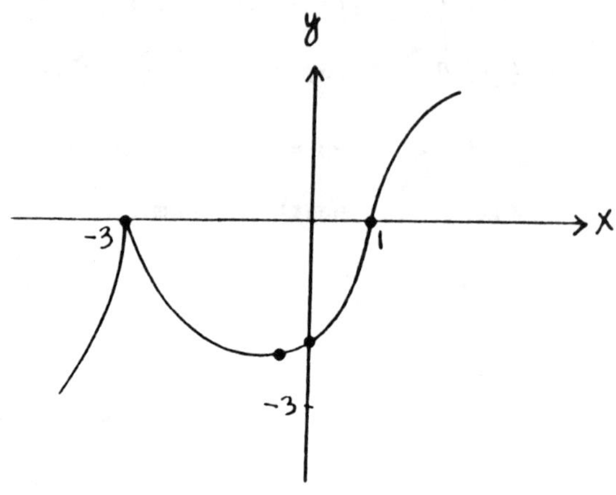

13. $y' = \frac{(x + 1)2x - x^2}{(x + 1)^2} = \frac{x(x + 2)}{(x + 1)^2}$,

$$y'' = \frac{(x + 1)^2(2x + 2) - (x^2 + 2x)2(x + 1)}{(x + 1)^4} = \frac{2}{(x + 1)^3}$$

$(0, 0)$	$(-2, -4)$
y'' is +	y'' is -
min.	max.

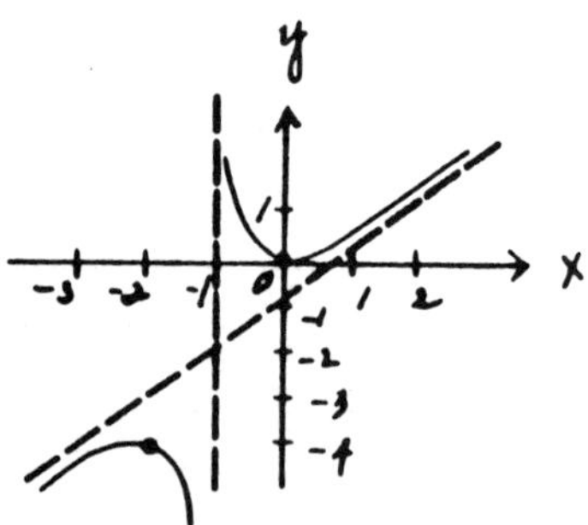

17. Equation Graph

Equation	Graph
(a)	(b)
(b)	(c)
(c)	(d)
(d)	(a)
(e)	(f)
(f)	(e)

CHAPTER 4, REVIEW (pp. 189-191)

1.

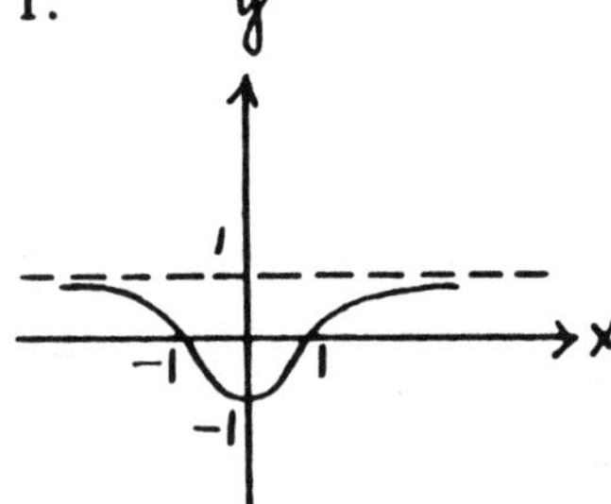

5. $y' = 6x^2 - 6x - 12 = 6(x - 2)(x + 1)$,

$y'' = 12x - 6 = 6(2x - 1)$

(-1, 15)	(2, -12)
$y'' < 0$	$y'' > 0$
rel. max.	rel. min.

9.

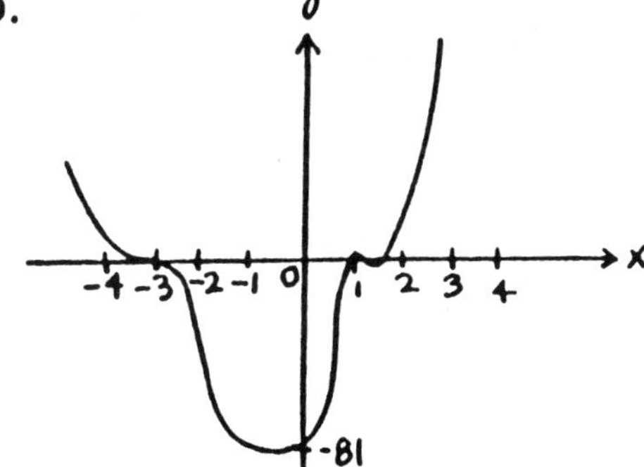

13.

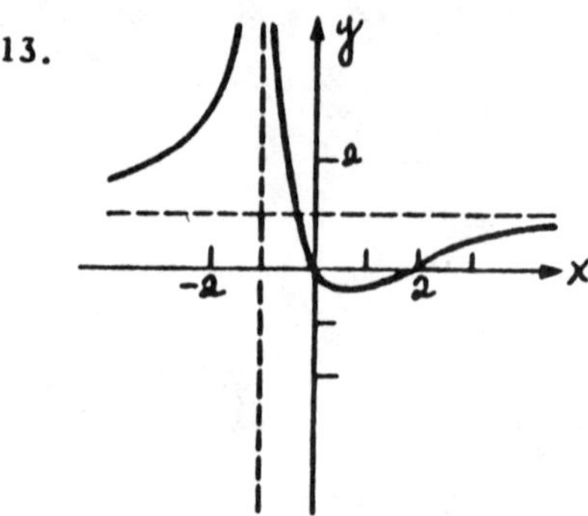

17. $y' = x^3 \cdot 2(3x - 1)3 + 3x^2(3x - 1)^2$

$= 3x^2(3x - 1)(5x - 1) = 3(15x^4 - 8x^3 + x^2)$,

$y'' = 3(60x^3 - 24x^2 + 2x) = 6x(30x^2 - 12x + 1)$

$(\frac{1}{3}, 0)$	$(\frac{1}{5}, \frac{4}{3125})$	$(0, 0)$
$y'' > 0$	$y'' < 0$	$y'' = 0$
rel. min.	rel. max.	Neither; since (0, 0) is an odd intercept, the graph crosses the x axis.

21. $y' = 2(x^2 - 9)2x = 4x(x^2 - 9)$,

$y'' = 4x \cdot 2x + 4(x^2 - 9) = 12(x^2 - 3)$

Critical points:	$(0, 81)$	$(\pm 3, 0)$
	$y'' < 0$	$y'' > 0$
	rel. max.	absol. min.

Points of inflection: $(\pm\sqrt{3}, 36)$.

25. $y' = \dfrac{(3x^2 - 1)6x - 3x^2\ 6x}{(3x^2 - 1)^2} = \dfrac{-6x}{(3x^2 - 1)^2},$

$$y'' = \frac{(3x^2 - 1)^2(-6) + 6x\ 2(3x^2 - 1)6x}{(3x^2 - 1)^4}$$

$$= \frac{-18x^2 + 6 + 72x^2}{(3x^2 - 1)^3}$$

$$= \frac{6(9x^2 + 1)}{(3x^2 - 1)^3}$$

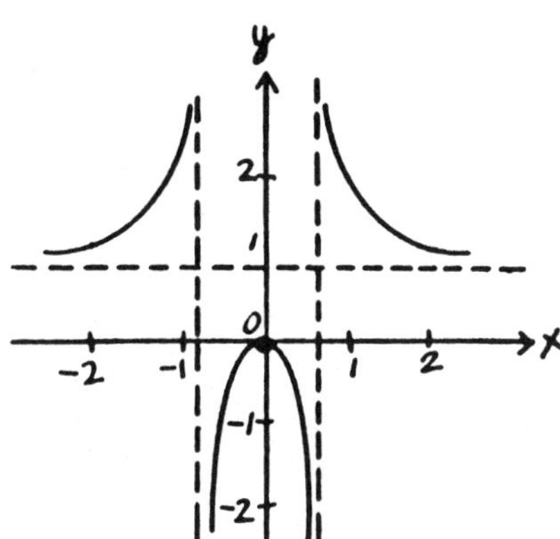

Critical point:

(0, 0)

$y'' < 0$

rel. max.

No point of inflection.

29. $f(-2) = f(2) = 0$. But $f(x)$ is not continuous on $[-2,2]$ nor does f' exist on $(-2,2)$ since f is not defined at $x = 0$. Therefore Rolle's theorem tells us nothing about this function.

33. $f(x) = (x - a)^m(x - b)^n, \quad m,n > 1,$

$$f'(x) = n(x - a)^m(x - b)^{n-1} + m(x - a)^{m-1}(x - b)^n$$

$$= (x - a)^{m-1}(x - b)^{n-1}[n(x - a) + m(x - b)]$$

$$= (x - a)^{m-1}(x - b)^{n-1}[(n + m)x - (na + mb)]$$

Critical points: $(a, 0)$, $(b, 0)$,

$$\left(\frac{na + mb}{m + n}, (-1)^n n^n m^m \left(\frac{b - a}{m + n}\right)^{m+n}\right)$$

$a < \dfrac{na + mb}{m + n}$	$\dfrac{na + mb}{m + n} < b$
$ma + na < na + mb$	$na + mb < mb + nb$
$a < b$	$a < b$

37. $y = x\sqrt{4 - x^2}$

$y^2 = x^2\sqrt{4 - x^2}$

$y = \pm|x|\sqrt{4 - x^2}$

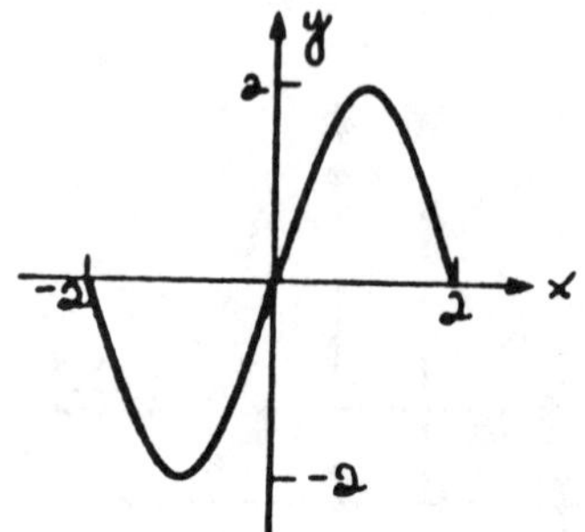

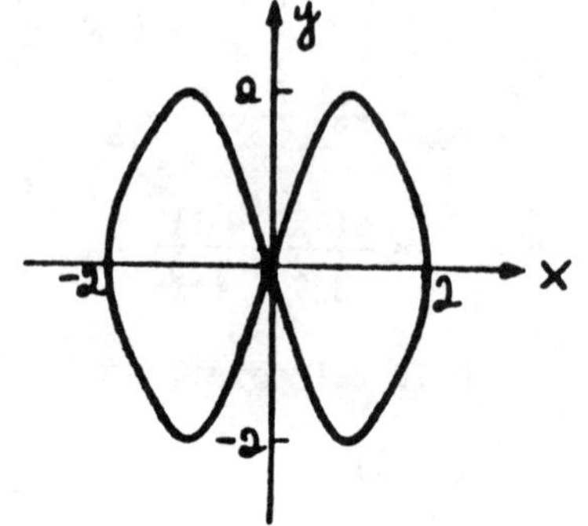

Chapter 5
Further Applications of the Derivative

CHAPTER 5, SECTION 1 (pp. 201-206)

1. $P = xy, \quad x + y = 48, \quad P = x(48 - x) = 48x - x^2$

   $\frac{dP}{dx} = 48 - 2x = 0, \quad x = 24, \; y = 24.$

5. $xy = 180{,}000, \quad y = \frac{180{,}000}{x}$

   $L = 4x + 2y = 4x + \frac{360{,}000}{x}$

   $L' = 4 - \frac{360{,}000}{x^2} = 0$

   $x^2 = 90{,}000$

   $x = 300$ ft, $\quad y = 600$ ft, $\quad L = 2400$ ft.

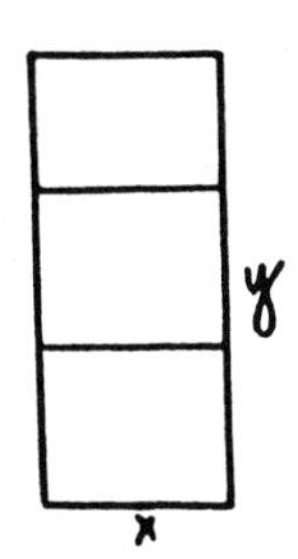

9. $4000 = x^2y, \quad y = \frac{4000}{x^2}$

   $A = 4xy + x^2 = \frac{16{,}000}{x} + x^2$

   $A' = -\frac{16{,}000}{x^2} + 2x = 0$

   $x^3 = 8000, \quad x = 20$ cm, $\quad y = 10$ cm.

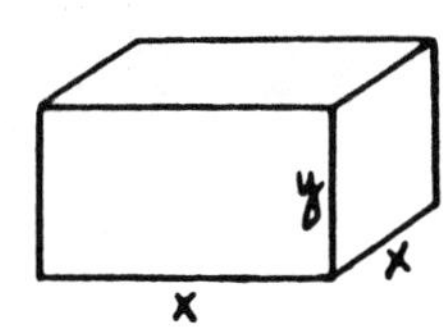

13. $x^2 + y^2 = R^2$

$$A = \frac{1}{2}\cdot 2x(R - y) = x(R - y)$$

$$= \sqrt{R^2 - y^2}(R - y)$$

$$A' = \sqrt{R^2 - y^2}(-1) + (R - y)\frac{-2y}{2\sqrt{R^2 - y^2}}$$

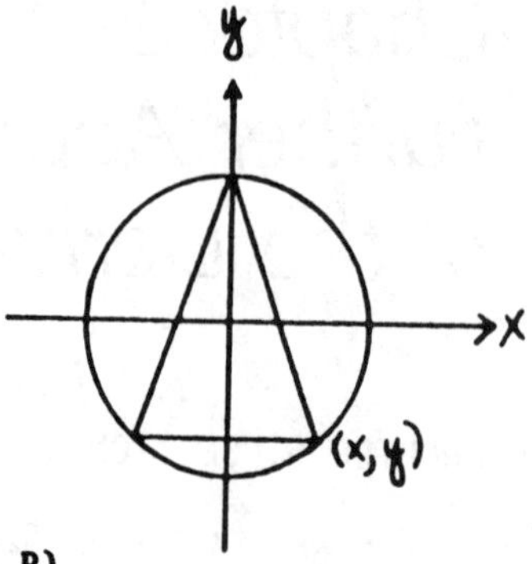

$$= \frac{2y^2 - Ry - R^2}{\sqrt{R^2 - y^2}} = \frac{(2y + R)(y - R)}{\sqrt{R^2 - y^2}}$$

$y = -\frac{R}{2}$, Base $= 2x = 2\sqrt{R^2 - y^2} = \sqrt{3}R$

Height $= R - y = \frac{3R}{2}$.

y

x

(x,y)

17. $\frac{9}{3} = \frac{h}{3 - r}$, $h = 9 - 3r$, $V = \pi r^2 h$

$$= 9\pi r^2 - 3\pi r^3$$

$V' = 18\pi r - 9\pi r^2 = 0$, $r = 2$

$h = 9 - 6 = 3$, $V = \pi\cdot 4\cdot 3 = 12\pi$.

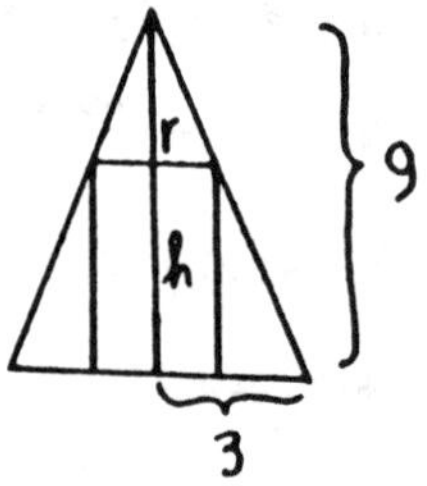

r

h

9

3

21. $V = \pi r^2 h = 2\pi$, $h = \frac{2}{r^2}$, $s = 2\pi rh + 2\pi r^2 = \frac{4\pi}{r} + 2\pi r^2$

$V' = -\frac{4\pi}{r^2} + 4\pi r = 0$, $r = 1$, $h = 2$.

25. $A = \frac{4b}{a} x\sqrt{a^2 - x^2}$, $0 \le x \le a$

$$A^2 = \frac{16b^2}{a^2} x^2(a^2 - x^2) = \frac{16b^2}{a^2}(a^2x^2 - x^4)$$

$$\frac{dA^2}{dx} = \frac{16b^2}{a^2}(2a^2x - 4x^3) = \frac{32b^2}{a^2} x(a^2 - 2x^2)$$

This is 0 if $x = 0$ or $x = \pm a/\sqrt{2}$. Since $x = -a/\sqrt{2}$ is meaningless and $x = 0$ gives $A = 0$, the maximum value of A occurs when $x = a/\sqrt{2}$.

29. $x = \dfrac{k(t + a)}{t^2 + b}$, $\dfrac{dx}{dt} = k\dfrac{t^2 + b - 2t(t + a)}{(t^2 + b)^2} = k\dfrac{b - 2at - t^2}{(t^2 + b)^2}$

$\dfrac{k(b - 2a - 1)}{(1 + b)^2} = 0$, $b - 2a - 1 = 0$, $b = 2a + 1$

When $t = 0$, $x = 20{,}000$; when $t = 1$, $x = 25{,}000$. This gives:

$\dfrac{ak}{b} = 20{,}000$ $\qquad\qquad$ $\dfrac{k(1 + a)}{1 + b} = 25{,}000$

$ak = 20{,}000b = 20{,}000(2a + 1)$ $\qquad$ $k(a + 1) = 25{,}000(b + 1)$

$= 25{,}000(2a + 2)$

$k = \dfrac{20{,}000(2a + 1)}{a} = \dfrac{25{,}000(2a + 2)}{a + 1}$

$\dfrac{8a + 4}{a} = \dfrac{10a + 10}{a + 1}$, $10a^2 + 10a = 8a^2 + 12a + 4$

$2a^2 - 2a - 4 = 0$, $a^2 - a - 2 = 0$, $(a - 2)(a + 1) = 0$

$a = 2$ or $a = -1$

$a \neq -1$, since then $\dfrac{k(1 + a)}{1 + b} \neq 25{,}000$. Therefore, $a = 2$, $b = 5$, $k = 50{,}000$; and the original formula is $x = 50{,}000\dfrac{t + 2}{t^2 + 5}$. When $t = 30$, $x = 50{,}000\dfrac{32}{905} = 1768$.

33. $t = \dfrac{d}{r}$, $t_1 = \dfrac{d_1}{r_1} = \dfrac{\sqrt{x^2 + 16}}{4}$

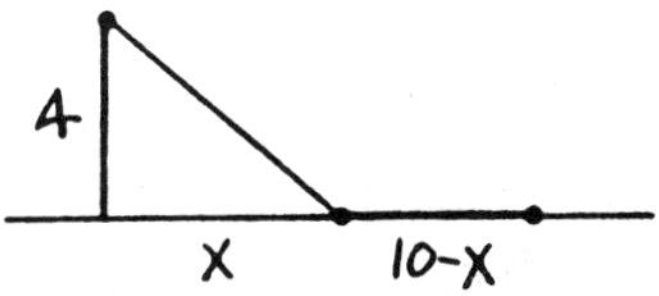

$t_2 = \dfrac{d_2}{r_2} = \dfrac{10 - x}{5}$

$t = t_1 + t_2 = \dfrac{\sqrt{x^2 + 16}}{4} + \dfrac{10 - x}{5}$

$\dfrac{dx}{dt} = \dfrac{x}{4\sqrt{x^2 + 16}} - \dfrac{1}{5}$, $5x = 4\sqrt{x^2 + 16}$

$25x^2 = 16x^2 + 256$, $x^2 = \dfrac{256}{9}$, $x = \dfrac{16}{3}$.

37. $I = \frac{E}{R + r}$, $P = RI^2 = \frac{RE^2}{(R + r)^2}$

$$\frac{dP}{dR} = \frac{(R + r)^2E^2 - RE^2\cdot 2(R + r)}{(R + r)^4} = \frac{E^2(r - R)}{(R + r)^3}, \quad R = r.$$

41. Number sold $= x = \frac{k}{p^2}$, Profit per widget $= p - 50$,

Total profit $= P = \frac{k}{p^2}(p - 50)$

$$\frac{dP}{dp} = k\,\frac{p^2 - (p - 50)2p}{p^4}, \quad k\,\frac{100 - p}{p^3} = 0, \quad p = \$1.00.$$

45. $x^2 = z^2 + (k - x)^2$, $z = \sqrt{2kx - k^2}$, $w = v - \sqrt{2kx - k^2}$,

$v^2 = w^2 + k^2 = v^2 - 2v\sqrt{2kx - k^2} + 2kx$, $kx = v\sqrt{2kx - k^2}$

$$v = \frac{kx}{\sqrt{2kx - k^2}}, \quad A = \frac{1}{2}xv = \frac{k}{2}\,\frac{x^2}{\sqrt{2kx - k^2}} \quad (\tfrac{k}{2} < x < k),$$

$$A^2 = \frac{k^2}{4}\,\frac{x^4}{2kx - k^2} = \frac{k}{4}\,\frac{x^4}{2x - k}$$

$$\frac{dA^2}{dx} = \frac{k}{4}\,\frac{2x^3(3x - 2k)}{(2x - k)^2}, \quad 3x - 2k = 0, \quad x = \frac{2k}{3}.$$

CHAPTER 5, SECTION 2 (pp. 209-212)

1. $V = xyz$

$$\frac{dV}{dt} = xy\,\frac{dz}{dt} + xz\,\frac{dy}{dt} + yz\,\frac{dx}{dt} = 10\cdot 10\cdot 2 + 10\cdot 10\cdot 7 + 10\cdot 10\cdot 5$$

$$= 1400 \text{ cm}^3/\text{min}.$$

$$s = 2xy + 2xz + 2yz$$

$$\frac{ds}{dt} = 2\left(x\,\frac{dy}{dt} + y\,\frac{dx}{dt} + x\,\frac{dz}{dt} + z\,\frac{dx}{dt} + y\,\frac{dz}{dt} + z\,\frac{dy}{dt}\right)$$

$$= 2(10\cdot 7 + 10\cdot 5 + 10\cdot 2 + 10\cdot 5 + 10\cdot 2 + 10\cdot 7)$$

$$= 560 \text{ cm}^2/\text{min}.$$

5. $x^2 = 10{,}000 + y^2$

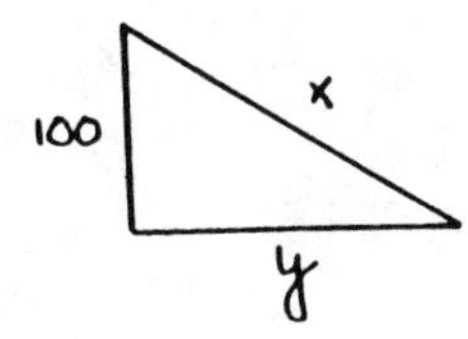

When $x = 260$,

$y = \sqrt{67{,}600 - 10{,}000} = \sqrt{57{,}600} = 240$

$2x\,\frac{dx}{dt} = 2y\,\frac{dy}{dt}$, $\frac{dy}{dt} = \frac{x}{y}\,\frac{dx}{dt} = \frac{260}{240}\cdot 5 = \frac{65}{12}$ ft/sec.

9. $\frac{dx}{dt} = -1$ ft/sec

$\frac{x}{y} = \frac{10 - x}{6}$

$6x = y(10 - x)$

$6\,\frac{dx}{dt} = -y\,\frac{dx}{dt} + (10 - x)\frac{dy}{dt}$

$6(-1) = -\frac{3}{2}(-1) + 8\,\frac{dy}{dt}$

$8\,\frac{dy}{dt} = -6 - \frac{3}{2} = -\frac{15}{2}$, $\frac{dy}{dt} = -\frac{15}{16}$ ft/sec.

13. $Pv = k$, $P\,\frac{dv}{dt} + v\,\frac{dP}{dt} = 0$

$1\,\frac{dv}{dt} + 1(0.1) = 0$, $\frac{dv}{dt} = -0.1$ ℓ/min.

17. $A = \frac{\sqrt{3}x^2}{4}$. When $A = 16$, $x = \frac{8}{\sqrt[4]{3}}$

$4 = \frac{\sqrt{3}}{2}\left(\frac{8}{\sqrt[4]{3}}\right)\frac{dx}{dt}$, $\frac{dx}{dt} = \frac{1}{\sqrt[4]{3}}$ cm/min.

21. $y = \sqrt{x}$, $y' = \frac{x^{-1/2}}{2}$, $\frac{dy'}{dt} = -\frac{1}{4x\sqrt{x}}\,\frac{dx}{dt}$

(a) $\frac{dy'}{dt} = -\frac{1}{4}\cdot 2 = -\frac{1}{2}$ unit/min.

(b) $\frac{dy'}{dt} = -\frac{1}{32}\cdot 2 = -\frac{1}{16}$ unit/min.

25. $V = \pi h^2(3 \cdot 4 - h)/3 = 4\pi h^2 - \dfrac{\pi h^3}{3}$

$\dfrac{dV}{dt} = (8\pi h - \pi h^2)\dfrac{dh}{dt}$, $3\pi = \pi(8 \cdot 3 - 3^2)\dfrac{dh}{dt}$,

$\dfrac{dh}{dt} = \dfrac{1}{5}$ m/min.

29. $E = 80x^2/2 = 40x^2$

$\dfrac{dE}{dt} = 80x\dfrac{dx}{dt} = 80 \cdot \dfrac{1}{4}(0.05) = 1$ N - m/sec

CHAPTER 5, SECTION 3 (pp. 218-220)

1. $dy = (3x^2 - 2x)dx.$

5. $dy = 8x(x^2 + 1)^3dx.$

9. $3x^2dx - 3x^2dy - 6xydx + 6xydy + 3y^2dx = 0$

$(x^2 - 2xy)dy = (x^2 - 2xy + y^2)dx$, $\quad dy = \dfrac{(x - y)^2}{x(x - 2y)}\,dx.$

13. $C = \pi D$, $\quad dC = \pi dD = \pm 0.03\pi$ cm.

$A = \dfrac{\pi D^2}{4}$, $\quad dA = \pi\,\dfrac{D}{2}\,dD = \dfrac{\pi(7.38)}{2}(.03) = \pm 0.1107\pi$ cm^2.

17. $R = 30\,\dfrac{50}{100 - 50} = 30$ ohms.

$dR = \dfrac{\ell r}{(\ell - x)^2}\,dx = \dfrac{100 \cdot 30}{(100 - 50)^2}(\pm 0.1) = \pm 0.12$ ohm.

21. $y = \sqrt[3]{x}$, $\quad x = 27$, $\quad y = 3$, $\quad dx = -2$,

$dy = \dfrac{dx}{3x^{2/3}} = \dfrac{-2}{3 \cdot 27^{2/3}} = -0.074$

$\sqrt[3]{25} \doteq y + dy = 2.926.$

25. $y = \frac{1}{\sqrt{x}}$, $x = 25$, $y = 0.2$, $dx = 2$

$$dy = -\frac{dx}{2x^{3/2}} = -\frac{2}{2\cdot 25^{3/2}} = -0.008$$

$$\frac{1}{\sqrt{27}} \doteq y + dy = 0.192.$$

29. $V = x^3$, $dV = 3x^2dx$

$$\frac{dV}{V} = \frac{3x^2dx}{x^3} = 3\,\frac{dx}{x}$$

$$0.03 = 3\,\frac{dx}{x},\quad \frac{dx}{x} = 0.01 = 1\%.$$

33. $kx = \frac{mv^2}{L + x}$, $kx^2 + kLx - mv^2 = 0$,

$480x^2 + 240x - 192 = 0$, $10x^2 + 5x - 4 = 0$,

$$x = \frac{-5 + \sqrt{25 + 160}}{20} = 0.43 \text{ m}.$$

$(2kx + kL)dx = 2mvdv$, $dx = 2 \text{ mm} = 0.02 \text{ m}$

$$dv = \frac{2kx + kL}{2mv}dx = \frac{2\cdot 480(0.93) + 480(0.5)}{2\cdot 3\cdot 8}(0.002)$$

$$= \frac{240}{48}[4(0.43) + 1](0.02) = (0.1)(2.72) \qquad = 0.047 \text{ m/sec}.$$

CHAPTER 5, SECTION 4 (pp. 224-226)

1. $C = 0.000\,008x^2 + 0.08x + 3200$, $Q = 0.000\,008x + 0.08 + \frac{3200}{x}$

$$Q' = 0.000\,008 - \frac{3200}{x^2} = 0,\quad \frac{3200}{x^2} = 0.000\,008$$

$$x^2 = \frac{3200}{0.000\,008} = 400{,}000{,}000$$

$x = 20{,}000.$

5. $P = px - C = 60x - 0.01x^2 - (0.004x^2 + 25x + 4000)$

$= -0.014x^2 + 35x - 4000$

$\frac{dP}{dx} = -0.028x + 35 = 0, \quad P'' = -0.028 < 0$

$0.028x = 35, \quad x = \frac{35}{0.028} = 1250$

$p = 60 - 0.01(1250) = \$47.50.$

9. $P = px - C = 70x - 0.1x^2 - 30x - 1000$

$= -0.1x^2 + 40x - 1000$

$P' = -0.2x + 40 = 0, \quad P'' = -0.2 < 0$

$0.2x = 40, \quad x = 200$

$p = 70 - 0.1(200) = \$50$

$P = -0.1(40{,}000) + 40(200) - 1000 = \$3000.$

13. $C = 100 + 12x, \quad R = px = 40x - 0.1x^2,$

$P = R - C = 28x - 0.1x^2 - 100$

$P' = 28 - 0.2x = 0, \quad x = 140$

$P = \$1{,}860.$

17. $P = R - C, \quad P' = R' - C' = 0, \quad R' = C'.$

CHAPTER 5, SECTION 5 (pp. 229-230)

1. $f(x) = 3x^4 - 2x^3 + 3x + C.$

5. $f(x) = \frac{3}{2}x^{2/3} + C, \quad f(1) = 2 = \frac{3}{2} + C.$

$C = \frac{1}{2}, \quad f(x) = \frac{3}{2}x^{2/3} + \frac{1}{2}.$

9. $f(x) = -\frac{1}{x} + C.$

13. $f(x) = \frac{1}{3}(x - 1)^3 + C.$

17. $f'(x) = 12x + C_1, \quad 12 + C_1 = 3, \quad C_1 = -9,$

$f'(x) = 12x - 9, \quad f(x) = 6x^2 - 9x + C_2,$

$-3 + C_2 = 4, \quad C_2 = 7, \quad f(x) = 6x^2 - 9x + 7.$

21. $f''(x) = 6x + C_1, \quad 6 = 6 + C_1,$

$C_1 = 0, \quad f''(x) = 6x$

$f'(x) = 3x^2 + C_2, \quad 4 = 3 + C_2,$

$C_2 = 1, \quad f'(x) = 3x^2 + 1,$

$f(x) = x^3 + x + C_3, \quad 0 = 1 + 1 + C_3,$

$C_3 = -2, \quad f(x) = x^3 + x - 2.$

25. $f'(x) = \frac{1}{x}, \quad f(1) = 0, \quad f'(1) = 1$

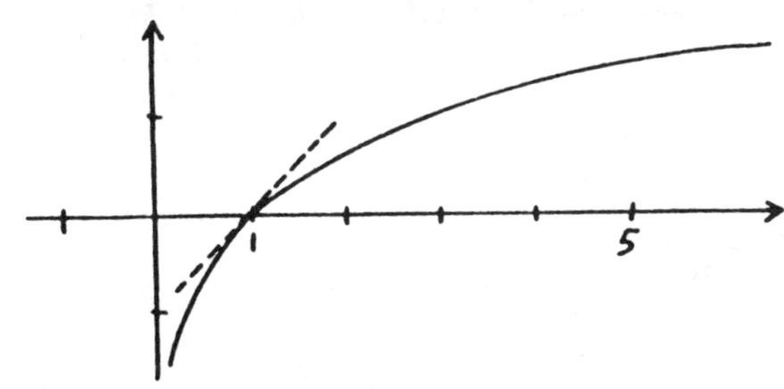

The slope of the graph is positive for $x > 0$. Thus the graph is increasing. For small values of x, $f'(x)$ is large; for large values of x, $f'(x)$ is small. Thus the graph is increasing rapidly when x is near 0, has slope 1 at (1, 0), and increases more and more slowly as x increases. (The discussion above does not dictate a vertical asymptote at $x = 0$, but the tangent must at least approach the vertical.)

CHAPTER 5, SECTION 6 (pp. 235-237)

1. $a = -9.8$, $v = -9.8t + C_1$, $30 = C_1$, $v = -9.8t + 30$, $x = -4.9t^2 + 30t + C_2$, $0 = C_2$, $x = -4.9t^2 + 30t$.

5. $v = 4t + 1$, $x = 2t^2 + t + C$, $C = 2$,

   $x = 2t^2 + t + 2$.

9. $v = at$, $15 = 6a$, $a = 2.5$ m/s^2

   $v = (2.5)(12) = 30$ m/s.

13. $x = -4.9^2 + 24.5t$

    $v = -9.8t + 24.5 = 0$, $t = 2.5$s

    $x(2.5) = -4.9(2.5^2) + 24.5(2.5) = 30.625$m.

    $-4.9t^2 + 24.5t = 20$

    $49t^2 - 245t + 200 = 0$

    $t = \dfrac{245 \pm \sqrt{245^2 - 49\cdot800}}{98} = 1.027$s or 3.973s.

17. $a = 1.8$, $v = 1.8t_1 = 12$, $t_1 = \frac{20}{3}$s.

    $x = 0.9t_1{}^2 = \frac{9}{10}(\frac{20}{3})^2 = 40$m

    $v = -3t_2 + 12 = 0$, $t_2 = 4$

    $x = -\frac{3}{2}t_2{}^2 + 12t_2 = -24 + 48 = 24$m

    $500 - 40 - 24 = 12t_3$, $t_3 = \frac{109}{3}$s.

    $t = t_1 + t_2 + t_3 = \frac{20}{3} + 4 + \frac{109}{3} = 47$s.

21. $v_1 = -at + 40$, $v_2 = 10$, $x_2(0) = 300$,

$x_1(0) = 0$

$x_1 = -\frac{1}{2}at^2 + 40t$,

$x_2 = 300 + 10t - \frac{1}{2}at^2 + 40t = 300 + 10t$,

$at^2 - 60t + 600 = 0$

$t = \frac{60 \pm \sqrt{3600 - 2400a}}{2}$.

No positive root t, hence no collision, if $3600 - 2400a < 0$,

$a > \frac{3}{2}$ m/s^2

CHAPTER 5, REVIEW (pp. 237-241)

1. $f(x) = \frac{x^3}{3} - \frac{1}{x} + C$, $1 = \frac{1}{3} - 1 + C$

$C = \frac{5}{3}$, $f(x) = \frac{x^3}{3} - \frac{1}{x} + \frac{5}{3} = \frac{x^4 + 5x - 3}{3x}$.

5. $V = \frac{4}{3}\pi r^3$, $\frac{dV}{dt} = 4\pi r^2 \frac{dr}{dt}$

$5 = 4\pi \cdot 9 \frac{dr}{dt}$, $\frac{dr}{dt} = \frac{5}{36\pi}$ m/min.

9. $y - 5 = m(x - 4)$

When $y = 0$, $x = 4 - \frac{5}{m} = \frac{4m - 5}{m}$.

When $x = 0$, $y = 5 - 4m$.

$A = \frac{1}{2}\frac{4m - 5}{m}(5 - 4m) = \frac{-16m^2 + 40m - 25}{2m} = -8m + 20 - \frac{25}{2m}$

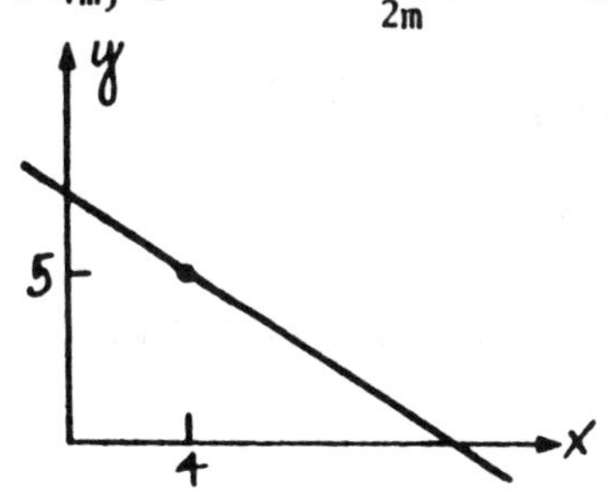

$$A' = -8 + \frac{25}{2m^2} = 0, \quad m^2 = \frac{25}{16}, \quad m = \pm\frac{5}{4}$$

(From the figure, $m = -\frac{5}{4}$.)

$$A = \frac{-16(5/4)^2 - 40(5/4) - 25}{2(-5/4)} = 40.$$

13. $P = x + 2y + \frac{\pi x}{2}, \quad y = \frac{2P - 2x - \pi x}{4}$

$$A = xy + \frac{\pi x^2}{8} = \frac{2Px - 2x^2 - \pi x^2}{4} + \frac{\pi x^2}{8} = \frac{4Px - 4x^2 - \pi x^2}{8}$$

$$A' = \frac{4P - 8x - 2\pi x}{8}$$

$$2P - 4x - \pi x = 0, \quad x = \frac{2P}{4 + \pi}, \quad y = \frac{2P - \frac{(2 + \pi)2P}{4 + \pi}}{4} = \frac{P}{4 + \pi}$$

$$y + \frac{x}{2} = \frac{2P}{4 + \pi}.$$

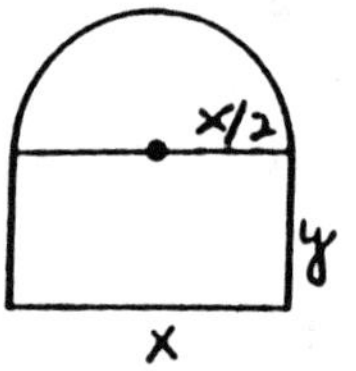

x/2

y

x

17. **If y is the distance in miles, n is the number of steps and ℓ is the length of a stride, then**

$$y = \frac{n\ell}{12 \cdot 5280}$$

When $\ell = 28$, $y = 10.3$.

$$10.3 = \frac{28n}{12 \cdot 5280}, \quad n = 23{,}300$$

$$y = \frac{23{,}300\ell}{12 \cdot 5280} = 0.368\ \ell$$

$$dy = 0.368\ d\ell = 0.368(2) = 0.7 \text{ mi.}$$

21. $E_K = [(1 - \frac{v^2}{c^2})^{-1/2} - 1]E_0$

$$E_K' = -\frac{E_o}{2}(1 - \frac{v^2}{c^2})^{-3/2}(\frac{-2vv'}{c^2})$$

$$= E_o(1 - \frac{(0.8)^2c^2}{c^2})^{-3/2}\frac{(0.8c)(0.01c)}{c^2}$$

$$= \frac{0.008E_o}{(0.36)^{3/2}} = \frac{E_o}{27}.$$

25. $\frac{a}{x} = \frac{y}{b}, \quad y = \frac{ab}{x}$

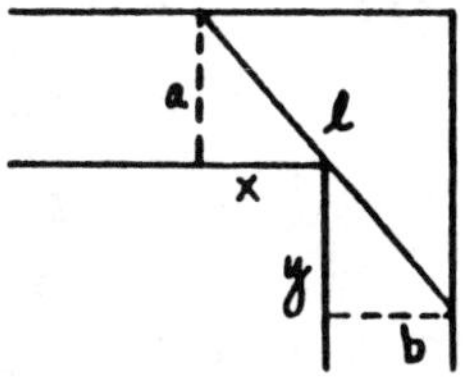

$$\ell = \sqrt{x^2 + a^2} + \sqrt{y^2 + b^2}$$

$$= \sqrt{x^2 + a^2} + \sqrt{\frac{a^2b^2}{x^2} + b^2}$$

$$= \sqrt{x^2 + a^2} + \frac{b}{x}\sqrt{a^2 + x^2} = \frac{(x + b)\sqrt{x^2 + a^2}}{x}$$

$$\ell' = \frac{x[(x + b)\frac{x}{\sqrt{x^2 + a^2}} + \sqrt{x^2 + a^2}] - (x + b)\sqrt{x^2 + a^2}}{x^2}$$

$$= \frac{x^2(x + b) + x(x^2 + a^2) - (x + b)(x^2 + a^2)}{x^2\sqrt{x^2 + a^2}} = \frac{x^3 - a^2b}{x^2\sqrt{x^2 + a^2}}$$

$$x^3 = a^2b, \quad x = a^{2/3}b^{1/3}$$

$$\ell = \frac{(a^{2/3}b^{1/3} + b)\sqrt{a^{4/3}b^{2/3} + a^2}}{a^{2/3}b^{1/3}}$$

$$= \frac{b^{1/3}(a^{2/3} + b^{2/3})a^{2/3}\sqrt{b^{2/3} + a^{2/3}}}{a^{2/3}b^{1/3}} = (a^{2/3} + b^{2/3})^{3/2}.$$

Chapter 6
The Integral

CHAPTER 6, SECTION 1 (pp. 248-250)

1. $\sum_{i=1}^{7} i^3 = 1^3 + 2^3 + 3^3 + 4^3 + 5^3 + 6^3 + 7^3$

$= 1 + 8 + 27 + 64 + 125 + 216 + 343$

5. $\sum_{i=0}^{7} (2i + 1) = (2\cdot 0 + 1) + (2\cdot 1 + 1) + (2\cdot 2 + 1) + (2\cdot 3 + 1)$

$+ (2\cdot 4 + 1) + (2\cdot 5 + 1) + (2\cdot 6 + 1) + (2\cdot 7 + 1)$

$= 1 + 3 + 5 + 7 + 9 + 11 + 13 + 15$

9. $\sum_{i=1}^{n} (2i + 1) = (2\cdot 1 + 1) + (2\cdot 2 + 1) + (2\cdot 3 + 1) + \cdots$

$+ (2\cdot n + 1)$

$= 3 + 5 + 7 + \cdots + (2n + 1)$

13. $2 + 4 + 6 + \cdots + 22 = \sum_{i=1}^{11} 2i$

10

17. $1 + 3 + 5 + \cdots + (2n - 1) = \sum_{i=1}^{n} (2i - 1)$

21. $\sum_{i=1}^{n} (2i - 1) = 1 + 3 + 5 + \cdots + (2n - 1)$

$\sum_{i=0}^{n-1} (2i + 1) = 1 + 3 + 5 + \cdots + (2n - 1)$

Equal.

25. $\sum_{i=0}^{n} (i^2 + 1) = 1 + 2 + 5 + \cdots + (n^2 + 1)$

$$\sum_{i=1}^{n+1} [(i - 1)^2 + 1] = 1 + 2 + 5 + \cdots + (n^2 + 1)$$

Equal.

29. $$\sum_{i=1}^{n} \left(\frac{i}{n}\right)^2 \frac{1}{n} = \frac{1}{n^3} \sum_{i=1}^{n} i^2 = \frac{1}{n^3} \frac{n(n + 1)(2n + 1)}{6}$$
$$= \frac{(n + 1)(2n + 1)}{6n^2}$$

33. $$\sum_{i=1}^{n} \left(\frac{i - 1}{n}\right)^2 \frac{1}{n} = \frac{1}{n^3} \sum_{i=1}^{n} (i - 1)^2$$
$$= \frac{1}{n^3} \sum_{i=0}^{n-1} i^2 = \frac{1}{n^3} \sum_{i=1}^{n-1} i^2$$
$$= \frac{1}{n^3} \frac{(n - 1)n(2n - 1)}{6} = \frac{(n - 1)(2n - 1)}{6n^2}$$

37. $$\sum_{i=1}^{n} 1 = \underbrace{1 + 1 + 1 + \cdots + 1}_{n \text{ terms}} = n$$

41.
$$\begin{array}{ccccccccccccc} & 1 & + & 2 & + & 3 & + \cdots + & (n-2) & + & (n-1) & + & n \\ & n & + & (n-1) & + & (n-2) & + \cdots + & 3 & + & 2 & + & 1 \\ \hline & (n+1) & + & (n+1) & + & (n+1) & + \cdots + & (n+1) & + & (n+1) & + & (n+1) \end{array}$$

Hence, $2 \sum_{i=1}^{n} i = n(n + 1)$, $\sum_{i=1}^{n} i = \frac{n(n + 1)}{2}$.

CHAPTER 6, SECTION 2 (pp. 258-259)

1. $A(R_1) = A(R_2)$

$ab = A(R_1 \cup R_2) = A(R_1) + A(R_2) = 2A(R_1)$

$A(R_1) = \frac{1}{2} ab$

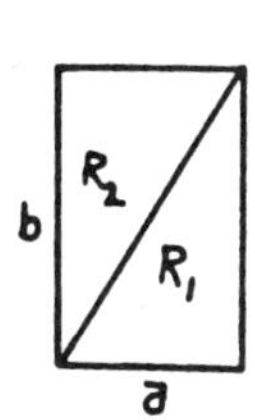

5. $x_0 = 0$ $x_1^* = \frac{1}{n}$

$x_1 = \frac{1}{n}$ $x_2^* = \frac{2}{n}$

$x_2 = \frac{2}{n}$

$\vdots$ $\vdots$

$x_n = \frac{n}{n}$ $x_n^* = \frac{n}{n}$

9. $x_0 = 1$ $x_1^* = 1 + \frac{1}{n}$

$x_1 = 1 + \frac{1}{n}$ $x_2^* = 1 + \frac{2}{n}$

$x_2 = 1 + \frac{2}{n}$

$\vdots$ $\vdots$

$x_n = 1 + \frac{n}{n}$ $x_n^* = 1 + \frac{n}{n}$

$$A = \lim_{n\to+\infty} \sum_{i=1}^{n} \frac{1}{n} f(1 + \frac{i}{n}) = \lim_{n\to+\infty} \sum_{i=1}^{n} \frac{1}{n}(1 + \frac{i}{n})^2$$

$$= \lim_{n\to+\infty} \sum_{i=1}^{n} (\frac{1}{n} + \frac{2i}{n^2} + \frac{i^2}{n^3})$$

$$= \lim_{n\to+\infty} [\frac{1}{n}\cdot n + \frac{2}{n^2}\frac{n(n+1)}{2} + \frac{1}{n^3}\frac{n(n+1)(2n+1)}{6}]$$

$$= \lim_{n\to+\infty} [1 + (1 + \frac{1}{n}) + \frac{(1 + 1/n)(2 + 1/n)}{6} = 1 + 1 + \frac{1}{3} = \frac{7}{3}.$$

13. $x_0 = 0$ $x_1^* = \frac{1}{n}$

$x_1 = \frac{1}{n}$ $x_2^* = \frac{2}{n}$

$x_2 = \frac{2}{n}$

$\vdots$ $\vdots$

$x_n = \frac{n}{n}$ $x_n^* = \frac{n}{n}$

$$A = \lim_{n\to+\infty} \sum_{i=1}^{n} \frac{1}{n} f(\frac{i}{n}) = \lim_{n\to+\infty} \sum_{i=1}^{n} \frac{1}{n}\frac{3i^2}{n^2}$$

$$= \lim_{n\to+\infty} \frac{3}{n^3} \frac{n(n + 1)(2n + 1)}{6} = \lim_{n\to+\infty} \frac{(1 + 1/n)(2 + 1/n)}{2} = 1.$$

17. $x_0 = 0$

$$x_1 = \frac{1}{n} \qquad x_1^* = \frac{1}{n}$$

$$x_2 = \frac{2}{n} \qquad x_2^* = \frac{2}{n}$$

$$\vdots \qquad\qquad \vdots$$

$$x_n = \frac{n}{n} \qquad x_n^* = \frac{n}{n}$$

$$A = \lim_{n\to+\infty} \sum_{i=1}^{n} \frac{1}{n} f(\frac{i}{n}) = \lim_{n\to+\infty} \sum_{i=1}^{n} \frac{1}{n}(\frac{i}{n} - \frac{i^2}{n^2})$$

$$= \lim_{n\to+\infty} [\frac{1}{n^2} \frac{n(n + 1)}{2} - \frac{1}{n^3} \frac{n(n + 1)(2n + 1)}{6}]$$

$$= \lim_{n\to+\infty} [\frac{1 + 1/n}{2} - \frac{(1 + 1/n)(2 + 1/n)}{6}] = \frac{1}{6}.$$

21. (a) $0 < 1 - \frac{i - 1}{n} = \frac{n + i - 1}{n}$, $i = 1, 2, \ldots, n$,

$$(\frac{n + i - 1}{n})^2 < \frac{(n + i - 1)(n + i)}{n^2},$$

$$1 + \frac{i - 1}{n} < \frac{1}{n} \sqrt{(n + i - 1)(n + i)} < \frac{1}{n} \sqrt{(n + i)^2}$$

$$= \frac{n + i}{n} = 1 + \frac{i}{n}.$$

(b) $x_0 = 1$

$$x_1 = 1 + \frac{1}{n} \qquad x_1^* = 1 \qquad x_1^* = 1 + \frac{1}{n}$$

$$x_2 = 1 + \frac{2}{n} \qquad x_2^* = 1 + \frac{1}{n} \qquad x_2^* = 1 + \frac{2}{n}$$

$$\vdots \qquad\qquad \vdots \qquad\qquad \vdots$$

$$x_n = 1 + \frac{n}{n} \qquad x_n^* = 1 + \frac{n - 1}{n} \qquad x_n^* = 1 + \frac{n}{n}$$

$$A = \lim_{n\to\infty} \sum_{i=1}^{n} \frac{1}{n} f(1 + \frac{i-1}{n}) = \lim_{n\to\infty} \sum_{i=1}^{n} \frac{1}{n}(1 + \frac{i-1}{n})^{-2}$$

$$\geq \lim_{n\to\infty} \sum_{i=1}^{n} \frac{1}{n}\left(\frac{\sqrt{(n+i-1)(n+i)}}{n}\right)^{-2}$$

$$= \lim_{n\to\infty} \sum_{i=1}^{n} \frac{1}{n} \frac{n^2}{(n+i-1)(n+i)}$$

$$= \lim_{n\to\infty} \sum_{i=1}^{n} n(\frac{1}{n+i+1} - \frac{1}{n+i})$$

$$= \lim_{n\to\infty} n \sum_{i=1}^{n} [(\frac{1}{n} - \frac{1}{n+1}) + (\frac{1}{n+1} - \frac{1}{n+2}) + \cdots$$

$$+ (\frac{1}{n+n-1} - \frac{1}{n+n})]$$

$$= \lim_{n\to\infty} n(\frac{1}{n} - \frac{1}{2n}) = 1 - \frac{1}{2} = \frac{1}{2}.$$

Using the right-hand endpoints,

$$A = \lim_{n\to\infty} \sum_{i=1}^{n} \frac{1}{n} f(1 + \frac{i}{n}) = \lim_{n\to\infty} \sum_{i=1}^{n} \frac{1}{n}(1 + \frac{i}{n})^{-2}$$

$$\leq \lim_{n\to\infty} \frac{1}{n} \sum_{i=1}^{n} \frac{n^2}{(n+i)(n+i-1)} = \frac{1}{2}.$$

Thus $\frac{1}{2} \leq A \leq \frac{1}{2}$, so $A = \frac{1}{2}$.

CHAPTER 6, SECTION 3 (pp. 266-267)

1. $x_0 = 0$

$x_1 = \frac{3}{n}$ $\qquad x_1^* = \frac{3}{n}$

$x_2 = \frac{6}{n}$ $\qquad x_2^* = \frac{6}{n}$

$\vdots \qquad \vdots$

$x_n = \frac{3n}{n}$ $\qquad x_n^* = \frac{3n}{n}$

$$\int_0^3 (x^2 - 1)dx = \lim_{n\to+\infty} \sum_{i=1}^{n} \frac{3}{n} f(\frac{3i}{n}) = \lim_{n\to+\infty} \sum_{i=1}^{n} \frac{3}{n} (\frac{9i^2}{n^2} - 1)$$

$$= \lim_{n\to+\infty} [\frac{27}{n^3} \frac{n(n + 1)(2n + 1)}{6} - \frac{3}{n} \cdot n]$$

$$= \lim_{n\to+\infty} [\frac{9(1 + 1/n)(2 + 1/n)}{2} - 3] = 6.$$

5. $x_0 = 0$

$x_1 = \frac{1}{n}$ $\quad x_1^* = \frac{1}{n}$

$x_2 = \frac{2}{n}$ $\quad x_2^* = \frac{2}{n}$

$\vdots$ $\quad \vdots$

$x_n = \frac{n}{n}$ $\quad x_n^* = \frac{n}{n}$

$$\int_0^1 x^3 dx = \lim_{n\to+\infty} \sum_{i=1}^{n} \frac{1}{n} \frac{i^3}{n^3} = \lim_{n\to+\infty} \frac{1}{n^4} \frac{n^2(n + 1)^2}{4}$$

$$= \lim_{n\to+\infty} \frac{(1 + 1/n)^2}{4} = \frac{1}{4}.$$

9. $x_0 = 0$

$x_1 = \frac{1}{n}$ $\quad x_1^* = \frac{1}{n}$

$x_2 = \frac{2}{n}$ $\quad x_2^* = \frac{2}{n}$

$\vdots$ $\quad \vdots$

$x_n = \frac{n}{n}$ $\quad x_n^* = \frac{n}{n}$

$$\int_0^1 (x^2 - x)dx = \lim_{n\to+\infty} \sum_{i=1}^{n} \frac{1}{n}(\frac{i^2}{n^2} - \frac{i}{n})$$

$$= \lim_{n\to+\infty} [\frac{1}{n^3} \frac{n(n + 1)(2n + 1)}{6} - \frac{1}{n^2} \frac{n(n + 1)}{2}]$$

$$= \lim_{n\to+\infty} [\frac{(1 + 1/n)(2 + 1/n)}{6} - \frac{1 + 1/n}{2}] = -\frac{1}{6}.$$

13. $x_0 = 0$ $\quad x_1 = \frac{1}{n}$ $\quad x_2 = \frac{2}{n}$ $\quad \vdots \quad x_n = \frac{n}{n}$

$x_1^* = \frac{1}{n}$ $\quad x_2^* = \frac{2}{n}$ $\quad \vdots \quad x_n^* = \frac{n}{n}$

$x_0 = 1$ $\quad x_1 = 1 + \frac{1}{n}$ $\quad x_2 = 1 + \frac{2}{n}$ $\quad \vdots \quad x_n = 1 + \frac{n}{n}$

$x_1^* = 1 + \frac{1}{n}$ $\quad x_2^* = 1 + \frac{2}{n}$ $\quad \vdots \quad x_n^* = 1 + \frac{n}{n}$

$$A = \int_0^2 |x^2 - x|dx = \int_0^1 (x - x^2)dx + \int_1^2 (x^2 - x)dx$$

$$\int_0^1 (x - x^2)dx = \lim_{n\to+\infty} \sum_{i=1}^{n} \frac{1}{n}\left(\frac{i}{n} - \frac{i^2}{n^2}\right)$$

$$= \lim_{n\to+\infty} \left[\frac{1}{n^2}\frac{n(n + 1)}{2} - \frac{1}{n^3}\frac{n(n + 1)(2n + 1)}{6}\right]$$

$$= \lim_{n\to+\infty} \left[\frac{(1 + 1/n)}{2} - \frac{(1 + 1/n)(2 + 1/n)}{6}\right] = \frac{1}{6}$$

$$\int_1^2 (x^2 - x)dx = \lim_{n\to+\infty} \sum_{i=1}^{n} \frac{1}{n}\left[\left(1 + \frac{i}{n}\right)^2 - \left(1 + \frac{i}{n}\right)\right]$$

$$= \lim_{n\to+\infty} \sum_{i=1}^{n} \frac{1}{n}\left(\frac{i}{n} + \frac{i^2}{n^2}\right)$$

$$= \lim_{n\to+\infty} \left[\frac{1}{n^2}\frac{n(n + 1)}{2} + \frac{1}{n^3}\frac{n(n + 1)(2n + 1)}{6}\right]$$

$$= \lim_{n\to+\infty} \left[\frac{(1 + 1/n)}{2} + \frac{(1 + 1/n)(2 + 1/n)}{6}\right] = \frac{5}{6}$$

$A = \frac{1}{6} + \frac{5}{6} = 1.$

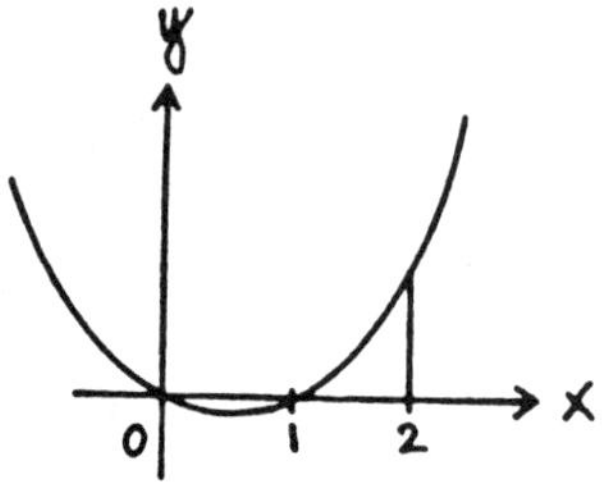

17. $A = \int_0^1 (x^2 - x^3)dx = \frac{1}{12}$ (from Problem 16).

21. The graph of an odd function is symmetric about the origin. Thus,

$$\int_{-a}^{0} f(x)dx = -\int_{0}^{a} f(x)\ dx$$

or

$$\int_{-a}^{a} f(x)dx = \int_{-a}^{0} f(x)dx + \int_{0}^{a} f(x)dx = 0.$$

The graph of an even function is symmetric about the y axis. Thus,

$$\int_{-a}^{0} f(x)dx = \int_{0}^{a} f(x)dx$$

or

$$\int_{-a}^{a} f(x)dx = \int_{-a}^{0} f(x)dx + \int_{0}^{a} f(x)dx = 2\int_{0}^{a} f(x)dx.$$

25. Let $S_1 = \{a = x_0 < x_1 < \cdots < x_m = c\}$ be a subdivision of $[a, c]$, $S_2 = \{c = x_m < x_{m+1} < \cdots < x_n = b\}$ be a subdivision of $[c, b]$, and $x_{i-1} \le x_i^* \le x_i$ $(i = 1,2,\ldots,n)$. Then $S = \{x_0, x_1, \ldots, x_m, \ldots, x_n\}$ is a subdivision of $[a, b]$, and

$$\begin{aligned}
\int_{a}^{b} f(x)dx &= \lim_{\|S\|\to 0} \sum_{i=1}^{n} f(x_i^*)\Delta x_i \\
&= \lim_{\|S\|\to 0} \left[\sum_{i=1}^{m} f(x_i^*)\Delta x_i + \sum_{i=m+1}^{n} f(x_i^*)\Delta x_i\right] \\
&= \lim_{\|S_1\|\to 0} \sum_{i=1}^{m} f(x_i^*)\Delta x_i + \lim_{\|S_2\|\to 0} \sum_{i=m+1}^{n} f(x_i^*)\Delta x_i \\
&= \int_{a}^{c} f(x)dx + \int_{c}^{b} f(x)dx.
\end{aligned}$$

CHAPTER 6, SECTION 4 (pp. 271-272)

1. $\int_0^2 x^2dx = \frac{x^3}{3}\Big|_0^2 = \frac{8}{3}.$

5. $\int_1^2 (3x^2 + 2x)dx = (x^3 + x^2)\Big|_1^2 = 8 + 4 - 2 = 10.$

9. $\int_1^2 x^2(x - 1)dx = \int_1^2 (x^3 - x^2)dx = (\frac{x^4}{4} - \frac{x^3}{3})\Big|_1^2$

$= (4 - \frac{8}{3}) - (\frac{1}{4} - \frac{1}{3}) = \frac{17}{12}.$

13. $\int_1^8 \sqrt[3]{x}\, dx = \frac{3x^{4/3}}{4}\Big|_1^8 = \frac{3}{4}(16 - 1) = \frac{45}{4}.$

17. $\int_1^2 \frac{1}{x^2} dx = -\frac{1}{x}\Big|_1^2 = -\frac{1}{2} + 1 = \frac{1}{2}.$

21. $A = \int_0^1 (x^2 + 1)dx$

$= (\frac{x^3}{3} + x)\Big|_0^1 = \frac{4}{3}.$

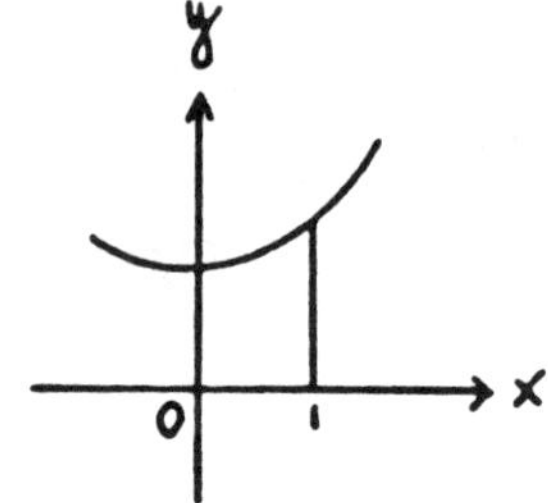

25. $A = \int_0^1 (x - 1)^2 dx$

$= \int_0^1 (x^2 - 2x + 1)dx$

$= (\frac{x^3}{3} - x^2 + x)\Big|_0^1 = \frac{1}{3} - 1 + 1 = \frac{1}{3}.$

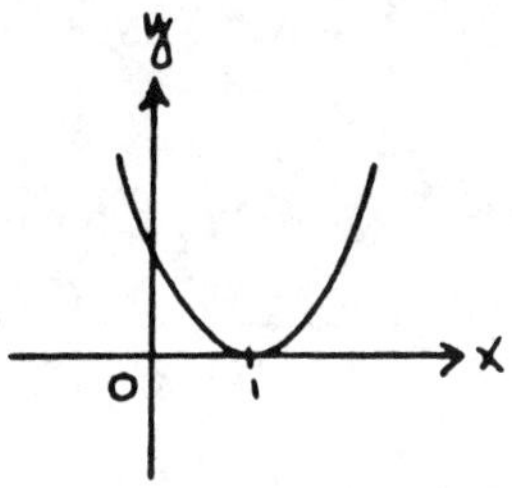

29. $A = \int_0^4 \sqrt{x}\, dx = \frac{2}{3} x^{3/2}\Big|_0^4$

$= \frac{2}{3} \cdot 8 = \frac{16}{3}.$

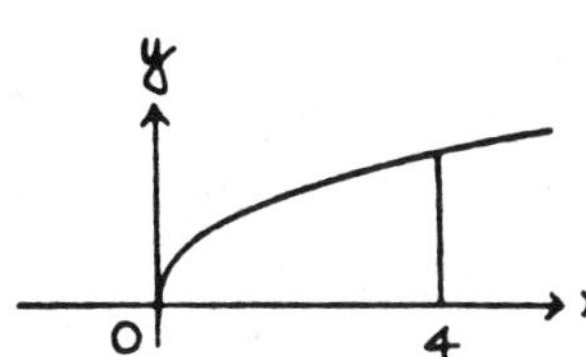

33. The mean-value theorem for derivatives:

$f(b) - f(a) = f'(x_0)(b - a)$. If f' is continuous on $[a, b]$, then

$$\int_a^b f'(x)dx = f(b) - f(a) = f'(x_0)(b - a) \text{ for some } x_0,$$

$a < x_0 < b$.

CHAPTER 6, SECTION 5 (pp. 278-279)

1. $\int 2(2x - 3)^2 dx = \frac{1}{3}(2x - 3)^3 + C.$

5. $\int \sqrt{4x - 3}\, dx = \frac{1}{4} \int 4\sqrt{4x - 3}\, dx = \frac{1}{4} \frac{(4x - 3)^{3/2}}{3/2} + C$

$= \frac{1}{6}(4x - 3)^{3/2} + C.$

9. $\int (x^2 + 4)^2 x^3 dx = \int (x^7 + 8x^5 + 16x^3)dx = \frac{x^8}{8} + \frac{4x^6}{3} + 4x^4 + C$

13. $\int \frac{(x^{1/3} - 1)^5}{x^{2/3}} dx = 3 \int \frac{1}{3} x^{-2/3}(x^{1/3} - 1)^5 dx$

$= \frac{1}{2}(x^{1/3} - 1)^6 + C.$

17. $\int x\sqrt[3]{3x^2 - 5}\,dx = \frac{1}{6}\int 6x\sqrt[3]{3x^2 - 5}\,dx = \frac{1}{6}\cdot\frac{3}{4}(3x^2 - 5)^{4/3} + C$

$= \frac{1}{8}(3x^2 - 5)^{4/3} + C.$

21. $\int \frac{x^3 + 8x^2 + 20x + 20}{(x + 2)^2}\,dx = \int (x + 4 + \frac{4}{(x + 2)^2})dx$

$= \frac{x^2}{2} + 4x - \frac{4}{x + 2} + C.$

25. $\int_1^2 (x^3 - x)^2(3x^2 - 1)dx = \frac{1}{3}(x^3 - x)^3\Big|_1^2 = \frac{1}{3}(6^3 - 0) = 72.$

29. $\int_1^2 (x^2 - 3)^3x\,dx = \frac{1}{2}\int_1^2 (x^2 - 3)^3 2x\,dx = \frac{1}{8}(x^2 - 3)^4\Big|_1^2$

$= \frac{1}{8}(1 - 16) = -\frac{15}{8}.$

33. $\int \frac{x^3 - 3x^2 + 3x}{(x - 1)^2}\,dx = \int (x - 1 + \frac{1}{(x - 1)^2})dx$

$= \frac{1}{2}x^2 - x - \frac{1}{x - 1} + C.$

CHAPTER 6, SECTION 6 (pp. 283-284)

1. $A = \int_{-2}^{2} (y_1 - y_2)dx$

$= \int_{-2}^{2} (2x - x^2 - 2x + 4)dx$

$= -\frac{x^3}{3} + 4x\Big|_{-2}^{2}$

$= (-\frac{8}{3} + 8) - (\frac{8}{3} - 8) = \frac{32}{3}.$

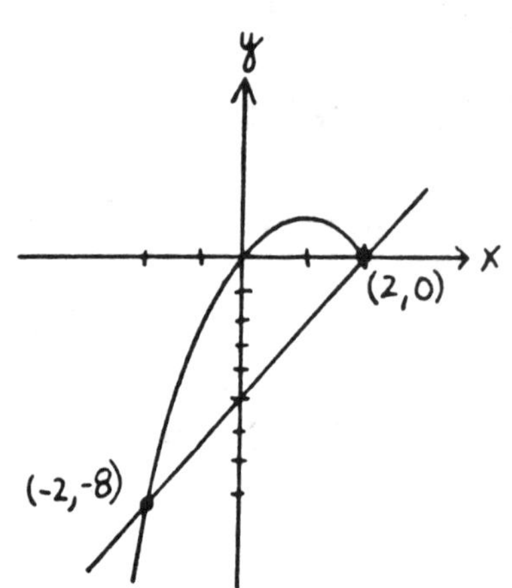

5. $A = \int_2^4 -x\,dy + \int_4^5 x\,dy$

$= \int_2^4 (4y - y^2)dy + \int_4^5 (y^2 - 4y)dy$

$= (2y^2 - \frac{y^3}{3})\Big|_2^4 + (\frac{y^3}{3} - 2y^2)\Big|_4^5$

$= (32 - \frac{64}{3}) - (8 - \frac{8}{3}) + (\frac{125}{3} - 50) - (\frac{64}{3} - 32) = \frac{23}{3}.$

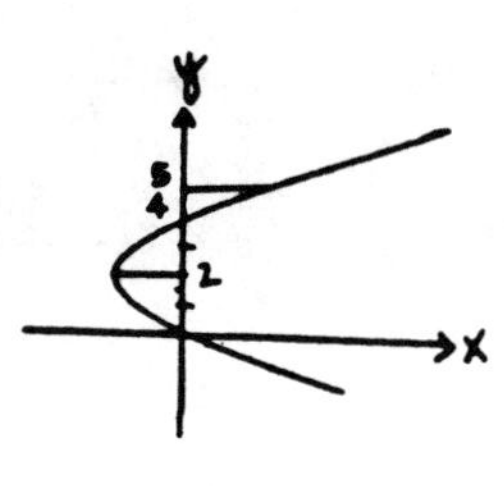

9. $A = \int_0^3 (y_1 - y_2)dx$

$= \int_0^3 (2x - x^2 + x)dx$

$= \int_0^3 (3x - x^2)dx = (\frac{3x^2}{2} - \frac{x^3}{3})\Big|_0^3$

$= \frac{27}{2} - 9 = \frac{9}{2}.$

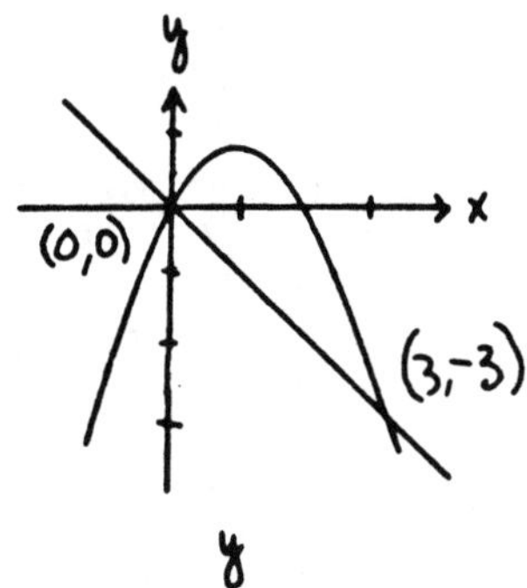

13. $y = \frac{4}{x^2}, \quad y = 7 - 3x$

$\frac{4}{x^2} = 7 - 3x, \quad 3x^3 - 7x^2 + 4 = 0,$

$(x - 1)(3x + 2)(x - 2) = 0$

$x = 1, -\frac{2}{3}, 2$

$A = \int_1^2 (y_1 - y_2)dx = \int_1^2 (7 - 3x - \frac{4}{x^2})dx$

$= (7x - \frac{3x^2}{2} + \frac{4}{x})\Big|_1^2 = (14 - 6 + 2) - (7 - \frac{3}{2} + 4) = \frac{1}{2}.$

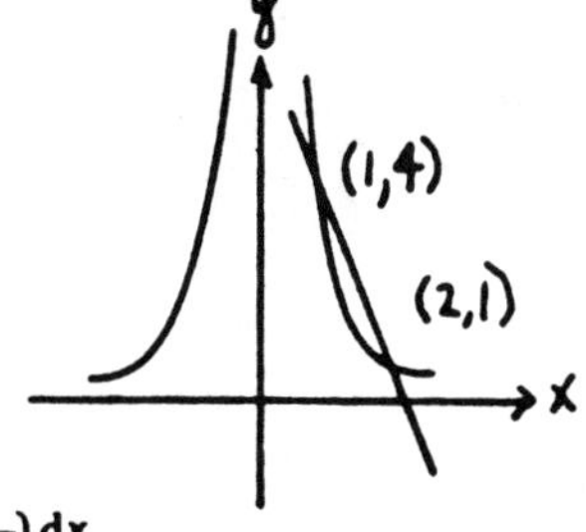

17. $A = \int_{-2}^{1/2} (y_1 - y_2)dx = \int_{-2}^{1/2} [(2 - x - x^2) - (x^2 + 2x)]dx$

$= \int_{-2}^{1/2} (2 - 3x - 2x^2)dx$

$$= (2x - \frac{3x^2}{2} - \frac{2x^3}{3})\Big|_{-2}^{1/2}$$

$$= (1 - \frac{3}{8} - \frac{1}{12}) - (-4 - 6 + \frac{16}{3})$$

$$= \frac{125}{24}.$$

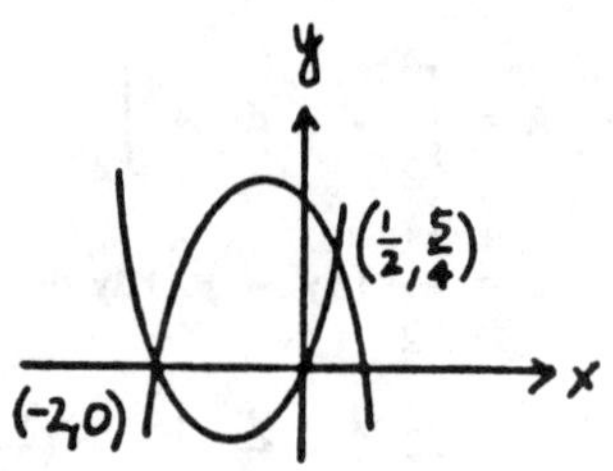

21. $A = \int_0^2 (y_1 - y_2)dx = \int_0^2 (2x - x^2)dx$

$$= (x^2 - \frac{x^3}{3})\Big|_0^2 = 4 - \frac{8}{3} = \frac{4}{3}.$$

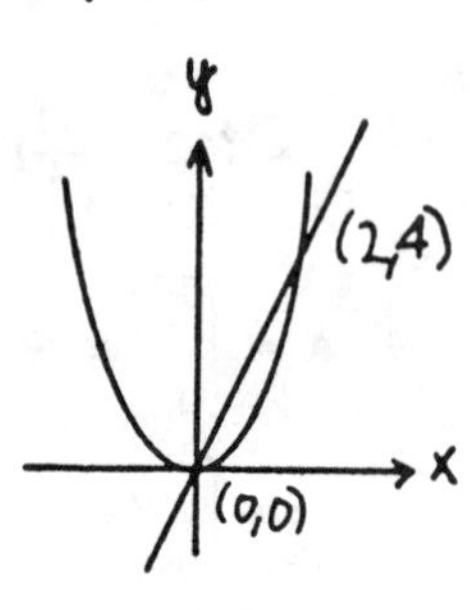

25. $A = \int_1^2 (y_1 - y_2)dx + \int_2^5 (y_1 - y_3)dx$

$$= \int_1^2 (-\frac{x}{2} + \frac{7}{2} + 5x - 8)dx + \int_2^5 (-\frac{x}{2} + \frac{7}{2} - x + 4)dx$$

$$= \int_1^2 (\frac{9x}{2} - \frac{9}{2})dx + \int_2^5 (-\frac{3x}{2} + \frac{15}{2})dx$$

$$= (\frac{9x^2}{4} - \frac{9x}{2})\Big|_1^2$$

$$+ (-\frac{3x^2}{4} + \frac{15x}{2})\Big|_2^5$$

$$= (9 - 9) - (\frac{9}{4} - \frac{9}{2})$$

$$+ (-\frac{75}{4} + \frac{75}{2})$$

$$- (-3 + 15) = 9.$$

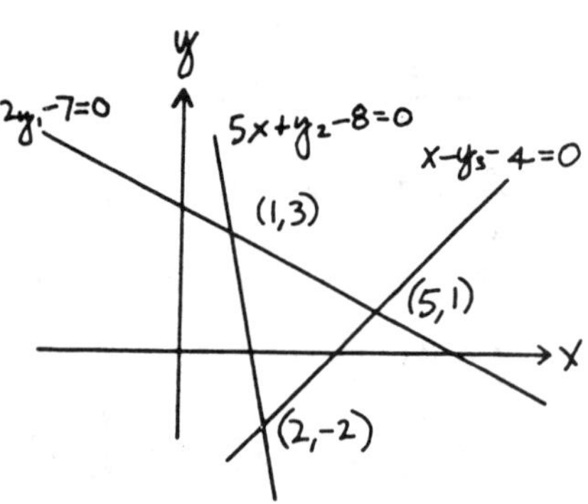

CHAPTER 6, SECTION 7 (pp. 291-293)

1. $W = \int_0^{12} 80x\,dx = 40x^2\Big|_0^{12} = 5760$ J.

5. $W = \int_0^3 9800 \cdot 4\pi(6 - x)dx = 4\pi \cdot 9800(6x - \frac{x^2}{2}\Big|_0^3 = 1.663 \times 10^6$

9. $F_1 = 0.1k,\ 60 + F_1 = 0.16k,\ 60 = 0.06k,\ k = 1000,\ F = 1000x$

$$W = \int_{0.1}^{0.16} 1000x\,dx = 500x^2\Big|_{0.1}^{0.16} = 7.8 \text{ J.}$$

13. $W = \int_{7.5}^{15} 20(2x - 15)dx = 20(x^2 - 15x)\Big|_{7.5}^{15} = 1125$ J.

17. $W = 9800 \cdot \frac{1}{3}\pi(1.5)^2(4.5) \cdot 10 = 1.039 \times 10^6$ J.

$$W = \int_0^{4.5} 9800\pi\frac{x^2}{9}(x + 4)\,dx = \frac{9800\pi}{9}\left(\frac{x^4}{4} + \frac{4x^3}{3}\right)\Big|_0^{4.5}$$

$$= 7.663 \times 10^5 \text{ J.}$$

21. $W_{anchor} = 8720 \cdot 20 + 10000 \cdot 4 = 214{,}400$ J

$$W_{chain} = \int_0^4 100x\,dx + \int_4^{24} [87.2(x - 4) + 100 \cdot 4]dx$$

$$= 50x^2\Big|_0^4 + \left(87.2\frac{x^2}{2} + 12.8 \cdot 4x\right)\Big|_4^{24}$$

$$= 800 + (25113.6 + 1228.8) - (697.6 + 204.8)$$

$$= 26240 \text{ J.}$$

$$W = W_{anchor} + W_{chain} = 214{,}400 + 26240 = 2.406 \times 10^5 \text{ J.}$$

25. $F = 21 = 10x - x^2,\ x = 3.$

$$W = \int_0^3 (10x - x^2)dx + 21(4 - 3)$$

$$= 5x^2 - \frac{x^3}{3}\Big|_0^3 + 21 = 45 - 9 + 21 = 57 \text{ J.}$$

CHAPTER 6, SECTION 8 (pp. 301-302)

1. $x_0 = 1 \qquad y_0 = 2$

   $x_1 = \frac{5}{4} \qquad y_1 = 2.5625$

   $x_2 = \frac{3}{2} \qquad y_2 = 3.25$

   $x_3 = \frac{7}{4} \qquad y_3 = 4.0625$

   $x_4 = 2 \qquad y_4 = 5$

   $$\int_1^2 (x^2 + 1)dx \doteq \frac{1}{8}(2 + 5.1250 + 6.5 + 8.1250 + 5) = 3.344$$

   $$\int_1^2 (x^2 + 1)dx \doteq \frac{1}{12}(2 + 10.25 + 6.5 + 16.25 + 5) = 3.333$$

   $$\int_1^2 (x^2 + 1)dx = (\frac{x^3}{3} + x)\Big|_1^2 = (\frac{8}{3} + 2) - (\frac{1}{3} + 1) = 3.333.$$

5. $x_0 = 0 \qquad y_0 = 1$

   $x_1 = \frac{1}{4} \qquad y_1 = 1.118$

   $x_2 = \frac{1}{2} \qquad y_2 = 1.2245$

   $x_3 = \frac{3}{4} \qquad y_3 = 1.323$

   $x_4 = 1 \qquad y_4 = 1.414$

   $$\int_0^1 \sqrt{x + 1}\, dx \doteq \frac{1}{8}(1 + 2.236 + 2.449 + 2.646 + 1.414) = 1.218$$

   $$\int_0^1 \sqrt{x + 1}\, dx \doteq \frac{1}{12}(1 + 4.472 + 2.449 + 5.292 + 1.414) = 1.219$$

   $$\int_0^1 \sqrt{x + 1}\, dx = \frac{2}{3}(x + 1)^{3/2}\Big|_0^1 = \frac{2}{3}(2\sqrt{2} - 1) = 1.219.$$

9. $x_0 = 1 \qquad y_0 = 2$

   $x_1 = \frac{5}{4} \qquad y_1 = 1.64$

   $x_2 = \frac{3}{2} \qquad y_2 = 1.4444$

$x_3 = \frac{7}{4}$ $\quad y_3 = 1.3265$

$x_4 = 2$ $\quad y_4 = 1.25$

$$\int_1^2 \frac{x^2+1}{x^2}\,dx \doteq \frac{1}{8}(2 + 3.28 + 2.8888 + 2.6530 + 1.25) = 1.509$$

$$\int_1^2 \frac{x^2+1}{x^2}\,dx \doteq \frac{1}{12}(2 + 6.56 + 2.8888 + 5.3060 + 1.25) = 1.500$$

$$\int_1^2 \frac{x^2+1}{x^2}\,dx = \int_1^2 (1 + x^{-2})dx = \left(x - \frac{1}{x}\right)\Big|_1^2$$

$$= \left(2 - \frac{1}{2}\right) - (1 - 1) = 1.500.$$

13. $x_0 = 1$ $\quad y_0 = 2$

$x_1 = 1.4$ $\quad y_1 = 1.7143$

$x_2 = 1.8$ $\quad y_2 = 1.5556$

$x_3 = 2.2$ $\quad y_3 = 1.4545$

$x_4 = 2.6$ $\quad y_4 = 1.3846$

$x_5 = 3$ $\quad y_5 = 1.3333$

$$\int_1^3 \frac{x+1}{x}\,dx \doteq \frac{1}{5}(2 + 3.4286 + 3.1112 + 2.9090 + 2.7692 + 1.3333) = 3.110$$

$f'(x) = -x^{-2}$, $\quad f''(x) = 2x^{-3}$,

$$E_T = -\frac{h^2}{12}(b - a)f''(c) = -\frac{(2/5)^2}{12}(2)\,\frac{2}{c^3}, \quad 1 \le c \le 3$$

$$|E_T| = \frac{4}{75c^3} \le \frac{4}{75} = 0.053, \quad 3.056 \le I \le 3.163.$$

17. $x_0 = 0$ $\quad y_0 = 1$

$x_1 = \frac{1}{4}$ $\quad y_1 = 0.9682$

$x_2 = \frac{1}{2}$ $\quad y_2 = 0.8660$

$x_3 = \frac{3}{4}$ $\quad y_3 = 0.6614$

$x_4 = 1$ $\quad y_4 = 0$

$$\int_0^1 \sqrt{1 - x^2}\, dx \doteq \frac{1}{12}(1 + 3.8728 + 1.7320 + 2.6456 + 0)$$
$$= 0.771$$

$f'(x) = -\dfrac{x}{\sqrt{1 - x^2}}$ does not exist at $x = 1$.

21. $x_0 = 0 \qquad y_0 = 0$

$x_1 = \frac{1}{4} \qquad y_1 = 0.2$

$x_2 = \frac{1}{2} \qquad y_2 = 0.3333$

$x_3 = \frac{3}{4} \qquad y_3 = 0.4286$

$x_4 = 1 \qquad y_4 = 0.5$

$$\int_0^1 \frac{x\,dx}{1 + x} \doteq \frac{1}{12}(0 + 0.8 + 0.6666 + 1.7144 + 0.5) = 0.307$$

$f'(x) = (1 + x)^{-2}, \quad f''(x) = -2(1 + x)^{-3},$

$f'''(x) = 6(1 + x)^{-4}, \quad f^{(4)}(x) = -\dfrac{24}{(1 + x)^5},$

$$E_s = -\frac{h^4}{180}(b - a)f^{(4)}(c) = -\frac{(1/4)^4}{180}(1)\ \frac{-24}{(1 + c)^5},\ 0 \le c \le 1$$

$$|E_s| = \frac{1}{1920(1 + c)^5} \le \frac{1}{1920} = 0.0005, \quad 0.3065 \le I \le 0.3075.$$

25. $x_0 = 1 \qquad y_0 = 0$

$x_1 = \frac{5}{4} \qquad y_1 = 0.45$

$x_2 = \frac{3}{2} \qquad y_2 = 0.8333$

$x_3 = \frac{7}{4} \qquad y_3 = 1.1786 \qquad x_6 = \frac{5}{2} \qquad y_6 = 2.1$

$x_4 = 2 \qquad y_4 = 1.5 \qquad x_7 = \frac{11}{4} \qquad y_7 = 2.3864$

$x_5 = \frac{9}{4} \qquad y_5 = 1.8056 \qquad x_8 = 3 \qquad y_8 = 2.6667$

$$\int_1^3 \frac{x^2 - 1}{x}\, dx \doteq \frac{1}{8}(0 + 0.9 + 1.6666 + 2.3572 + 3 + 3.6112$$
$$+ 4.2 + 4.7728 + 2.6667) = 2.897.$$

$$\int_1^3 \frac{x^2 - 1}{x}\,dx \doteq \frac{1}{12}(0 + 1.8 + 1.6666 + 4.7144 + 3 + 7.2224$$

$$+ 4.2 + 9.5456 + 2.6667) = 2.901.$$

29. $f'(x) = \dfrac{x}{\sqrt{x^2 + 4}}, \quad f''(x) = \dfrac{4}{(x^2 + 4)^{3/2}}, \quad h = \dfrac{1}{n}$

$$E_T = -\frac{h^2}{12}(b - a)f''(c) = -\frac{1/n^2}{12}(1)\,\frac{4}{(c^2 + 4)^{3/2}}, \quad 0 \le c \le 1$$

$$|E_T| = \frac{1}{3n^2(c^2 + 4)^{3/2}} \le \frac{1}{3n^2 \cdot 4^{3/2}} = \frac{1}{24n^2} < \frac{1}{1000}$$

$$n^2 > \frac{1000}{24} > 41, \quad n = 7.$$

33.
```
5     REM BASIC, TRAPEZOIDAL RULE
10    INPUT A,B,N
20    X=B
30    GOSUB 200: T=W
40    X=A
50    GOSUB 200: U=W
60    T=T+U
70    FOR I=1 to N-1
80    X=X+(B-A)/N
90    GOSUB 200: Y=W
100   T=T+2*Y
110   NEXT
120   T=(B-A)*T/(2*N)
130   PRINT T
140   END
200   W=1/X
210   RETURN
```

This program approximates $\int_a^b \frac{dx}{x}$ by the trapezoidal rule.

With $a = 1$, it approximates $\ln b$.

CHAPTER 6, REVIEW (pp. 302-304)

1. $\sum_{i=1}^{n} [(\frac{2i}{n})^2 - \frac{6i}{n}]\frac{2}{n}$

$$= \sum_{i=1}^{n} \left(\frac{8i^2}{n^3} - \frac{12i}{n^2}\right) = \frac{8}{n^3}\,\frac{n(n+1)(2n+1)}{6} - \frac{12}{n^2}\,\frac{n(n+1)}{2}$$

$$= \frac{4(n+1)(2n+1)}{3n^2} - \frac{6(n+1)}{n} = \frac{2(n+1)(4n+2-9n)}{3n^2}$$

$$= \frac{2(n+1)(2-5n)}{3n^2}.$$

5. $$\int_1^2 \frac{(x^2-2)^2}{x^2}\,dx = \int_1^2 \frac{x^4 - 4x^2 + 4}{x^2}\,dx = \int_1^2 (x^2 - 4 + 4x^{-2})\,dx$$

$$= \left(\frac{x^3}{3} - 4x - \frac{4}{x}\right)\Big|_1^2 = \left(\frac{8}{3} - 8 - 2\right) - \left(\frac{1}{3} - 4 - 4\right) = \frac{1}{3}.$$

9. $$A = \int_{-3}^{2} (x^3 - 4x + 15)\,dx$$

$$= \left(\frac{x^4}{4} - 2x^2 + 15x\right)\Big|_{-3}^{2}$$

$$= (4 - 8 + 30) - \left(\frac{81}{4} - 18 - 45\right)$$

$$= \frac{275}{4}.$$

13. $$A = \int_0^1 (y_1 - y_2)\,dx$$

$$= \int_0^1 (x^3 - x^4)\,dx$$

$$= \left(\frac{x^4}{4} - \frac{x^5}{5}\right)\Big|_0^1 = \frac{1}{4} - \frac{1}{5} = \frac{1}{20}.$$

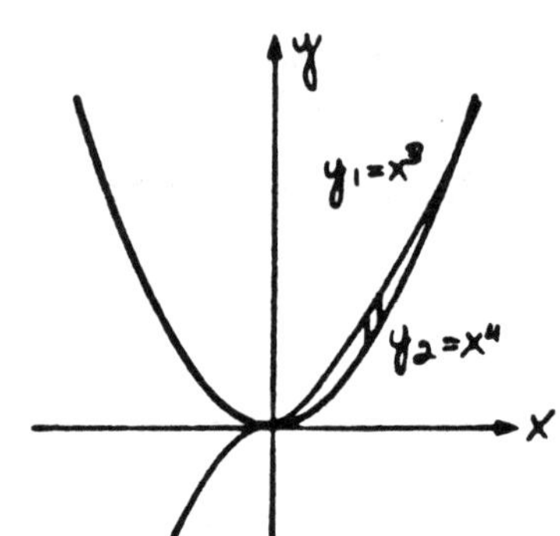

17. $F = kx,\ 3 = k(\frac{1}{10}),\ k = 30$

$$W = \int_0^{0.2} 30x\,dx = 15x^2\Big|_0^{0.2} = 0.6 \text{ J.}$$

$$W = \int_{0.2}^{0.4} 30x\,dx = 15x^2\Big|_{0.2}^{0.4} = 2.4 - 0.6 = 1.8 \text{ J.}$$

21. $x_0 = 0 \qquad y_0 = 1$

$x_1 = 1 \qquad y_1 = \sqrt{2} = 1.414$

$x_2 = 2 \qquad y_2 = 3$

$x_3 = 3 \qquad y_3 = \sqrt{28} = 5.292$

$x_4 = 4 \qquad y_4 = \sqrt{65} = 8.062$

$$\int_0^4 \sqrt{x^3 + 1}\,dx \doteq \frac{h}{2}(y_0 + 2y_1 + 2y_2 + 2y_3 + y_4)$$

$$= \frac{1}{2}(1 + 2.828 + 6 + 10.584 + 8.062) = 14.237.$$

$$\int_0^4 \sqrt{x^3 + 1}\,dx \doteq \frac{h}{3}(y_0 + 4y_1 + 2y_2 + 4y_3 + y_4)$$

$$= \frac{1}{3}(1 + 5.657 + 6 + 21.168 + 8.062) = 13.962.$$

25. $\int_1^2 (3x^2 - 4x + 1)dx$

$$= \lim_{n\to+\infty} \sum_{i=1}^{n} f(x_i^*)\Delta x_i = \lim_{n\to+\infty} \sum_{i=1}^{n} f(1 + \frac{i}{n})\frac{1}{n}$$

$$= \lim_{n\to+\infty} \sum_{i=1}^{n} [3(1 + \frac{i}{n})^2 - 4(1 + \frac{i}{n}) + 1]\frac{1}{n}$$

$$= \lim_{n\to+\infty} \frac{1}{n} \sum_{i=1}^{n} (\frac{2i}{n} + \frac{3i^2}{n^2})$$

$$= \lim_{n\to+\infty} \frac{1}{n}[\frac{2}{n}\,\frac{n(n + 1)}{2} + \frac{3}{n^2}\,\frac{n(n + 1)(2n + 1)}{6}]$$

$$= \lim_{n\to+\infty} \left[\left(1 + \frac{1}{n}\right) + \frac{(1 + 1/n)(2 + 1/n)}{2}\right] = 1 + \frac{1 \cdot 2}{2} = 2.$$

29. (a) $d = rt$ for constant speed r, so $d \doteq v(t_i^*)\Delta t$.

(b) Since $v(t) \geq 0$ there are no reversals of direction, so $D = \lim_{\|S\|\to 0} \sum_{i=1}^{n} v(t_i^*)\Delta t = \int_a^b v(t)dt$ gives the total distance traveled from $t = a$ to $t = b$. In general, $\int_a^b v(t)dt$ gives the net distance traveled while $\int_a^b |v(t)|dt$ gives the total distance traveled.

Chapter 7
Limits and Continuity: A Geometric Approach

CHAPTER 7, SECTION 1 (p. 310)

1. $P = (0, 0)$, $A = (0, 0)$.

5. $P = (2, 4)$, no A.

9. $P = (0, 0)$, $A = (0, 0)$.

13. Not a limit point.

17. Limit point.

21. Not a limit point.

25. Limit point.

CHAPTER 7, SECTION 2 (p. 314)

1. Not satisfied.

5. Satisfied.

9. $A = P = (0, 0)$.

   Clearly, 0 is a limit point of the domain. If α and β are two horizontal lines with P between them, let ℓ be any vertical line to the left of the y axis and m the vertical line through the point of intersection of α and $\mathscr{G}$. Thus every point of $\mathscr{G}$ - A between ℓ and m is also between α and β.

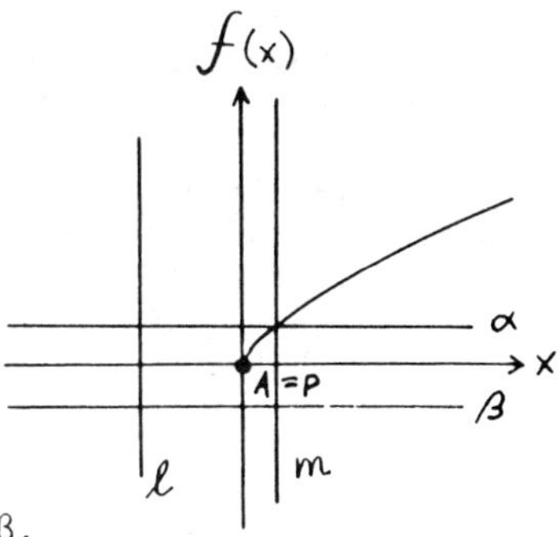

13. A = (1, 2), P = (1, 1).

Clearly, 1 is a limit point of the domain. If α and β are two horizontal lines with P between them, let ℓ and m be vertical lines through the points of

intersection of α and $\mathscr{G}$. Thus every point of $\mathscr{G}$ - A between ℓ and m is also between α and β.

17. A = P = (0, 3).

Clearly, 0 is a limit point of the domain. If α and β are two horizontal lines with P between them, let ℓ and m be the vertical lines through the points of intersection of α and β with $\mathscr{G}$. Thus every point of $\mathscr{G}$ - A between ℓ and m is also between α and β.

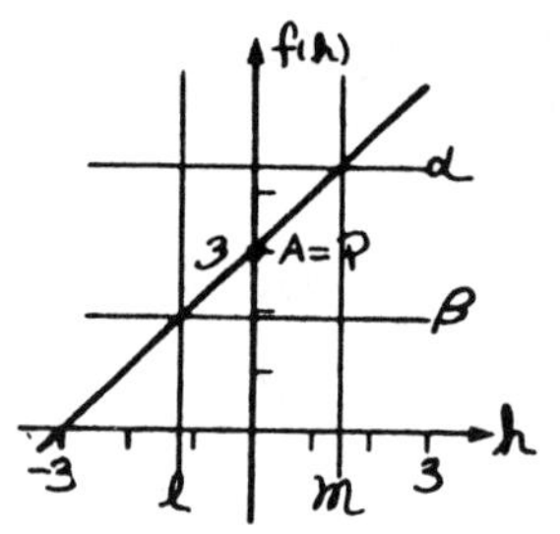

21. A = P = (a, k).

f(x) = k is clearly defined for all values of x; so the first part of the definition is satisfied. Let α and β be any pair of horizontal lines with P between them. Choose ℓ and m to be any pair of vertical lines with P between them. Since every point of $\mathscr{G}$ is between α and β, every point of $\mathscr{G}$ - A between ℓ and m is between α and β. Therefore, $\lim_{x \to a} k = k$.

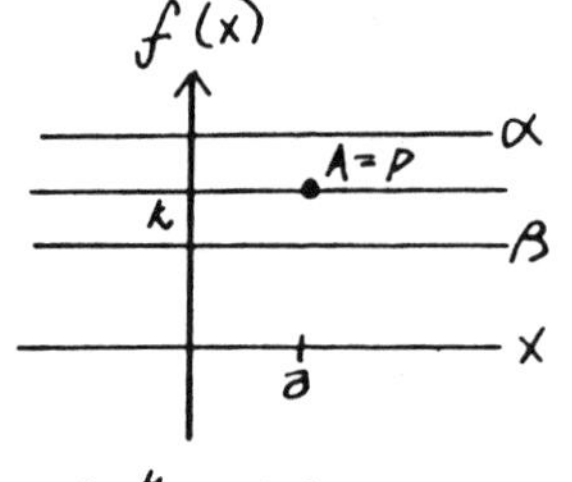

CHAPTER 7, SECTION 3 (pp. 318-319)

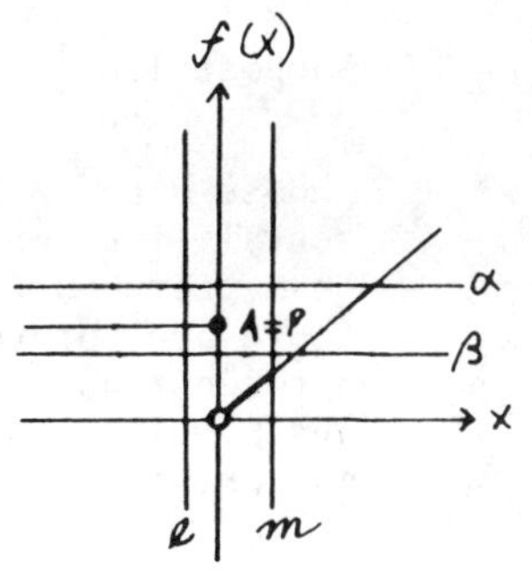

1. A = P = (0, 1).

   Choose α to be any horizontal line above y = 1 and β to be a horizontal line between y = 1 and the x axis. Now, no matter how ℓ and m are chosen, there are points of $\mathcal{G}$ - A between ℓ and m but not between α and β.

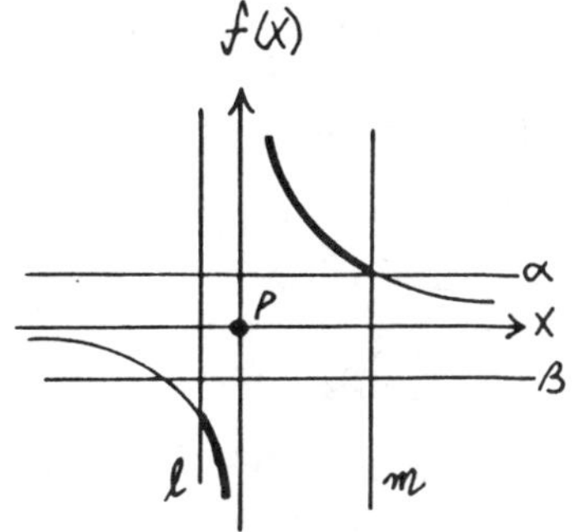

5. No A, P = (0, 0).

   Choose α and β to be any pair of horizontal lines with P between them. No matter how ℓ and m are chosen, there are points of $\mathcal{G}$ - A (= $\mathcal{G}$) between ℓ and m but not between α and β.

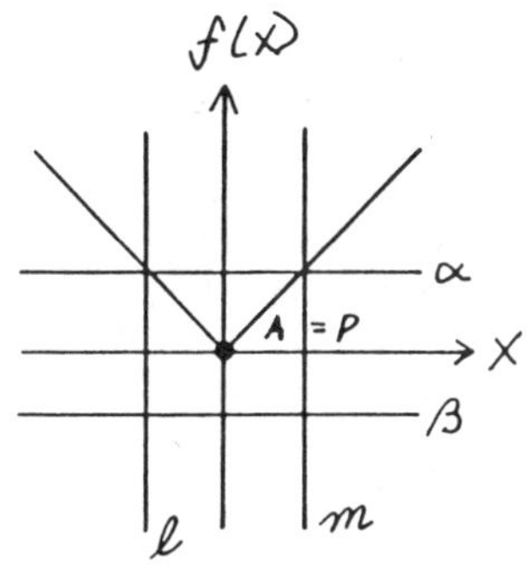

9. A = P = (0, 0).

   Clearly, 0 is a limit point of the domain. If α and β are two horizontal lines with P between them, let ℓ and m be the vertical lines through the points of intersection of α and $\mathcal{G}$. Now, every point of $\mathcal{G}$ - A between ℓ and m is also between α and β. The given limit statement is true.

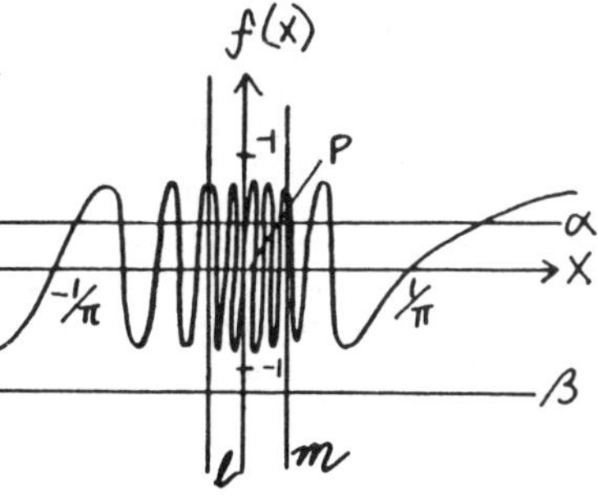

13. No A, P = (0, 0).

    Note that there are infinitely many waves between the y axis and any vertical line. Choose the vertical lines α and β so that either α is below y = 1 or β is above y = -1. Now, no matter how ℓ and m are chosen, there are points of $\mathcal{G}$ - A (= $\mathcal{G}$) between ℓ and m but not between α and β. The limit statement is false.

17. Suppose $\lim_{x \to 0} \frac{1}{x} = L$. No A, P = (0, L).

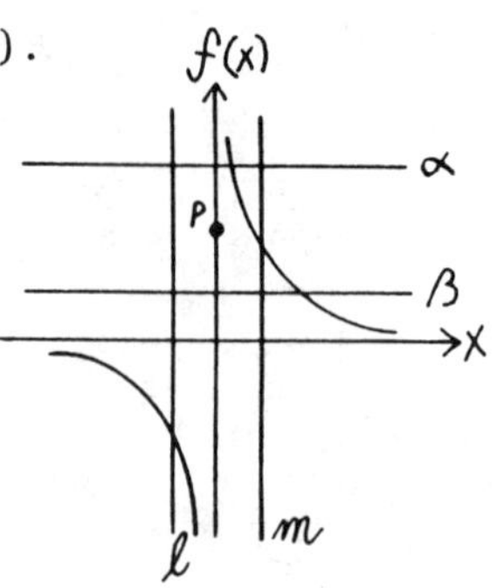

Choose α and β to be any two horizontal lines with P between them. No matter how ℓ and m are chosen, there are points of $\mathscr{G}$ - A (= $\mathscr{G}$) between ℓ and m but not between α and β. Thus the given limit does not exist.

CHAPTER 7, SECTION 4 (p. 323)

1. This means

$$\lim_{x \to 0} G(x) = 0, \text{ where } G(x) = \begin{cases} 1 & \text{if } x = 0, \\ x & \text{if } x > 0. \end{cases}$$

A = (0, 1), P = (0, 0).

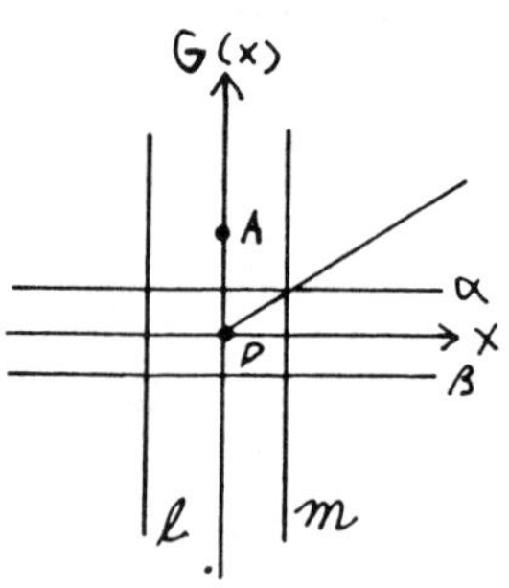

Clearly, 0 is a limit point of the domain of G. If α and β are horizontal lines with P between them, let ℓ be any vertical line to the left of the y axis and m the vertical line through the point of intersection of α and $\mathscr{G}$. Thus every point of $\mathscr{G}$ - A between ℓ and m is also between α and β.

5. This means

$$\lim_{x \to 0} G(x) = 0, \text{ where } G(x) = \begin{cases} 0 & \text{if } x = 0, \\ 1 & \text{if } x > 0. \end{cases}$$

A = P = (0, 0).

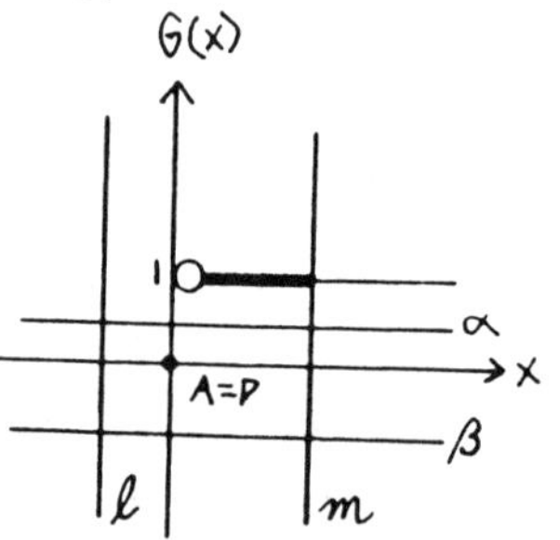

Choose α to be $y = \frac{1}{2}$ and β to be any horizontal line below the x axis. Now, no matter how ℓ and m are chosen, there are points of $\mathscr{G}$ - A between ℓ and m but not between α and β.

9. $\lim_{x \to 1^-} f(x) = 2$, $\lim_{x \to 1^+} f(x) = 0$, $\lim_{x \to 1} f(x)$ does not exist.

13. $\lim_{x \to 0^-} f(x)$ does not exist, $\lim_{x \to 0^+} f(x) = 0$, $\lim_{x \to 0} f(x)$ does not exist.

17. Suppose $\lim_{x \to 0^+} \log x = L$. Since log x is defined only for positive values of x, $\lim_{x \to 0} \log x = L$.

No A, P = (0, L).

Let α and β be two horizontal lines with P between them. No matter how ℓ and m are chosen, there are points of $\mathcal{G}$ - A (= $\mathcal{G}$) between ℓ and m but not between α and β.

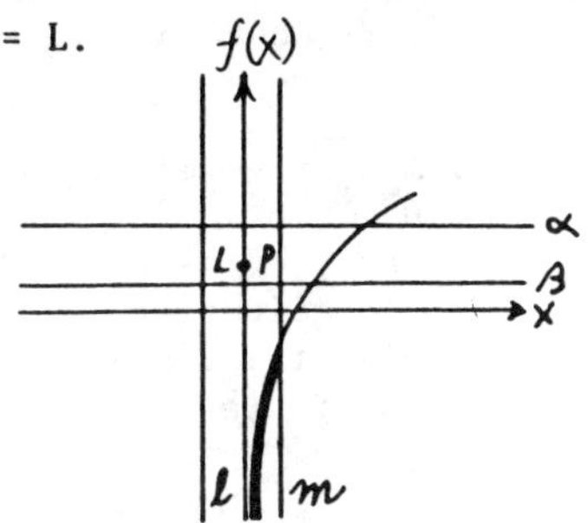

CHAPTER 7, SECTION 5 (p. 329)

1. No A.

Clearly, 1 is a limit point of the domain. Let α be a horizontal line. If α is on or below the x axis, every point of $\mathcal{G}$ is above α and the choice of ℓ and m is immaterial. If α is above the x axis, take ℓ and m to be the vertical lines through the points of intersection of α and $\mathcal{G}$. Now every point of $\mathcal{G}$ - A between ℓ and m is above α.

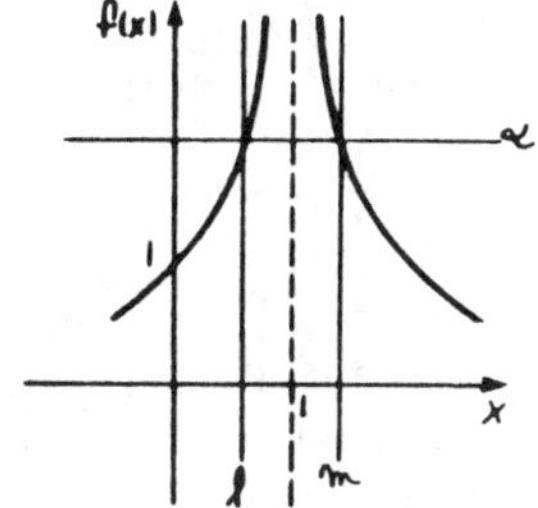

5. Clearly, there are points of $\mathcal{G}$ to the right of any vertical line. Suppose α and β are two horizontal lines with y = 0 between them. Let ℓ be the vertical line through the point of intersection of α and $\mathcal{G}$. Now every point of $\mathcal{G}$ to the right of ℓ is between α and β.

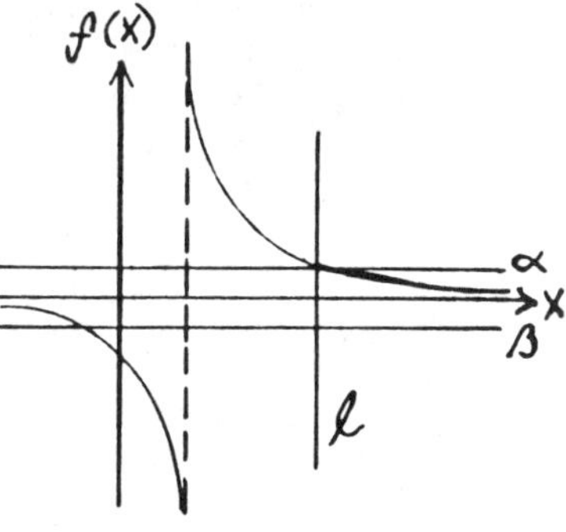

9. $\lim_{x\to+\infty} f(x) = +\infty$ means that

   (a) if h is a vertical line, then there is a point of $\mathcal{G}$ to the right of h, and

   (b) if α is a horizontal line, then there is a vertical line ℓ such that every point of $\mathcal{G}$ to the right of ℓ is above α.

13. $\lim_{x\to0^-} \frac{1}{x} = -\infty$ means $\lim_{x\to0} g(x) = -\infty$, where $g(x) = \frac{1}{x}$ when $x < 0$.
Clearly, 0 is a limit point of the domain of g. Let α be a horizontal line. If α is on or above the x axis, all of $\mathcal{G}$ is below α and the choice of ℓ and m is immaterial. If α is below the x axis, let ℓ be the vertical line through the point of intersection of α and $\mathcal{G}$, and let m be any vertical line to the right of the y axis. Now every point of $\mathcal{G}$ between ℓ and m is below α.

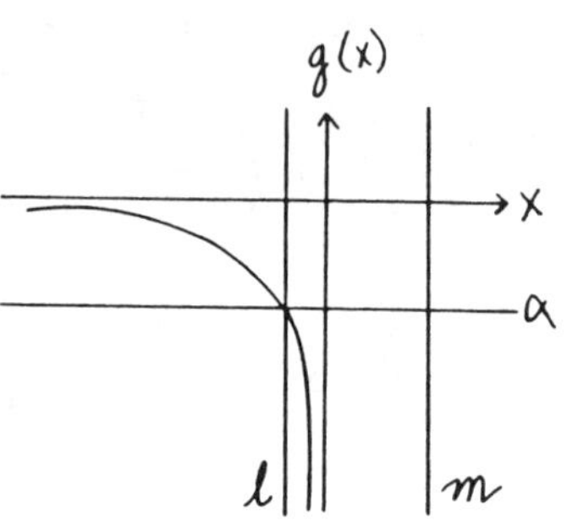

$\lim_{x\to0^+} \frac{1}{x} = +\infty$ means $\lim_{x\to0} G(x) = +\infty$, where $G(x) = \frac{1}{x}$, where $x > 0$.

Clearly, 0 is a limit point of the domain of G. Let α be a horizontal line. If α is on or below the x axis, all of $\mathcal{G}$ is below α and the choice of ℓ and m is immaterial. If α is above the x axis, let ℓ be any line to the left of the y axis and let m be the line through the point of intersection of α and $\mathcal{G}$. Now every point of $\mathcal{G}$ between ℓ and m is above α.

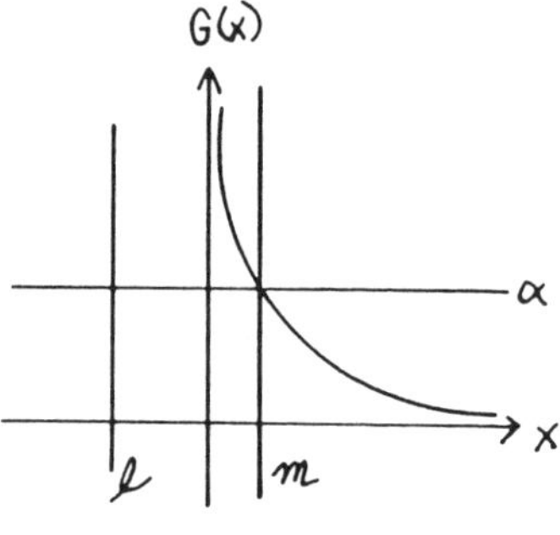

Let α be a horizontal line. No matter how ℓ and m are chosen with $x = 0$ between them, there are points of $\mathcal{G}$ to the left of the y axis which are below it, and points of $\mathcal{G}$ to the right of the y axis which are above it. Thus $\lim_{x\to0} \frac{1}{x}$ is neither $+\infty$ nor $-\infty$.

17. Clearly, there are points of $\mathcal{G}$ to the right of any vertical line. Suppose α is a horizontal line. If α is below the x axis, all of $\mathcal{G}$ is above α and the choice of ℓ is immaterial. If α is on or above the x axis, let ℓ be the vertical line through the point of intersection of α and $\mathcal{G}$. Now every point of $\mathcal{G}$ to the right of ℓ is above α. The limit statement is true.

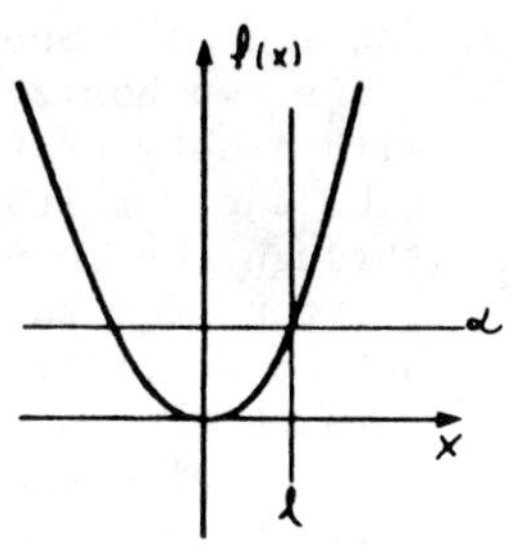

CHAPTER 7, SECTION 6 (pp. 333-334)

1. Continuous. Suppose α and β are two horizontal lines with A between them. If β does not intersect $\mathcal{G}$, let ℓ be the vertical line through the left-most intersection of α and $\mathcal{G}$. If β intersects $\mathcal{G}$, let ℓ be the vertical line through the right-most intersection of β and $\mathcal{G}$. In either case, let m be the vertical line through the right-most intersection of α and $\mathcal{G}$. Thus every point of $\mathcal{G}$ between ℓ and m is also between α and β.

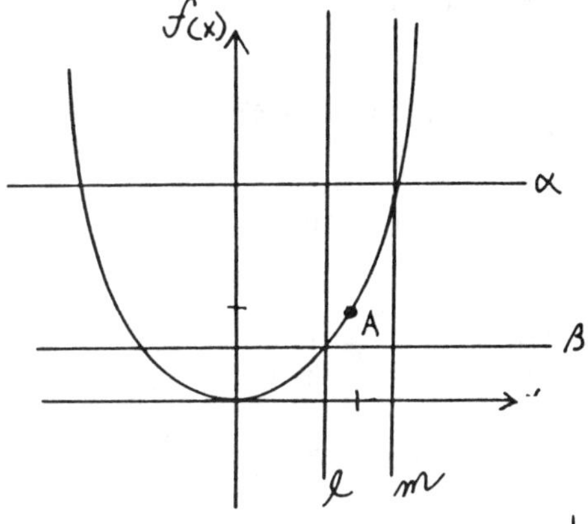

5. Continuous. Suppose α and β are two horizontal lines with A between them. Let ℓ be any vertical line to the left of A; let m be any vertical line between A and the y axis. Now every point of $\mathcal{G}$ between ℓ and m is also between α and β.

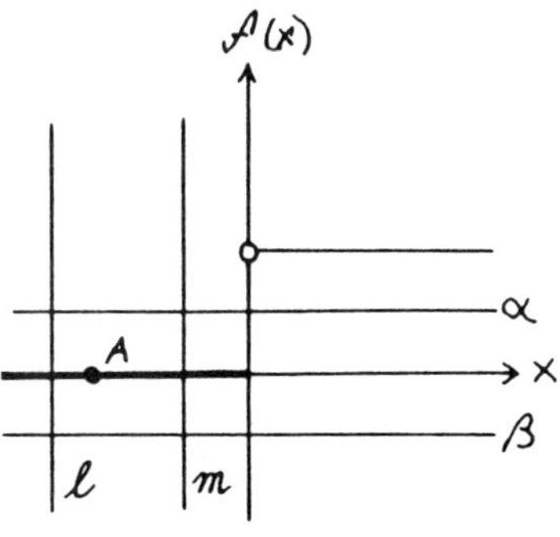

9. Continuous. Suppose α and β are two horizontal lines with A between them. Let ℓ be the vertical line through the point of intersection of β and 𝒟; let m be the vertical line through the point of intersection of α and 𝒟. Now every point of 𝒟 between ℓ and m is also between α and β.

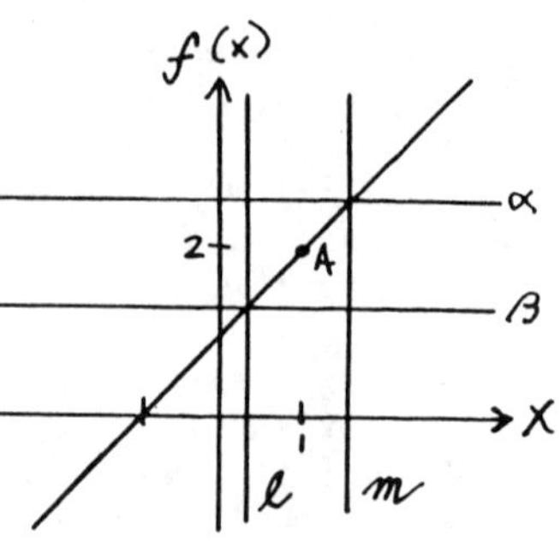

13. Not continuous on the set of real numbers at x = 3. Continuous on the domain.

17. Discontinuous at x = 1.

21. Continuous. Given α and β, the vertical lines ℓ and m are determined in the same way as in Problem 15, Section 10.3.

25. Continuous nowhere; discontinuous everywhere.

CHAPTER 7, SECTION 7 (pp. 344-345)

1. $\lim_{x \to 0} \frac{x^2 + x}{x^2 - x} = \lim_{x \to 0} \frac{x + 1}{x - 1}$ (by Theorem 4)

$= \frac{1}{-1} = -1$ (by Theorem 3).

5. $\lim_{h \to 0} \frac{(2 + h)^2 - 4}{h} = \lim_{h \to 0} \frac{4h + h^2}{h} = \lim_{h \to 0} (4 + h)$ (by Theorem 4)

$= 4$ (by Theorem 3).

9. $\lim_{h \to 0} \frac{\frac{x + h}{x + h + 1} - \frac{x}{x + 1}}{h} = \lim_{h \to 0} \frac{\frac{h}{(x + h + 1)(x + 1)}}{h}$

$= \lim_{h \to 0} \frac{1}{(x + h + 1)(x + 1)}$ (by Theorem 4)

$= \frac{1}{(x + 1)^2}$ (by Theorem 3).

13. $y = (x + 2)^{1/3}(x - 1)^{2/3}$ is continuous everywhere.

$$y' = (x + 2)^{1/3}\ \tfrac{2}{3}(x - 1)^{-1/3} + (x - 1)^{2/3}\ \tfrac{1}{3}(x + 2)^{-2/3}$$

$$= \frac{2(x + 2) + (x - 1)}{3(x - 1)^{1/3}(x + 2)^{2/3}} = \frac{x + 1}{(x - 1)^{1/3}(x + 2)^{2/3}}$$

There is no derivative at $x = 1$ or $x = -2$.

17. First, we show that $\lim_{x \to a} [-g(x)] = -c$. Let G be the graph of $-g(x)$, $P = (a, -c)$, and $A = (a, -g(a))$. Let G' be the graph of $g(x)$, $P' = (a, c)$, and $A' = (a, g(a))$. Suppose α and β are two horizontal lines with P between them and having equations

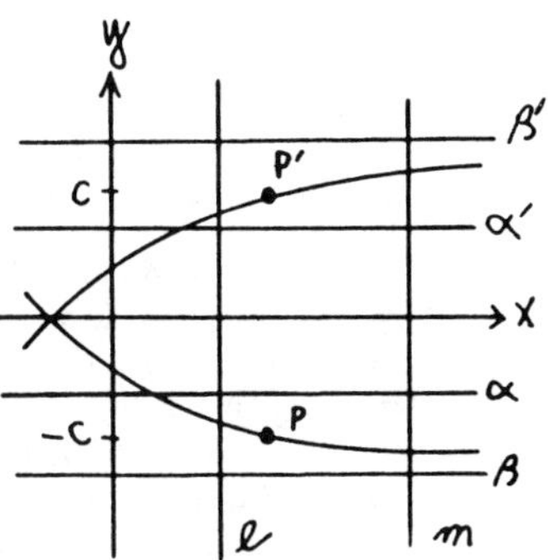

$y = k_1$ and $y = k_2$. Then $y = -k_1$ and $y = -k_2$ are two horizontal lines with P' between them (call them α' and β'). Then there are two vertical lines ℓ and m with P' between them such that every point of G' - A' between ℓ and m is also between α' and β'. But then every point of G - A between ℓ and m is between α and β. Therefore

$$\lim_{x \to a} [-g(x)] = -c.$$

Thus

$$\lim_{x \to a} [f(x) - g(x)] = \lim_{x \to a} \{f(x) + [-g(x)]\}$$

$$= \lim_{x \to a} f(x) + \lim_{x \to a} [-g(x)] = b - c.$$

CHAPTER 7, REVIEW (pp. 345-346)

1. Continuous for:

   $x \le -2$

   $-1 \le x < 0$

   $0 < x < 1$

   $x > 1$.

   Discontinuous for:

   $x = 0, \quad x = 1$.

5. True. $A = P = (2, 5)$.

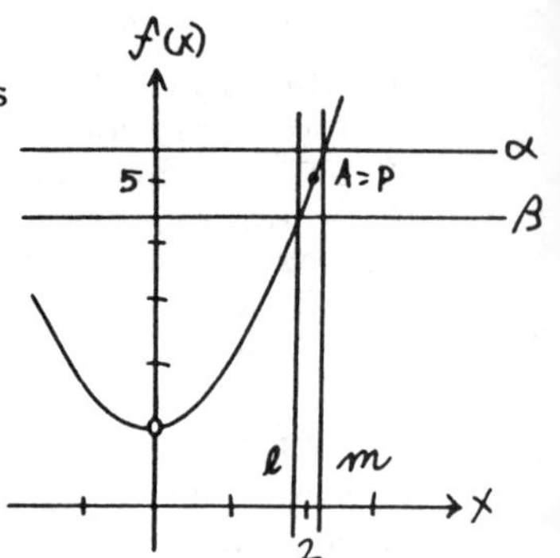

   Since f is defined for all values of x, 2 is a limit point of the domain of f. Let α and β be any pair of horizontal lines with P between them. Let ℓ and m be the vertical lines through the right-most point of intersection of $\mathcal{G}$ with α and β (if β does not intersect $\mathcal{G}$, let ℓ be $x = 0$). Then any point of $\mathcal{G}$ - A between ℓ and m is also between α and β.

9. True. $G(x) = \dfrac{1}{(x - 3)}$ if $x > 3$. No A.

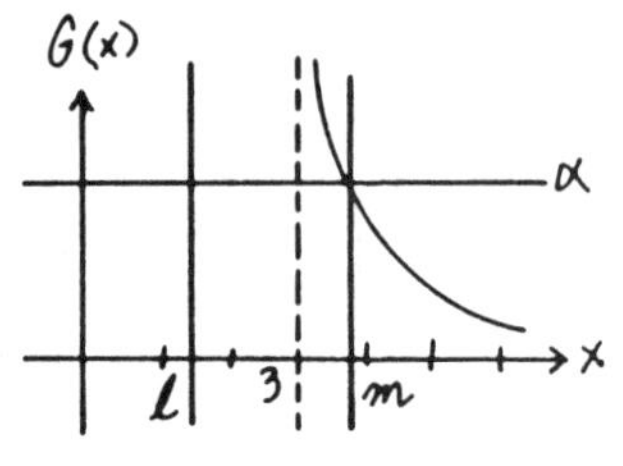

   Since G is defined for all $x > 3$, 3 is a limit point of the domain of G. Suppose α is a horizontal line. Let ℓ be any vertical line to the left of $x = 3$ and m the vertical line through the point of intersection of α and $\mathcal{G}$ (if there is no such point, let m be any vertical line to the right of $x = 3$). Then any point of $\mathcal{G}$ - A (= $\mathcal{G}$) between ℓ and m is also above α.

13. True. Since f is defined for all $x < 1$, the first part of the definition is satisfied. Let α be any horizontal line. If α lies above the x axis, let ℓ be any vertical line to the left of the y axis. If α lies on or below the x axis, let ℓ be the vertical line through the left-most point of intersection of α and $\mathcal{G}$. Then any point of $\mathcal{G}$ to the left of ℓ lies below α.

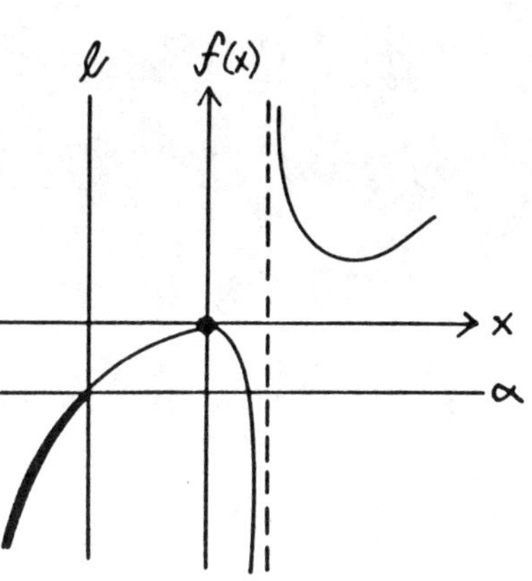

17. Continuous. Let α and β be any pair of horizontal lines with $A = (1, 1)$ between them. Let ℓ be the vertical line through the right-most point of intersection of $\mathcal{G}$ with β that is to the left of A (if there is no point of intersection of $\mathcal{G}$ with β which is to the left of A, let ℓ be $x = 0$), and let m be the vertical line through the point of intersection of $\mathcal{G}$ with β which is to the right of A. Then every point of $\mathcal{G}$ between ℓ and m is also between α and β.

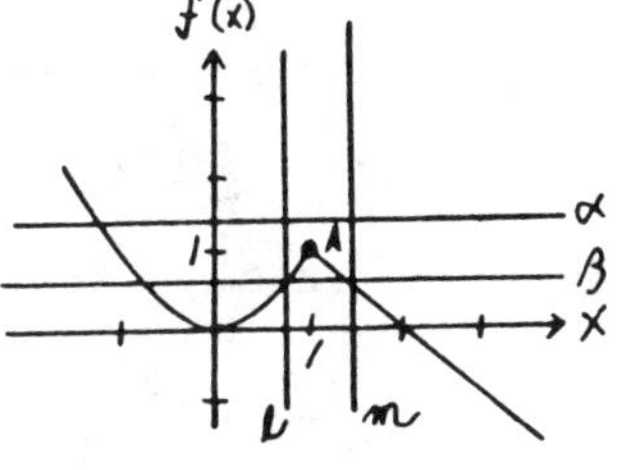

21. From the definition of $[x]$,

$$f(x) = \begin{cases} n - 1 & \text{if } n - 1 < x < n \\ n & \text{if } x = n \\ n + 1 & \text{if } n < x < n + 1. \end{cases}$$

Then, $\lim_{x \to n^-} f(x) = n - 1$ and $\lim_{x \to n^+} f(x) = n + 1$, so

$\lim_{x \to n^+} f(x) - \lim_{x \to n^-} f(x) = n + 1 - (n - 1) = 2.$

Chapter 8
Trigonometric Functions

CHAPTER 8, SECTION 1 (pp. 356-357)

1. $y' = 2\cos 2x.$

5. $y' = 2\sin 3x(\cos 3x \cdot 3) = 6\sin 3x\cos 3x.$

9. $y' = (\sec x\tan x)\tan x + \sec x(\sec^2 x)$

$= \sec^3 x + \sec x\tan^2 x.$

13. $y' = 2\csc x(-\csc x\cot x) - 2\cot x(-\csc^2 x)$

$= -2\csc^2 x\cot x + 2\csc^2 x\cot x = 0$, or

$y = \csc^2 x - (\csc^2 x - 1) = 1, \quad y' = 0.$

17. $y' = \dfrac{2x\sec^2 2x \cdot 2 - \tan 2x \cdot 2}{4x^2} = \dfrac{2x\sec^2 2x - \tan 2x}{2x^2}.$

21. $y' = \dfrac{2\cos 2x}{2\sqrt{\sin 2x}} = \dfrac{1}{0}$ (no value).

25. $y' = -\sin(\cos x)(-\sin x) = \sin x\sin(\cos x).$

29. $\sec^2(x + y)(1 + y') = y'$

$\sec^2(x + y) = y'[1 - \sec^2(x + y)]$

$y' = \dfrac{\sec^2(x + y)}{1 - \sec^2(x + y)} = \dfrac{\sec^2(x + y)}{-\tan^2(x + y)} = -\dfrac{1}{\sin^2(x + y)}$

$= -\csc^2(x + y).$

33. $y = 2\sqrt{3}\sin x + \cos 2x$

$y' = 2\sqrt{3}\cos x - 2\sin 2x$ $\quad\quad$ $y'' = -2\sqrt{3}\sin x - 4\cos 2x$

$2\sqrt{3}\cos x - 4\sin x\cos x = 0$

$2\cos x(\sqrt{3} - 2\sin x) = 0$

$\cos x = 0$		$\sin x = \sqrt{3}/2$	
$x = \frac{\pi}{2}$	$x = \frac{3\pi}{2}$	$x = \frac{\pi}{3}$	$x = \frac{2\pi}{3}$
$y = 2\sqrt{3} - 1$	$y = -2\sqrt{3} - 1$	$y = 2.5$	$y = 2.5$
y'' + min	y'' + absolute min.	y'' - max	y'' - max.

$y'' = 0$: $-\sqrt{3}\sin x - 2(1 - 2\sin^2 x) = 0$

$4\sin^2 x - \sqrt{3}\sin x - 2 = 0$

$$\sin x = \frac{\sqrt{3} \pm \sqrt{3 + 32}}{8} = \frac{\sqrt{3} \pm \sqrt{35}}{8}$$

$\sin x = 0.9560$ $\quad\quad$ $\sin x = -0.5230$

$x = 1.27,\ 1.87$ $\quad\quad$ $x = 3.692,\ 5.732$
$y = 2.49,\ 2.49$ $\quad\quad$ $y = -0.70,\ -1.36$

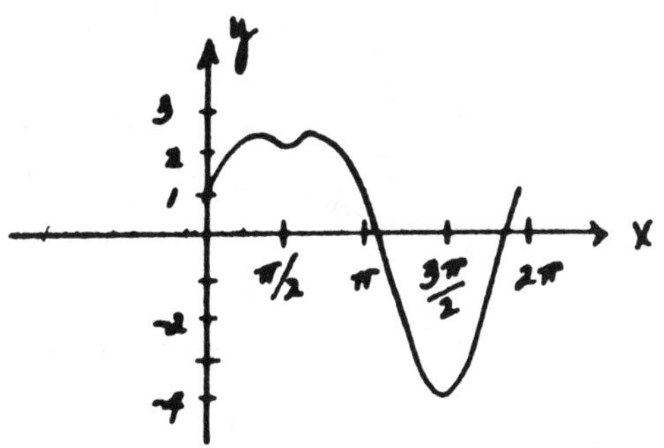

37. $y' = \cos x = \frac{1}{2}$, $\quad y - \frac{\sqrt{3}}{2} = \frac{1}{2}(x - \frac{\pi}{3})$

$3x - 6y + 3\sqrt{3} - \pi = 0.$

41. $\tan \alpha = \frac{x}{2}$

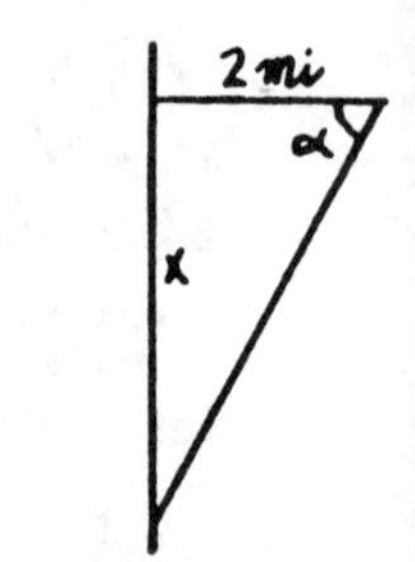

$x = 3.464, \quad \tan \alpha = 1.732, \quad \alpha = \frac{\pi}{3}$

$\frac{d\alpha}{dt} = 6\pi$ radians/min

$\frac{dx}{dt} = 2 \sec^2 \frac{\pi}{3} \cdot 6\pi = 48\pi$ mi/min.

45. $\sin(1/(2/\pi)) = \sin(\pi/2) = 1$, $\sin(1/(1/\pi)) = \sin \pi = 0$,
$\sin(1/(2/3\pi)) = \sin(3\pi/2) = -1$, $\sin(1/(1/2\pi))$
$= \sin 2\pi = 0, \cdot \cdot \cdot$. No limit exists.

CHAPTER 8, SECTION 2 (pp. 362-363)

1. $\int \sin 2\theta \, d\theta = \frac{1}{2} \int 2 \sin 2\theta \, d\theta = -\frac{1}{2} \cos 2\theta + C.$

5. $\int x^2 \sec x^3 \tan x^3 dx = \frac{1}{3} \int 3x^2 \sec x^3 \tan x^3 dx = \frac{1}{3} \sec x^3 + C.$

9. $\int_{-\pi/4}^{\pi/4} \tan^2 x \, dx = \int_{-\pi/4}^{\pi/4} (\sec^2 x - 1)dx = \tan x - x \Big|_{-\pi/4}^{\pi/4}$

$= (1 - \frac{\pi}{4}) - (-1 + \frac{\pi}{4}) = 2 - \frac{\pi}{2}.$

13. $\int (\sin^2 x - \cos^2 x)dx = \int -\cos 2x \, dx = -\frac{1}{2} \int 2 \cos 2x \, dx$

$= -\frac{1}{2} \sin 2x + C.$

17. $\int \frac{\tan x}{\sec^3 x} dx = \int \frac{\sin x}{\cos x} \cdot \cos^3 x \, dx = \int \cos^2 x \sin x \, dx$

$= -\int \cos^2 x(-\sin x)dx = -\frac{1}{3} \cos^3 x + C.$

21. $\int \cos^2 \theta \, d\theta = \frac{1}{2} \int (1 + \cos 2\theta)d\theta = \frac{1}{4} \int (2 + 2 \cos 2\theta)d\theta$

$= \frac{1}{4}(2\theta + \sin 2\theta) + C = \frac{\theta}{2} + \frac{1}{4} \sin 2\theta + C.$

25. $\int \sqrt{\cot x}\,\csc^2 x\,dx = -\int \sqrt{\cot x}(-\csc^2 x)dx = -\frac{2}{3}(\cot x)^{3/2} + C.$

29. From the figure, we have

$$A = \int_0^{\pi} \sin x\,dx = -\cos x\Big|_0^{\pi} = 2.$$

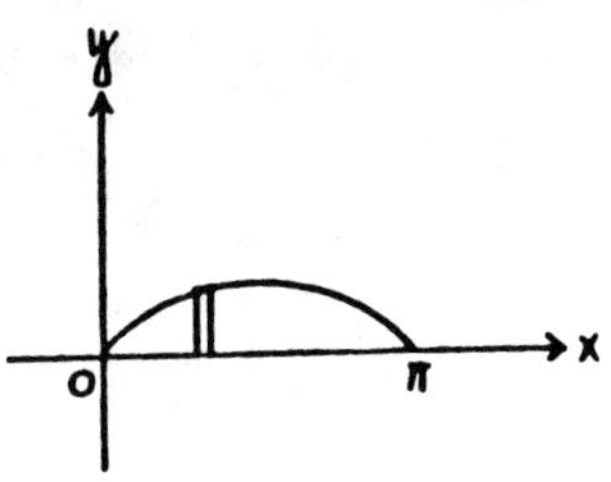

CHAPTER 8, SECTION 3 (pp. 372-373)

1. $y'/\csc y = \sin y\ y' = x$, $-\cos y = \frac{1}{2}x^2 + C.$

5. $\sec^2 y\ y' = \sin x$, $\tan y = -\cos x + C.$

9. $\frac{1}{2}y^2 = 2\sqrt{x} + C$, $8 = 2\sqrt{4} + C$, $C = 4$,

$y^2 = 4(\sqrt{x} + 2).$

13. $\omega = 6\pi$

$x = A\sin(6\pi t + \phi)$

$x(3.5) = 2\sqrt{2} = A\sin(21\pi + \phi) = -A\sin\phi$

$v(3.5) = 6\sqrt{2}\pi = 6\pi A\cos(21\pi + \phi) = -6\pi A\cos\phi$

$A = \sqrt{8 + 2} = \sqrt{10}$, $\tan\phi = 2$,

$\phi = 4.249$ (ϕ in third quadrant)

$x(t) = \sqrt{10}\sin(6\pi t + 4.249).$

17. $m = 1$

$2E = v^2 + kx^2$

$x(0) = 0.2$, $v(0) = 0.75$, $2E(0) = (0.75)^2 + k(0.2)^2$

$v = 1.5$ when $x = 0$, so also $2E = (1.5)^2$, Thus

$$(1.5)^2 = (0.75)^2 + (0.2)^2 k,$$

$$k = 42.1875$$

$$\omega = \sqrt{k/m} = 3.75\sqrt{3} \doteq 6.4952$$

$$x(t) = A \sin(3.75\sqrt{3}t + \phi)$$

$$0.2 = A \sin \phi$$

$$0.75 = 3.75\sqrt{3}A \cos \phi$$

$$A \sin \phi = 0.2,\ A \cos \phi = 0.2/\sqrt{3},\ \tan \phi = \sqrt{3},\ \phi = \pi/3$$

$$A = \sqrt{\frac{1}{25} + \frac{1}{75}} = \frac{2}{5\sqrt{3}} \doteq 0.231$$

$$x(t) = 0.231 \sin(6.495t + 1.047)$$

$$= \frac{2\sqrt{3}}{15} \sin(3.75\sqrt{3}t + \pi/3).$$

CHAPTER 8, SECTION 4 (pp. 381-382)

1. $0.$

5. $y' = \dfrac{2}{\sqrt{1 - 4x^2}}.$

9. $y' = \dfrac{2}{1 + 4x^2}.$

13. $y' = 1 - \dfrac{1}{1 + x^2} = \dfrac{x^2}{1 + x^2}.$

17. $y' = \dfrac{2}{\sqrt{1 - 4x^2}} = \dfrac{2}{\sqrt{1 - 1}} = \dfrac{2}{0}$; no value.

21. $y' = \dfrac{1}{\sqrt{1 - x^2}} - x \dfrac{-x}{\sqrt{1 - x^2}} - \sqrt{1 - x^2} = \dfrac{1 + x^2 - (1 - x^2)}{\sqrt{1 - x^2}}$

$$= \frac{2x^2}{\sqrt{1 - x^2}}.$$

25. $y' = \dfrac{-1/x^2}{\sqrt{1 - 1/x^2}} = \dfrac{-1}{|x|\sqrt{x^2 - 1}}.$

29. $y' = \dfrac{1}{1 + x^2} = \dfrac{1}{2}$ when $x = -1$.

$y + \frac{\pi}{4} = \frac{1}{2}(x + 1)$, $4y + \pi = 2x + 2$, $2x - 4y + (2 - \pi) = 0$.

33.

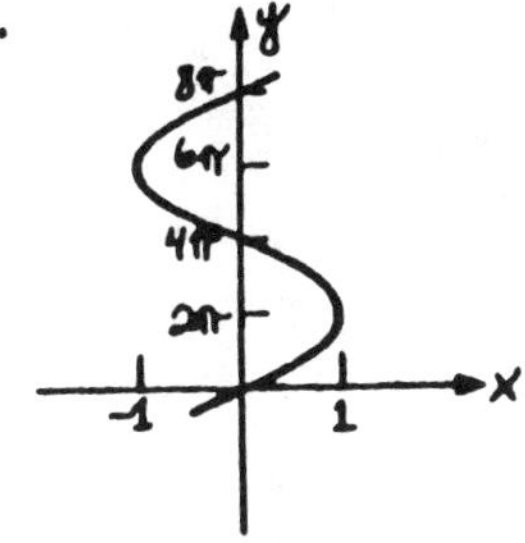

37. $y' = \dfrac{1}{1 + x^2} - 1$

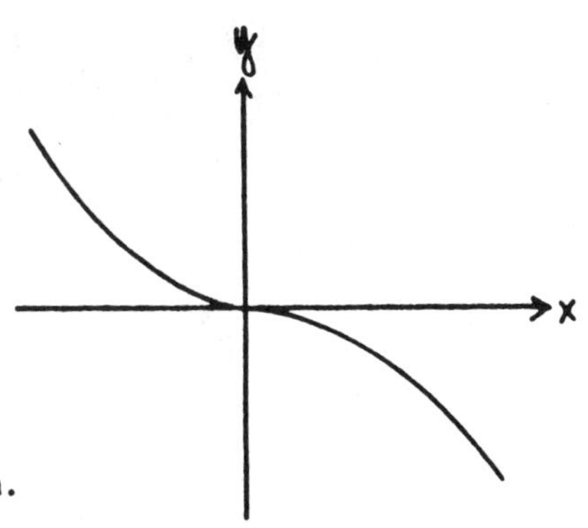

$y'' = \dfrac{-2x}{(1 + x^2)^2}$

$\dfrac{1}{1 + x^2} = 1$, $1 + x^2 = 1$

$x^2 = 0$, $x = 0$

(0, 0) neither max. nor min.

41. $y' = \dfrac{1}{1 + x^2} - \dfrac{1}{1 + x^2} = 0$, $y = C$.

Since this shows that y is constant, we can find the constant by determining y for any value of x. When $x = 1$, $y = \text{Arctan } 1 + \text{Arccot } 1 = \frac{\pi}{4} + \frac{\pi}{4} = \frac{\pi}{2}$.

CHAPTER 8, SECTION 5 (pp. 387-388)

1. $\displaystyle\int \frac{-x\ dx}{\sqrt{1 - x^2}} = \frac{1}{2}\int \frac{-2x\ dx}{\sqrt{1 - x^2}} = \sqrt{1 - x^2} + C.$

5. $$\int \frac{dx}{\sqrt{1 - 9x^2}} = \frac{1}{3}\int \frac{3\,dx}{\sqrt{1 - (3x)^2}} = \frac{1}{3}\text{ Arcsin } 3x + C.$$

9. $$\int \frac{dx}{\sqrt{1 - (2x + 1)^2}} = \frac{1}{2}\int \frac{2\,dx}{\sqrt{1 - (2x + 1)^2}}$$

$$= \frac{1}{2}\text{ Arcsin}(2x + 1) + C.$$

13. $$\int_0^1 \frac{dx}{1 + x^2} = \text{Arctan } x\Big|_0^1 = \frac{\pi}{4}.$$

17. $$\int \frac{dx}{(x - 2)\sqrt{x^2 - 4x + 3}} = \int \frac{dx}{(x - 2)\sqrt{(x - 2)^2 - 1}}$$

$$= \text{Arcsec}(x - 2) + C.$$

21. $$\int \frac{\cos x\,dx}{1 + \sin^2 x} = \text{Arctan } \sin x + C.$$

25. $$\int \frac{\text{Arctan } x}{1 + x^2}\,dx = \frac{1}{2}\text{ Arctan}^2 x + C.$$

29. $$\int \frac{u'\,dx}{a^2 + u^2} = \int \frac{u'/a^2}{1 + (u/a)^2}\,dx = \frac{1}{a}\int \frac{u'/a}{1 + (u/a)^2}\,dx$$

$$= \frac{1}{a}\text{ Arctan } \frac{u}{a} + C$$

$$\int \frac{dx}{4 + x^2} = \frac{1}{2}\text{ Arctan } \frac{x}{2} + C.$$

33. $\dfrac{1}{\sqrt{1 - x^2}} = 2$

$1 - x^2 = \dfrac{1}{4}$

$x^2 = \dfrac{3}{4}, \quad x = \pm\dfrac{\sqrt{3}}{2}$

$$A = 2\int_0^{\sqrt{3}/2} (2 - y)dx$$

$$= 2\int_0^{\sqrt{3}/2} \left(2 - \frac{1}{\sqrt{1 - x^2}}\right) dx = 2(2x - \text{Arcsin } x)\Big|_0^{\sqrt{3}/2}$$

$$= 2\sqrt{3} - 2 \text{ Arcsin } \frac{\sqrt{3}}{2}$$

$$= 2(\sqrt{3} - \frac{\pi}{3}).$$

37. $A = \displaystyle\int_0^k y\ dx = \int_0^k \frac{dx}{1 + x^2}$

$$= \text{Arctan } x\Big|_0^k$$

$$= \text{Arctan } k$$

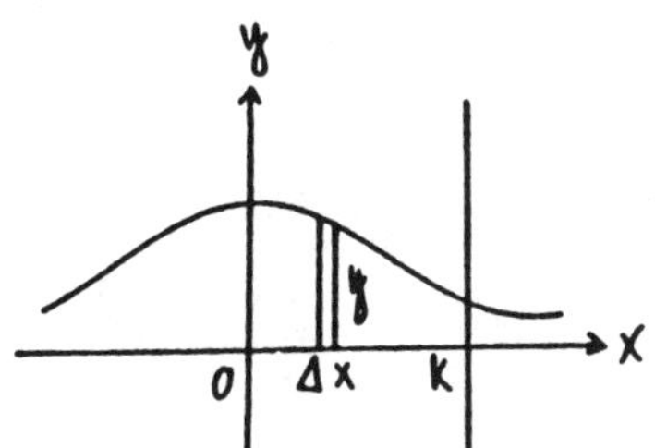

$\lim\limits_{k\to\infty} \text{Arctan } k = \frac{\pi}{2}$, the area of the unbounded region below $y = \dfrac{1}{1 + x^2}$, above the x axis, and to the right of $x = 0$.

CHAPTER 8, REVIEW (pp. 388-390)

1. $y' = \dfrac{\cos x \cdot \cos x - (1 + \sin x)(-\sin x)}{\text{cox}^2 x}$

$$= \frac{\cos^2 x + \sin x + \sin^2 x}{\cos^2 x} = \frac{1 + \sin x}{\cos^2 x}.$$

5. $y' = 9 - 3\csc^2 3x \cdot 3 - 3\cot^2 3x(-\csc^2 3x \cdot 3)$

$= 9 - 9\csc^2 3x + 9\csc^2 3x\cot^2 3x$

$= 9 - 9\csc^2 3x + 9\csc^2 3x(\csc^2 3x - 1)$

$= 9(1 - 2\csc^2 3x + \csc^4 3x) = 9(1 - \csc^2 3x)^2 = 9\cot^4 3x.$

9. $y' = 2 \text{ Arccos } 2x \cdot \dfrac{-2}{\sqrt{1 - 4x^2}} = \dfrac{-4 \text{ Arccos } 2x}{\sqrt{1 - 4x^2}}.$

13. $\displaystyle\int \cos^{5/2} 3x \tan 3x\, dx = \int \cos^{5/2} 3x \frac{\sin 3x}{\cos 3x} dx$

$\displaystyle= -\frac{1}{3}\int \cos^{3/2} 3x(-3\sin 3x)dx$

$\displaystyle= -\frac{1}{3}\cdot\frac{2}{5}\cos^{5/2} 3x + C$

$\displaystyle= -\frac{2}{15}\cos^{5/2} 3x + C.$

17. $\displaystyle\int \frac{dx}{(x + 3)\sqrt{x^2 + 6x + 8}} = \int \frac{dx}{(x + 3)\sqrt{(x + 3)^2 - 1}}$

$= \text{Arcsec}(x + 3) + C.$

21. $\displaystyle A = \int_0^2 y\, dx = \int_0^2 \frac{dx}{1 + x^2}$

$\displaystyle= \text{Arctan } x\Big|_0^2 = \text{Arctan } 2.$

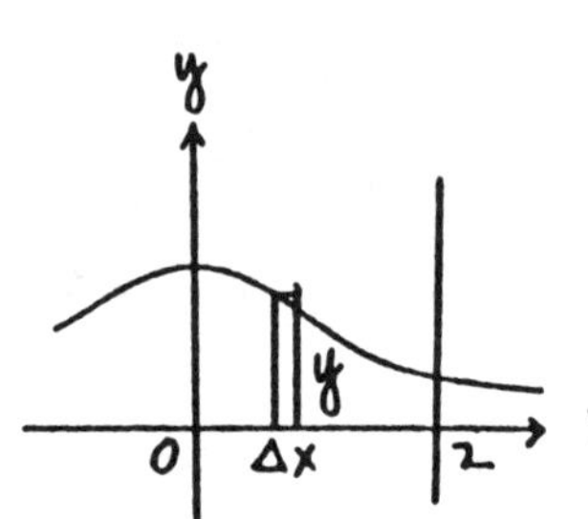

25. $k = 36$, $m = 4$, $\omega = \sqrt{k/m} = 3$.

$x = A\sin(3t + \phi)$, $x(0) = 0 = A\sin\phi$, $\phi = 0$.

$x = A\sin 3t$, $v = 3A\cos 3t$, $v(0.5) = 2 = 3A\cos 1.5$,

$A = \frac{2}{3}\sec 1.5$, $v(t) = 3A\cos 3t = 2\sec 1.5\cos 3t$.

29. $y' = \frac{1}{2} - \sin x$, $y'' = -\cos x$

Critical points: $\sin x = \frac{1}{2}$, $x = \frac{\pi}{6}, \frac{5\pi}{6}$

$\left(\frac{\pi}{6}, \frac{\pi + 6\sqrt{3}}{12}\right)$, $\left(\frac{5\pi}{6}, \frac{5\pi - 6\sqrt{3}}{12}\right)$

$y'' < 0$	$y'' > 0$
max	min

Points of inflection:

$\cos x = 0$

$x = \frac{\pi}{2}, \frac{3\pi}{2}$

$\left(\frac{\pi}{2}, \frac{\pi}{4}\right)$, $\left(\frac{3\pi}{2}, \frac{3\pi}{4}\right)$

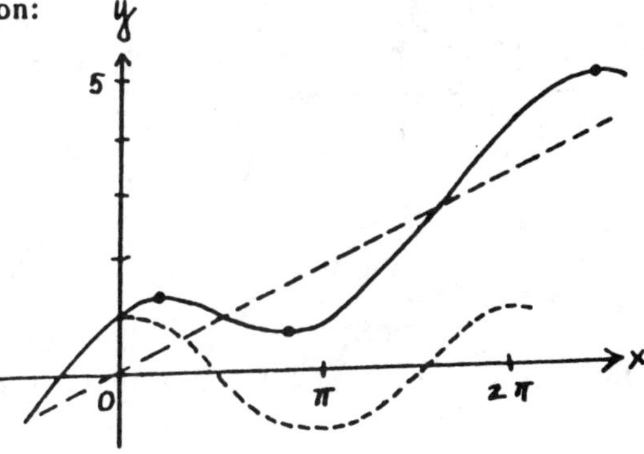

Chapter 9
Exponents, Logarithms, and Hyperbolic Functions

CHAPTER 9, SECTION 1 (pp. 397-398)

1. $y = 2^{x+1} = 2 \cdot 2^x$.

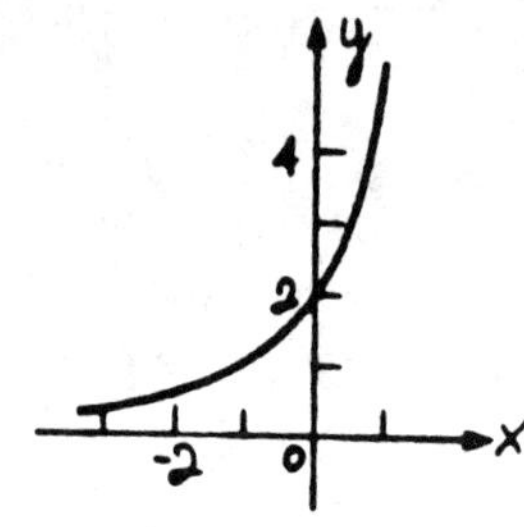

5.

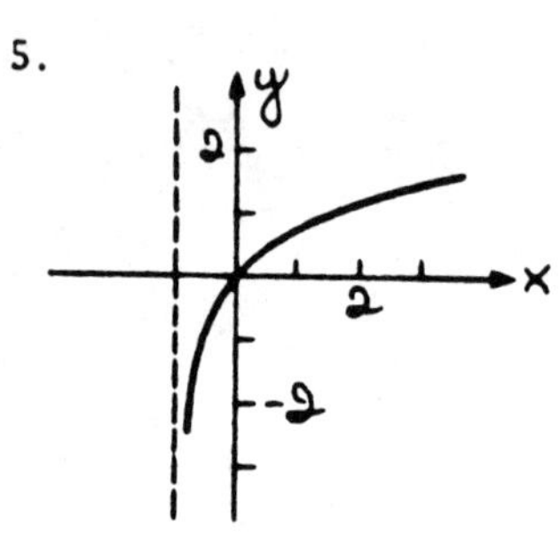

9. $y = e^x$

$x = \ln y$.

13. $\log x^2(x + 2) = 2 \log x + \log(x + 2)$.

17.

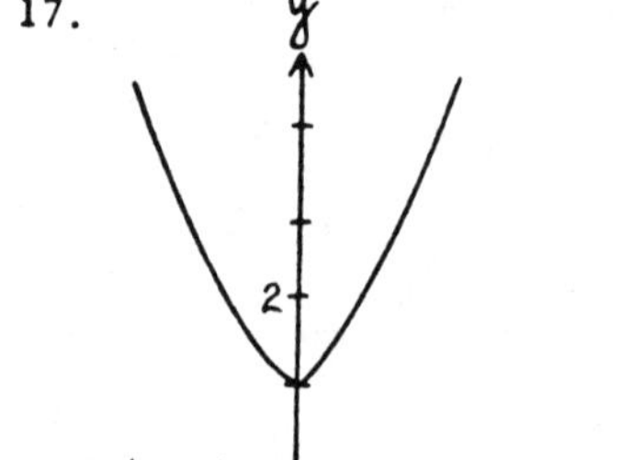

21.

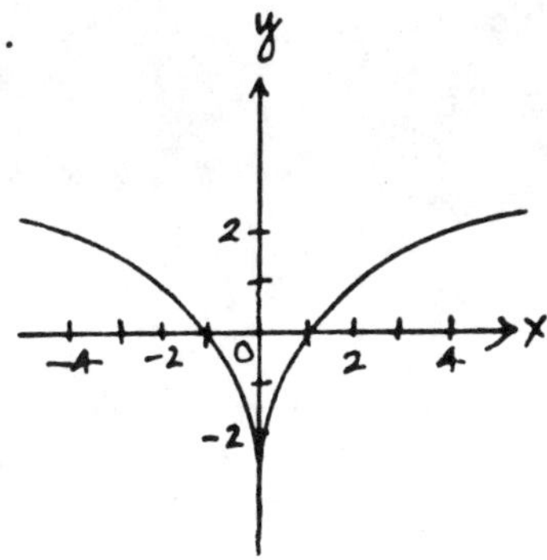

25. $y = \log_3 3^x$

$x = y.$

29. $\log_3 x = \frac{1}{2}\log_3 5 + \log_3 4 - 2\log_3 3$

$\log_3 x = \log_3 \frac{4\sqrt{5}}{3^2}, \quad x = \frac{4\sqrt{5}}{9}.$

33. $\log_5 \frac{x^2(x + 3)}{x - 1} = 2\log_5 x + \log_5(x + 3) - \log_5(x - 1).$

37. I. Let $a^{\log_a n} = x$. Then, by the definition of a logarithm, $\log_a x = \log_a n$ and $x = n$.

II. Let $\log_a a^n = y$. Then, by the definition of a logarithm, $a^n = a^y$ and $y = n$.

41. $F = P(1 + \frac{r}{n})^n = P[(1 + \frac{r}{n})^{n/r}]^r$

$\lim_{n\to+\infty} F = \lim_{r/n\to 0} P[(1 + \frac{r}{n})^{n/r}]^r = Pe^r$

(a) $F = P(1 + \frac{r}{n})^n = 1000(1.06) = \$1060.00$

(b) $F = P(1 + \frac{r}{n})^n = 1000(1.03)^2 = \$1060.90$

(c) $F = Pe^r = 1000\cdot e^{0.06} = 1000(1.0618) = \$1061.84.$

CHAPTER 9, SECTION 2 (pp. 404-406)

1. $y = \log 4 + \log x \qquad y' = \frac{\log e}{x}.$

5. $y = -\ln x, \quad y' = -\frac{1}{x}.$

9. $y = \frac{2}{3}[\log_5(3x+1) - \log_5(3x-1)$

$$y' = \frac{2}{3}\left(\frac{3}{3x+1} - \frac{3}{3x-1}\right)\log_5 e = \frac{-4\log_5 e}{9x^2 - 1} = \frac{4\log_5 e}{1 - 9x^2}.$$

13. $y' = \frac{x \cdot 1/x - \ln x}{x^2} = \frac{1 - \ln x}{x^2}.$

17. $y' = x \cdot \frac{1}{x} + \ln x = 1 + \ln 1 = 1.$

21. $y' = \frac{\cos \ln x}{x}.$

25. $y' = \frac{1/x}{\ln x} = \frac{1}{x \ln x}.$

29. $\frac{y'}{y} = \cos x, \quad y' = y \cos x.$

33. $2x + 2yy' = \frac{1 + y'}{x + y}, \quad 2x^2 + 2xy + 2xyy' + 2y^2y' = 1 + y'$

$$y'(2xy + 2y^2 - 1) = 1 - 2x^2 - 2xy, \quad y' = \frac{1 - 2x^2 - 2xy}{2xy + 2y^2 - 1}.$$

37. $\ln y = \frac{2}{3}\ln x + \frac{4}{3}\ln(x+1) - \frac{1}{3}\ln(x-5)$

$$\frac{y'}{y} = \frac{1}{3}\left(\frac{2}{x} + \frac{4}{x+1} - \frac{1}{x-5}\right), \quad y' = \frac{y}{3}\left(\frac{2}{x} + \frac{4}{x+1} - \frac{1}{x-5}\right).$$

41. $y' = \frac{-\sin x}{\cos x} = -1$ when $x = \frac{\pi}{4}.$

$$y + \frac{\ln 2}{2} = -\left(x - \frac{\pi}{4}\right), \quad 4y + 2\ln 2 = -4x + \pi$$

$$4x + 4y + (2\ln 2 - \pi) = 0.$$

45. $S = kx^2 \ln \frac{1}{x} = -kx^2 \ln x$

$$\frac{dS}{dx} = -k\left(x^2 \frac{1}{x} + 2x \ln x\right) = -kx(1 + 2\ln x)$$

$$1 + 2\ln x = 0, \quad \ln x = -\frac{1}{2}, \quad x = e^{-1/2} = \frac{1}{\sqrt{e}}.$$

CHAPTER 9, SECTION 3 (pp. 411-412)

1. $y' = 3^x \ln 3$.

5. $y' = 2e^{2x+2}$.

9. $y' = 1 + e^x$.

13. $y' = 3^x \cdot 3x^2 + 3^x(x^3 - 1)\ln 3 = 3^x[3x^2 + (x^3 - 1)\ln 3]$.

17. $y' = (\sec^2 x)(e^{\tan x})$.

21. $\ln y = x \ln \sin x$

$$\frac{y'}{y} = x\frac{\cos x}{\sin x} + \ln \sin x$$

$$y' = (\sin x)^x(x \cot x + \ln \sin x).$$

25. $\dfrac{1/x}{\ln x} = e^y \cdot y',\quad y' = \dfrac{1}{xe^y \ln x} = \dfrac{1}{x \ln x \ln \ln x}$.

29. $y' = e^{\sin x}\cos x = e \cdot 0 = 0$.

33. $y' = \dfrac{xe^x - e^x}{x^2} = 0,\quad y - e = 0$.

37. $y' = -e^x \sin e^x$

$\sin e^x = 0,\quad e^x = n\pi,\quad x = \ln(n\pi),\quad y = \pm 1$

$(\ln(n\pi), 1)$ max if n is even

$(\ln(n\pi), -1)$ min if n is odd.

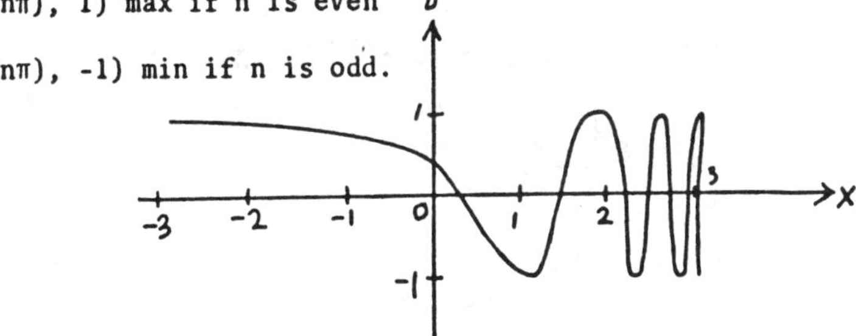

41. $y = f(x) = x^n$, $\ln y = n \ln x$, $\frac{y'}{y}$

$$= \frac{n}{x} y,\ y' = n \cdot \frac{x^n}{x} = nx^{n-1}.$$

CHAPTER 9, SECTION 4 (pp. 416-417)

1. $\int \frac{dx}{2x + 1} = \frac{1}{2} \int \frac{2\ dx}{2x + 1} = \frac{1}{2} \ln|2x + 1| + C.$

5. $\int \frac{\sin x}{\cos x} dx = -\int \frac{-\sin x}{\cos x} dx = -\ln|\cos x| + C$

$$= \ln|\cos x|^{-1} + C = \ln|\sec x| + C.$$

9. By synthetic division:

1	3	2	1
	1	4	
1	4	6	

so $\int \frac{x^2 + 3x + 2}{x - 1} dx = \int (x + 4 + \frac{6}{x - 1})dx$

$$= \frac{x^2}{2} + 4x + 6 \ln|x - 1| + C.$$

13. $\int \frac{2x + 1}{\sqrt{x^2 + x}} dx = \int (x^2 + x)^{-1/2}(2x + 1)dx = 2\sqrt{x^2 + x} + C.$

17. $\int (e^x - e^{-x})dx = e^x + e^{-x} + C.$

21. $\int (e^{\tan x})(\sec^2 x\ dx) = e^{\tan x} + C.$

25. $\int \frac{x^3 + 3x^2 + 5x + 3}{(x + 1)^2}$

$$= \int \left(x + 1 + \frac{2x + 2}{x^2 + 2x + 1}\right) dx = \frac{x^2}{2} + x + \ln(x^2 + 2x + 1) + C$$

$$= \frac{x^2}{2} + x + 2 \ln|x + 1| + C.$$

29. $\displaystyle\int_{\pi/6}^{\pi/2} \frac{\cos x}{\sin x}\,dx = \ln \sin x \Big|_{\pi/6}^{\pi/2} = \ln 1 - \ln \frac{1}{2} = \ln 2.$

33. $\displaystyle A = \int_0^{\ln 2} (2 - x)\,dy = \int_0^{\ln 2} (2 - e^y)\,dy = (2y - e^y)\Big|_0^{\ln 2}$

$$= 2 \ln 2 - 2 + 1 = 2 \ln 2 - 1.$$

37. $\displaystyle\int \frac{1}{e^x + 1}\,dx = \int \frac{e^{-x}}{1 + e^{-x}}\,dx = -\int \frac{-e^{-x}\,dx}{1 + e^{-x}}$

$$= -\ln(1 + e^{-x}) + C_1 = \ln \frac{C}{1 + e^{-x}} = \ln \frac{Ce^x}{e^x + 1},$$

if $C_1 = \ln C$. Also, $-\ln(1 + e^{-x}) + C_1$

$1 + e^x)] + C_1 = -\ln e^{-x} - \ln(1 + e^x) + C_1$

$= x - \ln(1 + e^x) + C_1.$

CHAPTER 9, SECTION 5 (pp. 426-428)

1. $x(t) = x(0)e^{kt}$, k min.

$2x(0) = x(0)e^{30k}$, $e^{30k} = 2$, $30k = \ln 2$, $k = \dfrac{\ln 2}{30} \doteq 0.023.$

5. $\dfrac{dx}{dt} = kx$, $\dfrac{x'}{x} = k$, $\ln|x| = kt + C$, $x = Ke^{kt}$.

Since $x = x_0$ when $t = 0$, $K = x_0$.

Since $x = 3x_0$ when $t = 1$,

$3x_0 = x_0e^k$, $e^k = 3$, $x = x_0 \cdot 3^t$.

If $x = 100x_0$, then $100x_0 = x_0 \cdot 3^t$

$3^t = 100$, $t \ln 3 = \ln 100$

$t = \dfrac{\ln 100}{\ln 3} = \dfrac{4.605}{1.099} = 4.192$ hr $= 4$ hr 11 min 31 sec.

9. $\frac{dx}{dt} = kx, \quad x = x_0e^{kt}$. Since $x = \frac{x_0}{2}$ when $t = 50$,

$\frac{x_0}{2} = x_0e^{50k}, \quad e^{50k} = 0.5, \quad 50k = \ln 0.5 = -0.6931,$

$k = -0.0139, \quad x = x_0e^{-0.0139t}$. When $x = 0.05x_0$,

$0.05x_0 = x_0e^{-0.0139t}, \quad e^{-0.0139t} = 0.05,$

$-0.0139t = \ln 0.05 = -2.9958, \quad t = 216$ hrs.

13. $T_1 = 8, T_2 = 50$

$k_1 = \frac{\ln 2}{8}, k_2 = \frac{\ln 2}{50}$. From Example 3,

$$t = \frac{\ln(k_2/k_1)}{k_2 - k_1} = \frac{200 \ln(0.16)}{\ln 2(4 - 25)} = 25.2 \text{ min.}$$

17. $\frac{dx}{dt} = kx^2, \quad \frac{dx}{x^2} = k\,dt, \quad -\frac{1}{x} = kt + C.$

Since $x = 250$ when $t = 0$,

$C = -\frac{1}{250}, \quad -\frac{1}{x} = kt - \frac{1}{250}.$

Since $x = 200$ when $t = 30$,

$\frac{1}{250} - \frac{1}{200} = 30k, \quad k = -\frac{1}{30,000}, \quad \frac{1}{250} - \frac{1}{x} = -\frac{t}{30,000}.$

When $x = 150$,

$\frac{1}{250} - \frac{1}{150} = -\frac{t}{30,000}, \quad t = 80$ min $= 1$ hr 20 min.

21. $k_1 = \frac{\ln 2}{8} = 0.0866, \; k_2 = \frac{\ln 2}{50} = 0.01386$

$x_1(t) = x_1(0)e^{-k_1t}, \; x_2(t) = \frac{k_1}{k_2 - k_1}x(0)[e^{-k_1t} - e^{-k_2t}]$

$x_1(t) = x_2(t),$

$$e^{-k_1t} = \frac{k_1}{k_2 - k_1}[e^{-k_1t} - e^{-k_2t}]$$

$$t = \frac{1}{k_1 - k_2} \ln \frac{2k_1 - k_2}{k_1} \quad (2k_1 > k_2)$$

$$t = \frac{1}{0.0727} \ln \frac{0.1593}{0.0866} = 8.378 \text{ min.}$$

$$\frac{x_1(8.378)}{x_1(0)} = e^{-\ln 2(8.378/8)} \times 100$$

$$= 48.39\%.$$

CHAPTER 9, SECTION 6 (pp. 436-437)

1\. (a) $\frac{15,000,000}{61,900} = 242$ yr, $1972 + 242 = 2214$.

(b) $A = \frac{x_0}{k}(e^{kt} - 1)$

$$15,000,000 = \frac{61,900}{0.065}(e^{0.065t} - 1)$$

$$e^{0.065t} - 1 = 15.75, \quad e^{0.065t} = 16.75$$

$$0.065t = \ln 16.75 = 2.818$$

$$t = 43, \quad 1972 + 43 = 2015.$$

5\. $A = \frac{x_0}{k}(e^{kt} - 1)$

$$913,000 = \frac{2,500}{0.067}(e^{0.067t} - 1)$$

$$e^{0.067t} - 1 = 24.5, \quad e^{0.067t} = 25.5$$

$$0.067t = \ln 25.5 = 3.2371$$

$$t = 48, \quad 1972 + 48 = 2020.$$

9\. $x = 1.4x - 4 \times 10^{-5}x^2$

$$0.4x - 4 \times 10^{-5}x^2 = 0$$

$$0.4x(1 - 10^{-4}x) = 0$$

$$x = 10^4 = 10,000.$$

$$y' = 1.4 - 8 \times 10^{-5}x = 1,$$

$8 \times 10^{-5}x = 0.4$

$x = 5{,}000, \quad y = 6{,}000,$

$y - x = 1{,}000.$

13. $f'(x) = 1 + i, \quad 1.4 - 8 \times 10^{-5}x = 1.08$

$8 \times 10^{-5}x = 0.32, \quad x = 4000.$

17. We want the maximum value of y'.

$y' = 1 + 7.5 \times 10^{-5}x - 3.75 \times 10^{-9}x^2$

$y'' = 7.5 \times 10^{-5} - 7.5 \times 10^{-9}x = 0$

$x = 10{,}000, \quad y' = 1.375 = 1 + i,$

$i = 0.375 = 37.5\%.$

CHAPTER 9, SECTION 7 (pp. 442-444)

1. $\cosh^2 x - \sinh^2 x = 1,$

$\cosh^2 x - \frac{16}{9} = 1,$

$\cosh^2 x = \frac{25}{9},$

$\cosh x = \frac{5}{3},$

$\tanh x = \frac{\sinh x}{\cosh x} = \frac{4/3}{5/3} = \frac{4}{5},$

$\coth x = \frac{1}{\tanh x} = \frac{5}{4},$

$\operatorname{sech} x = \frac{1}{\cosh x} = \frac{3}{5},$

$\operatorname{csch} x = \frac{1}{\sinh x} = \frac{3}{4}.$

5. $1 - \tanh^2 x = \operatorname{sech}^2 x$

$1 - \frac{25}{169} = \operatorname{sech}^2 x$

$\frac{144}{169} = \operatorname{sech}^2 x$

$\operatorname{sech} x = \frac{12}{13}$

$\coth x = \frac{1}{\tanh x} = -\frac{13}{5},$

$\cosh x = \frac{1}{\operatorname{sech} x} = \frac{13}{12},$

$\tanh x = \frac{\sinh x}{\cosh x},$

$-\frac{5}{13} = \frac{\sinh x}{13/12},$

$\sinh x = -\frac{5}{12},$

$\operatorname{csch} x = \frac{1}{\sinh x} = -\frac{12}{5}.$

9. $y' = 2x \cosh x^2$.

13. $y' = \dfrac{-\text{sech}\sqrt{x}\ \tanh\sqrt{x}}{2\sqrt{x}}$.

17. $y' = 2 \sinh x \cosh x + 2 \cosh x \sinh x = 4 \sinh x \cosh x$.

21. $y' = \cosh x = \cosh 0 = 1$, $y - 0 = (1)(x - 0)$, $x - y = 0$.

25. $\cosh^2 x - \sinh^2 x = 1$, $\dfrac{\cosh^2 x}{\cosh^2 x} - \dfrac{\sinh^2 x}{\cosh^2 x} = \dfrac{1}{\cosh^2 x}$,

$1 - \tanh^2 x = \text{sech}^2 x$.

29. $\sinh x \cosh y - \cosh x \sinh y$

$$= \frac{e^x - e^{-x}}{2}\,\frac{e^y + e^{-y}}{2} - \frac{e^x + e^{-x}}{2}\,\frac{e^y - e^{-y}}{2}$$

$$= \frac{e^{x+y} - e^{-x+y} + e^{x-y} - e^{-x-y} - e^{x+y} + e^{x-y} - e^{-x+y} + e^{-x-y}}{4}$$

$$= \frac{2e^{x-y} - 2e^{-x+y}}{4} = \frac{e^{x-y} - e^{-(x-y)}}{2} = \sinh(x - y).$$

33. $\cosh 2y = 2 \cosh^2 y - 1$, $\cosh^2 y = \dfrac{\cosh 2y + 1}{2}$,

$\cosh y = \sqrt{\dfrac{\cosh 2y + 1}{2}}$ (cosh y is never negative).

Letting $y = \dfrac{x}{2}$, $\cosh \dfrac{x}{2} = \sqrt{\dfrac{\cosh x + 1}{2}}$.

37. $$\frac{d}{dx} \tanh u = \frac{d}{dx}\frac{\sinh u}{\cosh u} = \frac{\cosh u \cosh u \cdot u' - \sinh u \sinh u \cdot u'}{\cosh^2 u}$$

$$= \frac{\cosh^2 u - \sinh^2 u}{\cosh^2 u} \cdot u' = \frac{1}{\cosh^2 u} \cdot u' = \text{sech}^2 u \cdot u'.$$

41. $y = \text{gd } x = \text{Arctan} \sinh x$

$$y' = \frac{\cosh x}{1 + \sinh^2 x} = \frac{\cosh x}{\cosh^2 x} = \frac{1}{\cosh x} = \text{sech } x$$

$$\lim_{x \to \pm\infty} \sinh x = \pm\infty$$

$$\lim_{x\to\pm\infty} \text{gd } x = \lim_{x\to\pm\infty} \text{Arctan sinh } x = \pm\frac{\pi}{2}.$$

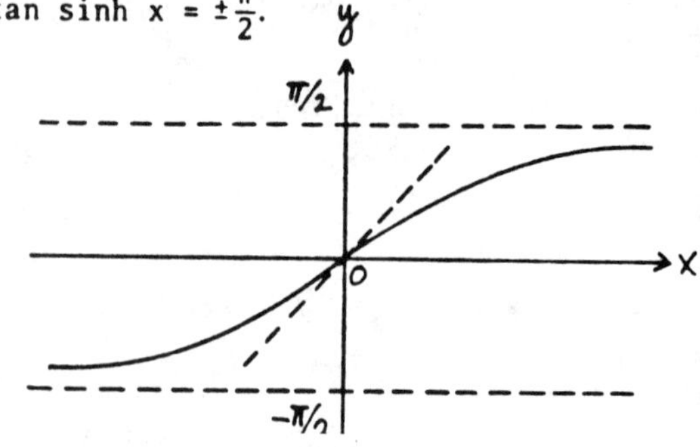

CHAPTER 9, SECTION 8 (pp. 446-477)

1. $\int \cosh(2x + 1)dx = \frac{1}{2}\int 2 \cosh(2x + 1)dx = \frac{1}{2} \sinh(2x + 1) + C$

5. $\int \sinh^3 x \cosh x \, dx = \frac{1}{4} \sinh^4 x + C.$

9. $\int \frac{\sinh x \, dx}{1 + \cosh^2 x} = \text{Arctan} \cosh x + C.$

13. $A = \int_0^1 y \, dx = \int_0^1 \cosh x \, dx = \sinh x\Big|_0^1 = \sinh 1 - \sinh 0 = \text{si}$

CHAPTER 9, SECTION 9 (pp. 452-454)

1. $\sinh^{-1}3 = \ln(3 + \sqrt{9 + 1}) = \ln(3 + \sqrt{10}) = \ln(3 + 3.162)$

$= \ln 6.162 = 1.818.$

5. $\coth^{-1}3 = \frac{1}{2} \ln \frac{4}{2} = \frac{1}{2} \ln 2 = \frac{1}{2}(0.6931) = 0.347.$

9. $y' = \frac{1/2\sqrt{x}}{\sqrt{1 + x}} = \frac{1}{2\sqrt{x^2 + x}}.$

13. $y' = \frac{e^x}{1 - e^{2x}} = \frac{1}{e^{-x} - e^x} = -\frac{1}{2} \text{csch } x.$

17. $y' = 2 \sinh^{-1}x \cdot \frac{1}{\sqrt{1 + x^2}} = \frac{2 \sinh^{-1}x}{\sqrt{1 + x^2}}.$

21. $y = \frac{1}{2}\ln(x^2 - 1) + x\tanh^{-1}x$

$$y' = \frac{x}{x^2 - 1} + \tanh^{-1}x + x \cdot \frac{1}{1 - x^2}$$

$$= \tanh^{-1}x.$$

25. $y = \operatorname{sech}^{-1}x,\quad x = \operatorname{sech} y = \frac{1}{\cosh y} = \frac{2}{e^y + e^{-y}}$

$$xe^y + xe^{-y} = 2,\quad xe^{2y} - 2e^y + x = 0$$

$$e^y = \frac{2 \pm \sqrt{4 - 4x^2}}{2x} = \frac{1 \pm \sqrt{1 - x^2}}{x}$$

If $0 < x \leq 1$, then $\frac{1 + \sqrt{1 - x^2}}{x} \geq 1$ and $\frac{1 - \sqrt{1 - x^2}}{x} < 1$.

By definition, $\operatorname{sech}^{-1}x \geq 0$, so $e^y \geq 1$. Thus,

$$y = \ln\frac{1 + \sqrt{1 - x^2}}{x}.$$

29. $y = \coth^{-1}u,\quad u = \coth y,\quad u' = -\operatorname{csch}^2 y \cdot y',$

$$y' = -\frac{u'}{\operatorname{csch}^2 y} = -\frac{u'}{\coth^2 y - 1} = \frac{u'}{1 - u^2} \quad \text{or}$$

$$y = \coth^{-1}u = \frac{1}{2}\ln\frac{u + 1}{u - 1} = \frac{1}{2}[\ln(u + 1) - \ln(u - 1)]$$

$$y' = \frac{1}{2}\left(\frac{u'}{u + 1} - \frac{u'}{u - 1}\right) = \frac{1}{2}\,\frac{u - 1 - u - 1}{u^2 - 1}\,u' = \frac{u'}{1 - u^2}.$$

33. $100\frac{dv}{dt} = 40v^2 - 980 = 40(v^2 - 24.5)$

$$\frac{5}{2}\,\frac{v'}{v^2 - 24.5} = 1,\quad \frac{5}{2}\cdot\frac{1}{24.5}\,\frac{v'}{1 - \frac{v^2}{24.5}} = -1$$

$$\frac{5}{2\sqrt{24.5}} \cdot \frac{\frac{1}{\sqrt{24.5}}\,v'}{1 - \left(\frac{v}{\sqrt{24.5}}\right)^2} = -1$$

$$\frac{5}{2\sqrt{24.5}}\tanh^{-1}\frac{v}{\sqrt{24.5}} = -t + C$$

$v = 0$ when $t = 0$, so $C = 0$

$$v = \sqrt{24.5}\tanh\left(-\frac{2\sqrt{24.5}t}{5}\right) = -\sqrt{24.5}\tanh\frac{2\sqrt{24.5}t}{5}.$$

$$\lim_{t\to+\infty} v(t) = -\sqrt{24.5} \text{ m/sec.}$$

$$100\frac{dv}{dt} = 600v^2 - 980$$

$$\lim_{t\to+\infty}\frac{dv}{dt} = 0, \text{ so}$$

$$\lim_{t\to+\infty} 100\frac{dv}{dt} = 0 = 600[\lim_{t\to+\infty} v(t)^2] - 980$$

$$\lim_{t\to+\infty} v(t) = -\sqrt{\frac{980}{600}} = -\frac{7}{\sqrt{30}} \text{ m/sec.}$$

CHAPTER 9, REVIEW (pp. 454-456)

1. $y' = \dfrac{e^x \sin x - e^x \cos x}{\sin^2 x} = \dfrac{(\sin x - \cos x)e^x}{\sin^2 x}.$

5. $y = \ln\dfrac{x^2(x-1)}{\sqrt{x+1}} = 2\ln x + \ln(x-1) - \dfrac{1}{2}\ln(x+1)$

$$y' = \frac{2}{x} + \frac{1}{x-1} - \frac{1}{2(x+1)}$$

$$= \frac{4(x^2-1) + 2(x^2+x) - (x^2-x)}{2x(x^2-1)} = \frac{5x^2+3x-4}{2x(x^2-1)}.$$

9. $y' = -e^x \operatorname{sech} e^x \tanh e^x.$

13. $\displaystyle\int \frac{(x-2)^2}{x^2+4}\,dx = \int \frac{x^2-4x+4}{x^2+4}\,dx$

$$= \int \left(1 - 2\frac{2x}{x^2+4}\right)dx = x - 2\ln(x^2+4) + C.$$

17. $y' = \dfrac{2\ln x}{x} = \dfrac{2}{e}$ when $x = e$.

$y - 1 = \frac{2}{e}(x - e)$, $ey - e = 2x - 2e$, $2x - ey - e = 0$.

21. $y' = \dfrac{x^2 \frac{1}{x} - 2x \ln x}{x^4} = \dfrac{x - 2x \ln x}{x^4} = \dfrac{1 - 2 \ln x}{x^3}$,

$$y'' = \frac{x^3(-\frac{2}{x}) - (1 - 2 \ln x)3x^2}{x^6} = \frac{-2 -3(1 - 2 \ln x)}{x^4}$$
$$= \frac{6 \ln x - 5}{x^4}.$$

Critical points:

$1 - 2 \ln x = 0, \quad \ln x = \frac{1}{2}$,

$x = e^{1/2} = \sqrt{e}, \quad y = \frac{1}{2e}$,

$y'' < 0, \quad (\sqrt{e}, \frac{1}{2e})$ max.

Point of inflection:

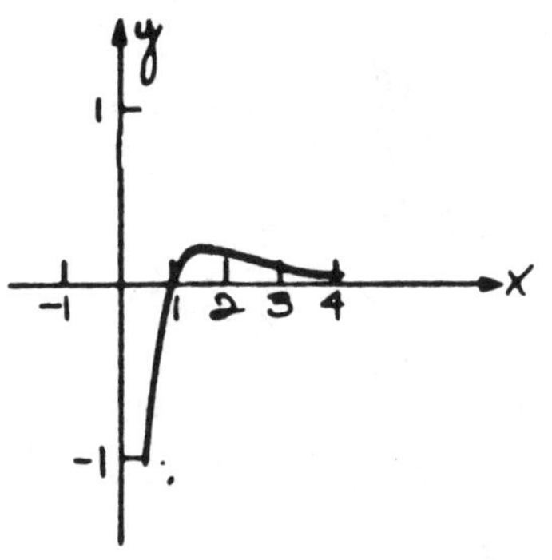

$6 \ln x - 5 = 0$

$\ln x = \frac{5}{6}; \quad x = e^{5/6}$,

$y = \dfrac{5}{6e^{5/3}}$

$(e^{5/6}, \dfrac{5}{6e^{5/3}})$.

25. $A = \int_0^1 (e^{2x} - e^x)dx$

$= (\frac{1}{2} e^{2x} - e^x)\Big|_0^1$

$= \frac{1}{2} e^2 - e - (\frac{1}{2} - 1)$

$= \dfrac{(e - 1)^2}{2}$.

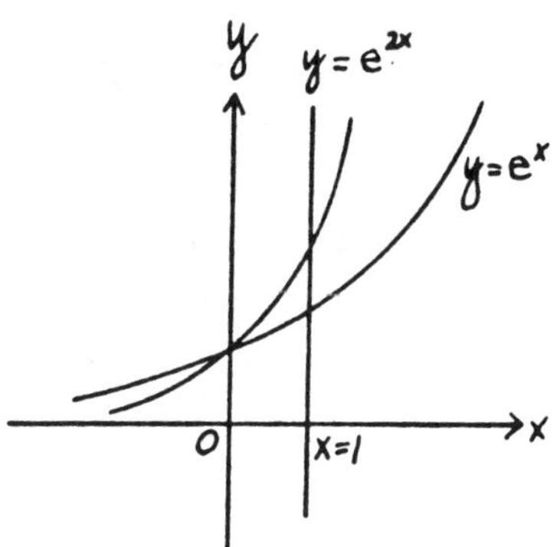

29. $x = \cosh t, \quad y = \sinh t$

$x^2 - y^2 = \cosh^2 t - \sinh^2 t = 1$.

33. $\frac{dx}{dt} = kx, \quad x = x_0 e^{kt} = 912 \times 10^6 e^{kt}$.

Since $x = 1590 \times 10^6$ when $t = 100$,

$1590 \times 10^6 = 912 \times 10^6 e^{100k}, \quad e^{100k} = 1.7434$

$100k = \ln 1.74 = 0.5558, \quad k = 0.005558$

$x = 912 \times 10^6 e^{0.005558t}$.

In year 2000: $t = 200$

$x = 912 \times 10^6 e^{1.1116} = 912 \times 10^6 (3.0395) = 2{,}772{,}000{,}000$.

In year 2500: $t = 700$

$x = 912 \times 10^6 e^{3.891} = 912 \times 10^6 (48.94) = 44{,}630{,}000{,}000$.

37. (a) $\frac{340{,}000}{7100} = 48$ yr, $1972 + 48 = 2020$.

(b) $A = \frac{x_0}{k}(e^{kt} - 1)$

$340{,}000 = \frac{7100}{0.04}(e^{0.04t} - 1)$

$e^{0.04t} - 1 = 1.9155, \quad e^{0.04t} = 2.9155$

$0.04t = \ln 2.9155 = 1.0702$

$t = 27$ yr, $1972 + 27 = 1999$.

41. $y' = 1 + 18.75 \times 10^{-5}x - 46.875 \times 10^{-9}x^2 = 1.09$

$46.875x^2 - 18.75 \times 10^4 x + 9 \times 10^7 = 0$

$x^2 - 4000x + 1{,}920{,}000 = 0$

$$(x - 2000)^2 = 4{,}000{,}000 - 1{,}920{,}000 = 2{,}080{,}000$$

$x = 2000 \pm 1442, \quad x = 3442$

$y - x = 3915 - 3442 = 473$

Chapter 10
Methods of Integration

CHAPTER 10, SECTION 1 (pp. 461-463)

1. $\int \frac{x^2 + 2x - 6}{x}\,dx = \int (x + 2 - \frac{6}{x})dx = \frac{x^2}{2} + 2x - 6\ \ln|x| + C.$

5. $\int \frac{u^3 + u}{u - 1}\,du = \int (u^2 + u + 2 + \frac{2}{u-1})du$

$= \frac{u^3}{3} + \frac{u^2}{2} + 2u + 2\ \ln|u - 1| + C.$

9. $\int (x + 1)(x^2 + 2x)^{2/3}\,dx = \frac{1}{2}\int (2x + 2)(x^2 + 2x)^{2/3}\,dx$

$= \frac{1}{2}\,\frac{(x^2 + 2x)^{5/3}}{5/3} + C = \frac{3}{10}(x^2 + 2x)^{5/3} + C.$

13. $-\int \frac{-\cos x}{1-\sin x}\,dx = -\ \ln|1 - \sin x| + C.$

17. $\int \frac{\sec^2 u}{1 + \tan u}\,du = \ln|1 + \tan u| + C.$

21. $\int \frac{\operatorname{csch}^2 x}{1 - \coth x}\,dx = \ln|1 - \coth x| + C.$

25. $\int \frac{dx}{\sqrt{2 - x^2}} = \int \frac{1/\sqrt{2}}{\sqrt{1 - x^2/2}}\,dx = \text{Arcsin}\ \frac{x}{\sqrt{2}} + C.$

29. $\int (1 - \sqrt{x})^2 dx = \int (1 - 2\sqrt{x} + x)dx = x - 2\,\frac{x^{3/2}}{3/2} + \frac{x^2}{2} + C$

$= x - \frac{4}{3}\,x^{3/2} + \frac{1}{2}\,x^2 + C.$

33. $\int \frac{e^x}{e^{2x}+1}\,dx = \int \frac{e^x}{(e^x)^2+1}\,dx = \text{Arctan } e^x + C.$

37. $\int \frac{e^{1/x}}{x^2}\,dx = -\int -\frac{1}{x^2} e^{1/x}\,dx = -e^{1/x} + C.$

41. $\int \frac{e^x - 1}{e^x + 1}\,dx = \int (-1 + \frac{2e^x}{e^x+1})dx = -x + 2\ln(e^x+1) + C,$

45. $\int \csc u \cdot u'dx = \int \frac{(\csc^2 u + \csc u \cot u)u'dx}{\csc u + \cot u}$

$= -\int \frac{(-\csc u \cot u - \csc^2 u)u'dx}{\csc u + \cot u} = -\ln|\csc u + \cot u| + C$

$= -\ln\left|\frac{1+\cos u}{\sin u}\right| + C = \ln\left|\frac{\sin u}{1+\cos u}\right| + C$

$= \ln\left|\frac{\sin u(1-\cos u)}{1-\cos^2 u}\right| + C = \ln\left|\frac{\sin u(1-\cos u)}{\sin^2 u}\right| + C$

$= \ln|\csc u - \cot u| + C.$

CHAPTER 10, SECTION 2 (pp. 468-469)

1. $u = \sqrt{x+3}$, $x = u^2 - 3$, $dx = 2u\,du$,

$\int x\sqrt{x+3}\,dx = \int (u^2-3)u\cdot 2u\,du = 2\int (u^4 - 3u^2)du$

$= 2(\frac{u^5}{5} - u^3) + C = \frac{2}{5}u^3(u^2-5) + C$

$= \frac{2}{5}(x+3)^{3/2}[(x+3)-5] + C = \frac{2}{5}(x+3)^{3/2}(x-2) + C.$

5. $u = \sqrt{2x+1}$, $x = \frac{u^2-1}{2}$, $dx = u\,du$,

$\int x\sqrt{2x+1}\,dx = \int \frac{u^2-1}{2}u\cdot u\,du = \frac{1}{2}\int (u^4 - u^2)du$

$$= \frac{1}{2}\left(\frac{u^5}{5} - \frac{u^3}{3}\right) + C = \frac{1}{30} u^3(3u^2 - 5) + C$$

$$= \frac{1}{30}(2x + 1)^{3/2}[3(2x + 1) - 5] = \frac{1}{15}(2x + 1)^{3/2}(3x - 1) + C.$$

9. $u = \sqrt{x - 2}$, $x = u^2 + 2$, $dx = 2u\ du$,

$$\int \frac{x^2}{\sqrt{x - 2}}\, dx = \int \frac{(u^2 + 2)^2}{u} \cdot 2u\ du = 2 \int (u^4 + 4u^2 + 4)du$$

$$= 2\left(\frac{u^5}{5} + \frac{4u^3}{3} + 4u\right) + C = \frac{2}{15} u(3u^4 + 20u^2 + 60) + C$$

$$= \frac{2}{15}\sqrt{x - 2}\ [3(x - 2)^2 + 20(x - 2) + 60] + C$$

$$= \frac{2}{15}\sqrt{x - 2}\ (3x^2 + 8x + 32) + C.$$

13. $u = \sqrt{x - 4}$, $x = u^2 + 4$, $dx = 2u\ du$,

$$\int_5^7 \frac{x}{\sqrt{x - 4}}\, dx = \int_1^{\sqrt{3}} \frac{u^2 + 4}{u}\ 2u\ du = 2 \int_1^{\sqrt{3}} (u^2 + 4)du$$

$$= 2\left(\frac{u^3}{3} + 4u\right)\Bigg|_1^{\sqrt{3}} = \frac{2}{3} u(u^2 + 12)\Bigg|_1^{\sqrt{3}} = \frac{2(15\sqrt{3} - 13)}{3}.$$

17. $u = \sqrt{2x + 1}$, $x = \frac{u^2 - 1}{2}$, $dx = u\ du$,

$$\int_0^4 \frac{3x - 2}{\sqrt{2x + 1}}\, dx = \int_1^3 \frac{(3u^2 - 3)/2 - 2}{u}\ u\ du = \int_1^3 \left(\frac{3}{2}u^2 - \frac{7}{2}\right)du$$

$$= \frac{1}{2}(u^3 - 7u)\Bigg|_1^3 = \frac{1}{2}[(27 - 21) - (1 - 7)] = 6.$$

21. $u = \sqrt[3]{x + 1}$, $x = u^3 - 1$, $dx = 3u^2 du$,

$$\int x \sqrt[3]{x + 1}\ dx = \int (u^3 - 1)u \cdot 3u^2 du = 3 \int (u^6 - u^3)du$$

$$= 3\left(\frac{u^7}{7} - \frac{u^4}{4}\right) + C = \frac{3}{28} u^4 (4u^3 - 7) + C$$

$$= \frac{3}{28}(x+1)^{4/3}[4(x+1) - 7] + C = \frac{3}{28}(x+1)^{4/3}(4x-3) + C.$$

25. $u = \sqrt{2x+1}$, $x = \frac{u^2 - 1}{2}$, $dx = u\,du$,

$$\int (2x+2)\sqrt{2x+1}\,dx = \int (u^2+1)u \cdot u\,du = \int (u^4 + u^2)du$$

$$= \frac{u^5}{5} + \frac{u^3}{3} + C = \frac{1}{15}u^3(3u^2+5) + C$$

$$= \frac{1}{15}(2x+1)^{3/2}[3(2x+1)+5] + C = \frac{1}{15}(2x+1)^{3/2}(6x+8) + C.$$

29. $u = \sqrt{2x-3}$, $x = \frac{u^2+3}{2}$, $dx = u\,du$,

$$\int_2^4 x^2\sqrt{2x-3}\,dx = \int_1^{\sqrt{5}} \frac{(u^2+3)^2}{4}\, u \cdot u\,du$$

$$= \frac{1}{4}\int_1^{\sqrt{5}} (u^6 + 6u^4 + 9u^2)du$$

$$= \frac{1}{4}\left(\frac{u^7}{7} + \frac{6u^5}{5} + 3u^3\right)\Bigg|_1^{\sqrt{5}} = \frac{1}{140}u^3(5u^4 + 42u^2 + 105)\Bigg|_1^{\sqrt{5}}$$

$$= \frac{1}{140}[5\sqrt{5}(125 + 210 + 105) - (5 + 42 + 105)]$$

$$= \frac{2200\sqrt{5} - 152}{140} = \frac{550\sqrt{5} - 38}{35}.$$

33. Let $F(x) = \int f(x)dx$. Then $F'(x) = f(x)$ and

$$\frac{d}{du}F(\phi(u)) = F'(\phi(u))\frac{d\phi(u)}{du} = f(\phi(u))\frac{d\phi(u)}{du},$$

so $F(\phi(u)) = \int f(\phi(u))\frac{d\phi(u)}{du}du.$

CHAPTER 10, SECTION 3 (pp. 472-473)

1. $\int \sin^2 x \cos x \, dx = \frac{\sin^3 x}{3} + C$.

5. $\int \sin^3 3x \cos^2 3x \, dx = -\frac{1}{3}\int (-3 \sin 3x)(1 - \cos^2 3x)\cos^2 3x \, dx$

$= \frac{1}{3}\int (-3 \sin 3x)(\cos^4 3x - \cos^2 3x)dx$

$= \frac{1}{3}(\frac{\cos^5 3x}{5} - \frac{\cos^3 3x}{3}) + C = \frac{1}{45}\cos^3 3x(3 \cos^2 3x - 5) + C$.

9. $\int \sin^2 2x \cos^2 2x \, dx = \int \frac{1 - \cos 4x}{2} \cdot \frac{1 + \cos 4x}{2} dx$

$= \frac{1}{4}\int (1 - \cos^2 4x)dx = \frac{1}{4}\int (1 - \frac{1 + \cos 8x}{2})dx$

$= \frac{1}{4}\int (\frac{1}{2} - \frac{1}{2}\cos 8x)dx = \frac{1}{4}\int (\frac{1}{2} - \frac{1}{16} \cdot 8 \cos 8x)dx$

$= \frac{1}{4}(\frac{x}{2} - \frac{\sin 8x}{16}) + C = \frac{x}{8} - \frac{\sin 8x}{64} + C$.

13. $\int_{-\pi/4}^{0} \sin^4 2\theta \, d\theta = \int_{-\pi/4}^{0} \frac{(1 - \cos 4\theta)^2}{4} d\theta$

$= \frac{1}{4}\int_{-\pi/4}^{0} (1 - 2 \cos 4\theta + \cos^2 4\theta)d\theta$

$= \frac{1}{4}\int_{-\pi/4}^{0} (1 - 2 \cos 4\theta + \frac{1 + \cos 8\theta}{2})d\theta$

$= \frac{1}{4}\int_{-\pi/4}^{0} (\frac{3}{2} - \frac{1}{2} \cdot 4 \cos 4\theta + \frac{1}{16} \cdot 8 \cos 8\theta)d\theta$

$= \frac{1}{4}(\frac{3\theta}{2} - \frac{\sin 4\theta}{2} + \frac{\sin 8\theta}{16})\Big|_{-\pi/4}^{0} = -\frac{1}{4}(-\frac{3\pi}{8}) = \frac{3\pi}{32}$.

17. $\displaystyle\int \cos^{3/2}x \sin^3 x\, dx = \int \cos^{3/2}x(1 - \cos^2 x)\sin x\, dx$

$\displaystyle = -\int (\cos^{3/2}x - \cos^{7/2}x)(-\sin x)\, dx$

$\displaystyle = -\frac{2}{5}\cos^{5/2}x + \frac{2}{9}\cos^{9/2}x + C\,.$

21. $\displaystyle\int \frac{dx}{\csc^2 x \cot x} = \int \sin^2 x \frac{\sin x}{\cos x}\, dx = \int (1 - \cos^2 x)\frac{\sin x}{\cos x}\, dx$

$\displaystyle = \int \left(\frac{\sin x}{\cos x} - \sin x \cos x\right)dx = -\ln|\cos x| - \frac{\sin^2 x}{2} + C\,.$

25. $\displaystyle\int \cos^3 t\, \sqrt{\sin t}\, dt = \int (1 - \sin^2 t)\, \sqrt{\sin t}\, \cos t\, dt$

$\displaystyle = \int (\sin^{1/2}t - \sin^{5/2}t)\cos t\, dt = \frac{2}{3}\sin^{3/2}t - \frac{2}{7}\sin^{7/2}t + C$

$\displaystyle = -\frac{2}{21}\sin^{3/2}t(3\sin^2 t - 7) + C\,.$

29. $\displaystyle\int \sin^m u \cdot \cos^n u \cdot u'dx = \int \sin^m u \cos^{n-1}u \cdot \cos u \cdot u'dx$

$\displaystyle = \int \sin^m u(1 - \sin^2 u)^{(n-1)/2}\cos u \cdot u'dx\,.$

Since n is even, $n - 1$ is odd and $(n - 1)/2$ is not a whole number. Thus $\sin^m u(1 - \sin^2 u)^{(n-1)/2}$ cannot be written as a polynomial in sin u. A similar problem arises when we try to express $\sin^m u$ as $\sin^{m-1}u \cdot \sin u$.

CHAPTER 10, SECTION 4 (pp. 478-479)

1. $\displaystyle\int \sec^2 x \tan^2 x\, dx = \frac{\tan^3 x}{3} + C\,.$

5. $\displaystyle\int \csc\theta \cot^3\theta \, d\theta = -\int(\csc^2\theta - 1)(-\csc\theta \cot\theta)d\theta$

$\displaystyle = -\frac{\csc^3\theta}{3} + \csc\theta + C .$

9. $\displaystyle\int \sec^6 x \, dx = \int(1 + \tan^2 x)^2 \sec^2 x \, dx$

$\displaystyle = \int(1 + 2\tan^2 x + \tan^4 x)\sec^2 x \, dx$

$\displaystyle = \tan x + \frac{2\tan^3 x}{3} + \frac{\tan^5 x}{5} + C .$

13. $\displaystyle\int_0^{\pi/3} \tan^2 x \, dx = \int_0^{\pi/3} (\sec^2 x - 1)dx$

$\displaystyle = (\tan x - x)\Big|_0^{\pi/3} = \sqrt{3} - \frac{\pi}{3} .$

17. $\displaystyle\int \tan^4 3x \, dx = \int(\sec^2 3x - 1)\tan^2 3x \, dx$

$\displaystyle = \int(\sec^2 3x \tan^2 3x - \tan^2 3x)dx$

$\displaystyle = \int(\sec^2 3x \tan^2 3x - \sec^2 3x + 1)dx$

$\displaystyle = \int(\frac{1}{3}\tan^2 3x \cdot 3\sec^2 3x - \frac{1}{3} \cdot 3\sec^2 3x + 1)dx$

$\displaystyle = \frac{\tan^3 3x}{9} - \frac{\tan 3x}{3} + x + C .$

21. $\displaystyle\int \frac{\tan\theta}{1 - \tan^2\theta} d\theta = \frac{1}{2}\int \tan 2\theta \, d\theta = -\frac{1}{4}\ln|\cos 2\theta| + C .$

25. $\displaystyle\int \frac{\sin\theta \cos\theta}{\sin^2\theta - \cos^2\theta} d\theta = -\frac{1}{2}\int \frac{2\sin\theta \cos\theta}{\cos^2\theta - \sin^2\theta} d\theta$

$\displaystyle = -\frac{1}{2}\int \frac{\sin 2\theta}{\cos 2\theta} d\theta = \frac{1}{4}\ln|\cos 2\theta| + C .$

29. $\displaystyle\int \frac{\tan^2 x}{\sec^3 x}\,dx = \int \sin^2 x \cos x\,dx = \frac{\sin^3 x}{3} + C\,.$

33. $\displaystyle\int_{-\pi/4}^{0} \frac{\sec^2\theta \tan\theta}{\sqrt{3 + \sec^2\theta}}\,d\theta = \int_{-\pi 4}^{0} \frac{\tan\theta}{\sqrt{4 + \tan^2\theta}}\sec^2\theta\,d\theta$

$\displaystyle = \sqrt{4 + \tan^2\theta}\Big|_{-\pi/4}^{0} = \sqrt{4} - \sqrt{5} = 2 - \sqrt{5}.$

CHAPTER 10, SECTION 5 (pp. 486-487)

1. $x = \sin\theta\,,\quad dx = \cos\theta\,d\theta\,,$

$\displaystyle\int \frac{\sqrt{1 - x^2}}{x^2}\,dx = \int \frac{\sqrt{1 - \sin^2\theta}}{\sin^2\theta}\cdot\cos\theta\,d\theta = \int \frac{\cos^2\theta}{\sin^2\theta}\,d\theta$

$\displaystyle = \int \cot^2\theta\,d\theta = \int(\csc^2\theta - 1)d\theta = -\cot\theta - \theta + C$

$\displaystyle = -\frac{\sqrt{1 - x^2}}{x} - \text{Arcsin}\,x + C\,.$

5. $x = 3\sin\theta\,,\quad dx = 3\cos\theta\,d\theta\,,$

$\displaystyle\int \frac{x^3 dx}{\sqrt{9 - x^2}} = \int \frac{27\sin^3\theta\cdot 3\cos\theta\,d\theta}{\sqrt{9 - 9\sin^2\theta}} = \int 27\sin^3\theta\,d\theta$

$\displaystyle = 27\int(1 - \cos^2\theta)\sin\theta\,d\theta = 27(-\cos\theta + \frac{\cos^3\theta}{3}) + C$

$\displaystyle = 9\cos\theta(\cos^2\theta - 3) + C = 9\cdot\frac{\sqrt{9 - x^2}}{3}(\frac{9 - x^2}{9} - 3) + C$

$\displaystyle = \frac{1}{3}\sqrt{9 - x^2}\,(9 - x^2 - 27) + C = -\frac{1}{3}\sqrt{9 - x^2}\,(x^2 + 18) + C\,.$

9. $x = 2\sin\theta\,,\quad dx = 2\cos\theta\,d\theta\,,$

$\displaystyle\int \frac{\sqrt{4 - x^2}}{x}\,dx = \int \frac{2\cos\theta\cdot 2\cos\theta\,d\theta}{2\sin\theta} = 2\int \frac{1 - \sin^2\theta}{\sin\theta}\,d\theta$

$$= 2\int(\csc\theta - \sin\theta)d\theta = 2\ln|\csc\theta - \cot\theta| + 2\cos\theta + C$$

$$= 2\ln\left|\frac{\sqrt{4-x^2}}{x} - \frac{2}{x}\right| + 2\cdot\frac{\sqrt{4-x^2}}{2} + C$$

$$= 2\ln\left|\frac{\sqrt{4-x^2}-2}{x}\right| + \sqrt{4-x^2} + C.$$

13. $x = 3\sin\theta$, $dx = 3\cos\theta\, d\theta$,

$$\int(9-x^2)^{3/2}dx = \int(9-9\sin^2\theta)^{3/2}\cdot 3\cos\theta\, d\theta$$

$$= \int 27\cos^3\theta\cdot 3\cos\theta\, d\theta = 81\int\cos^4\theta\, d\theta$$

$$= \frac{81}{4}\int(1+\cos 2\theta)^2 d\theta = \frac{81}{4}\int(1+2\cos 2\theta+\cos^2 2\theta)d\theta$$

$$= \frac{81}{4}\int(1+2\cos 2\theta+\frac{1+\cos 4\theta}{2})d\theta$$

$$= \frac{81}{4}(\frac{3\theta}{2}+\sin 2\theta+\frac{1}{8}\sin 4\theta) + C$$

$$= \frac{243}{8}\theta + \frac{81}{2}\sin\theta\cos\theta + \frac{81}{16}\sin 2\theta\cos 2\theta + C$$

$$= \frac{243}{8}\theta + \frac{81}{2}\sin\theta\cos\theta + \frac{81}{8}\sin\theta\cos\theta(\cos^2\theta - \sin^2\theta) + C$$

$$= \frac{243}{8}\text{Arcsin}\,\frac{x}{3} + \frac{81}{2}\cdot\frac{x}{3}\cdot\frac{\sqrt{9-x^2}}{3} + \frac{81}{8}\cdot\frac{x}{3}\cdot\frac{\sqrt{9-x^2}}{3}(\frac{9-x^2}{9} - \frac{x^2}{9}) + C$$

$$= \frac{243}{8}\text{Arcsin}\,\frac{x}{3} + \frac{9}{2}x\sqrt{9-x^2} + \frac{x}{8}\sqrt{9-x^2}\,(9-2x^2) + C.$$

17. $x = \frac{\sqrt{2}}{\sqrt{3}}\tan\theta$, $dx = \frac{\sqrt{2}}{\sqrt{3}}\sec^2\theta\, d\theta$,

$$\int\frac{dx}{(3x^2+2)^{3/2}} = \int\frac{(\sqrt{2}\sec^2\theta)/\sqrt{3}}{(2\tan^2\theta+2)^{3/2}}\,d\theta = \int\frac{\sqrt{2}\sec^2\theta\, d\theta}{\sqrt{3}\cdot 2\sqrt{2}\sec^3\theta}$$

$$= \frac{1}{2\sqrt{3}} \int \cos\theta \, d\theta = \frac{\sin\theta}{2\sqrt{3}} + C = \frac{\sqrt{3}\, x}{2\sqrt{3}\sqrt{3x^2 + 2}} + C$$

$$= \frac{x}{2\sqrt{3x^2 + 2}} + C .$$

21. $x = \text{Arcsin } e^y , \quad dx = \dfrac{e^y}{\sqrt{1 - e^{2y}}} dy$

$$s = \int_{-1}^{-1/2} \sqrt{1 + \frac{e^{2y}}{1 - e^{2y}}} \, dy = \int_{-1}^{-1/2} \frac{1}{\sqrt{1 - e^{2y}}} dy$$

$$e^y = \sin\theta , \quad y = \ln \sin\theta , \quad dy = \frac{\cos\theta}{\sin\theta} d\theta$$

$$= \int_{\text{Arcsin } 1/e}^{\text{Arcsin } 1/\sqrt{e}} \frac{\cos\theta \, d\theta}{\sin\theta \cos\theta} = \int_{\text{Arcsin } 1/e}^{\text{Arcsin } 1/\sqrt{e}} \csc\theta \, d\theta$$

$$= \ln\left|\csc\theta - \cot\theta\right| \Big|_{\text{Arcsin } 1/e}^{\text{Arcsin } 1/\sqrt{e}}$$

$$= \ln(\sqrt{e} - \sqrt{e - 1}) - \ln(e - \sqrt{e^2 - 1}) = \ln \frac{\sqrt{e} - \sqrt{e - 1}}{e - \sqrt{e^2 - 1}} .$$

CHAPTER 10, SECTION 6 (pp. 490-491)

1. $\displaystyle\int \frac{dx}{x^2 + 2x + 2} = \int \frac{dx}{(x + 1)^2 + 1} = \text{Arctan}(x + 1) + C .$

5. $x - 1 = \sin\theta , \quad dx = \cos\theta \, d\theta ,$

$$\int \sqrt{8x - 4x^2} \, dx = \int 2\sqrt{1 - (x - 1)^2} \, dx$$

$$= 2 \int \sqrt{1 - \sin^2\theta} \cos\theta \, d\theta = 2 \int \cos^2\theta \, d\theta$$

$$= \int (1 + \cos 2\theta) d\theta = \theta + \frac{1}{2} \sin 2\theta + C$$

$$= \theta + \sin\theta \cos \theta + C = \text{Arcsin}(x - 1) + (x-1)\sqrt{2x - x^2} + C.$$

9. $x - 5 = 2 \sin \theta$, $dx = 2 \cos\theta \, d\theta$,

$$\int \frac{\sqrt{-x^2 + 10x - 21}}{x - 5} dx = \int \frac{\sqrt{4 - (x - 5)^2}}{x - 5} dx$$

$$= \int \frac{\sqrt{4 - 4 \sin^2\theta} \; 2 \cos\theta}{2 \sin \theta} d\theta = 2 \int \frac{\cos^2\theta}{\sin \theta} d\theta$$

$$= 2 \int \frac{1 - \sin^2\theta}{\sin \theta} d\theta = 2 \int (\csc \theta - \sin \theta) d\theta$$

$$= 2 \ln|\csc \theta - \cot \theta| + 2 \cos \theta + C$$

$$= 2 \ln\left|\frac{2}{x - 5} - \frac{\sqrt{-x^2 + 10x - 21}}{x - 5}\right| + 2 \frac{\sqrt{-x^2 + 10x - 21}}{2} + C$$

$$= 2 \ln\left|\frac{2 - \sqrt{-x^2 + 10x - 21}}{x - 5}\right| + \sqrt{-x^2 + 10x - 21} + C .$$

13. $x + 1 = \tan \theta$, $x = \tan \theta - 1$, $dx = \sec^2\theta \, d\theta$,

$$\int \frac{x^3 + 1}{x^2 + 2x + 2} dx = \int \frac{x^3 + 1}{(x + 1)^2 + 1} dx$$

$$= \int \frac{[(\tan \theta - 1)^3 + 1]\sec^2\theta \, d\theta}{\tan^2\theta + 1}$$

$$= \int (\tan^3\theta - 3 \tan^2\theta + 3 \tan \theta) d\theta$$

$$= \int [\tan^2\theta(\tan \theta - 3) + 3 \tan \theta] d\theta$$

$$= \int [(\sec^2\theta - 1)(\tan \theta - 3) + 3 \tan \theta] d\theta$$

$$= \int (\sec^2\theta \tan \theta - 3 \sec^2\theta + 2 \tan \theta + 3) d\theta$$

$$= \frac{1}{2}\tan^2\theta - 3\tan\theta - 2\ln|\cos\theta| + 3\theta + C$$

$$= \frac{1}{2}(x+1)^2 - 3(x+1) - 2\ln\frac{1}{\sqrt{x^2+2x+2}} + 3\,\text{Arctan}(x+1) + C$$

$$= \frac{1}{2}(x+1)(x-5) + \ln(x^2+2x+2) + 3\,\text{Arctan}(x+1) + C\ .$$

17. $x + 1 = 3\tan\theta\ ,\quad dx = 3\sec^2\theta\, d\theta\ ,$

$$\int_{-1}^{2} \frac{dx}{x^2+2x+10} = \int_{-1}^{2} \frac{dx}{(x+1)^2+9}$$

$$= \int_{0}^{\pi/4} \frac{3\sec^2\theta\, d\theta}{9\tan^2\theta + 9} = \frac{1}{3}\int_{0}^{\pi/4} d\theta = \frac{\pi}{12}\ .$$

21. $x + 2 = \sin\theta,\ x = \sin\theta - 2,\ dx = \cos\theta\, d\theta$

$$\int_{-5/2}^{-2} \frac{x^2+1}{(-x^2-4x-3)^{3/2}}\,dx$$

$$= \int_{-5/2}^{-2} \frac{x^2+1}{[1-(x+2)^2]^{3/2}}\,dx$$

$$= \int_{-\pi/6}^{0} \frac{(\sin\theta-2)^2+1}{(1-\sin^2\theta)^{3/2}}\cos\theta\, d\theta$$

$$= \int_{-\pi/6}^{0} \frac{\sin^2\theta - 4\sin\theta + 5}{\cos^2\theta}\,d\theta$$

$$= \int_{-\pi/6}^{0} \left(6\sec^2\theta - 1 + 4\frac{-\sin\theta}{\cos^2\theta}\right)d\theta$$

$$= 6\tan\theta - \theta - 4\sec\theta\Big|_{-\pi/6}^{0}$$

$$= -4 + \frac{6}{\sqrt{3}} - \frac{\pi}{6} + 4\frac{2}{\sqrt{3}} = \frac{14}{\sqrt{3}} - 4 - \frac{\pi}{6}.$$

CHAPTER 10, SECTION 7 (pp. 497-498)

1. $u = x \qquad v' = \cos x$

   $u' = 1 \qquad v = \sin x$

   $$\int x \cos x \, dx = x \sin x - \int \sin x \, dx = x \sin x + \cos x + C.$$

5. $u = \ln x^2 = 2 \ln x \qquad v' = 1$

   $u' = \frac{2}{x} \qquad v = x$

   $$\int \ln x^2 dx = x \ln x^2 - \int 2 \, dx = x \ln x^2 - 2x + C .$$

9. $u = \text{Arcsin } x \qquad v' = 1$

   $u' = \frac{1}{\sqrt{1 - x^2}} \qquad v = x$

   $$\int \text{Arcsin } x \, dx = x \text{ Arcsin } x - \int \frac{x \, dx}{\sqrt{1 - x^2}}$$

   $$= x \text{ Arcsin } x + \frac{1}{2} \int \frac{-2x \, dx}{\sqrt{1 - x^2}} = x \text{ Arcsin } x + \sqrt{1 - x^2} + C .$$

13. $u = \sinh^{-1} x \qquad v' = 1$

    $u' = \frac{1}{\sqrt{1 + x^2}} \qquad v = x$

    $$\int \sinh^{-1} x \, dx = x \sinh^{-1} x - \int \frac{x \, dx}{\sqrt{1 + x^2}}$$

    $$= x \sinh^{-1} x - \frac{1}{2} \int \frac{2x \, dx}{\sqrt{1 + x^2}} = x \sinh^{-1} x - \sqrt{1 + x^2} + C .$$

17. $u = x^3 \qquad v' = x^2 \sqrt{x^3 + 3} = \frac{1}{3} \cdot 3x^2 \sqrt{x^3 + 3}$

    $u' = 3x^2 \qquad v = \frac{2}{9} (x^3 + 3)^{3/2}$

$$\int x^5 \sqrt{x^3+3}\, dx = \frac{2}{9} x^3(x^3+3)^{3/2} - \int \frac{2}{9} \cdot 3x^2(x^3+3)^{3/2} dx$$

$$= \frac{2}{9} x^3(x^3+3)^{3/2} - \frac{4}{45} (x^3+3)^{5/2} + C .$$

21. $u = x^n$ $\qquad v' = x^{n-1}\sqrt{x^n+3} = \frac{1}{n} nx^{n-1} \sqrt{x^n+3}$

$u' = nx^{n-1}$ $\qquad v = \frac{2}{3n} (x^n+3)^{3/2}$

$$\int x^{2n-1} \sqrt{x^n+3}\, dx$$

$$= \frac{2x^n}{3n} (x^n+3)^{3/2} - \int \frac{2}{3n} nx^{n-1}(x^n+3)^{3/2} dx$$

$$= \frac{2x^n}{3n} (x^n+3)^{3/2} - \frac{4}{15n} (x^n+3)^{5/2} + C .$$

25.

u	v'
x^4	$\sin x$
$4x^3$	$-\cos x$
$12x^2$	$-\sin x$
$24x$	$\cos x$
24	$\sin x$
0	$-\cos x$

$$\int x^4 \sin x\, dx = -x^4\cos x + 4x^3\sin x + 12x^2\cos x - 24x \sin x$$

$$- 24 \cos x + C .$$

29. $u = \ln^2 x$ $\qquad v' = x^3$

$u' = \frac{2 \ln x}{x}$ $\qquad v = \frac{x^4}{4}$

$$\int x^3\ln^2 x\, dx = \frac{x^4}{4} \ln^2 x - \int \frac{1}{2} x^3 \ln x\, dx$$

$u = \ln x \qquad v' = \frac{x^3}{2}$

$u' = \frac{1}{x} \qquad v = \frac{x^4}{8}$

$$= \frac{x^4}{4} \ln^2 x - \frac{x^4}{8} \ln x + \int \frac{x^3}{8}\, dx$$

$$= \frac{x^4}{4} \ln^2 x - \frac{x^4}{8} \ln x + \frac{x^4}{32} + C$$

$$= \frac{x^4}{32} (8 \ln^2 x - 4 \ln x + 1) + C .$$

33. $u = \sec^3\theta \qquad v' = \sec^2\theta$

$u' = 3 \sec^3\theta \tan \theta \qquad v = \tan \theta$

$$\int \sec^5\theta\, d\theta = \sec^3\theta \tan \theta - \int 3 \sec^3\theta \tan^2\theta\, d\theta$$

$$= \sec^3\theta \tan \theta - \int 3 \sec^3\theta(\sec^2\theta - 1)d\theta$$

$$= \sec^3\theta \tan \theta + 3 \int \sec^3\theta\, d\theta - 3 \int \sec^5\theta\, d\theta$$

Example 8

$$4 \int \sec^5\theta\, d\theta = \sec^3\theta \tan \theta + \frac{3}{2}(\sec \theta \tan \theta + \ln|\sec \theta + \tan \theta|)$$

$$\int \sec^5\theta\, d\theta = \frac{1}{8}(2 \sec^3\theta \tan \theta + 3 \sec \theta \tan \theta + 3 \ln| \sec \theta + \tan \theta|) + C .$$

37. $u = \sin x \qquad v' = \sin 5x$

$u' = \cos x \qquad v = -\frac{1}{5} \cos 5x$

$$\int \sin x \sin 5x\, dx = -\frac{1}{5} \sin x \cos 5x + \int \frac{1}{5} \cos x \cos 5x\, dx$$

$u = \cos x \qquad v' = \frac{1}{5} \cos 5x$

$u' = -\sin x \qquad v = \frac{1}{25} \sin 5x$

$$\int \sin x \sin 5x \, dx = -\frac{1}{5} \sin x \cos 5x + \frac{1}{25} \cos x \sin 5x$$

$$+ \int \frac{1}{25} \sin x \sin 5x \, dx$$

$$\frac{24}{25} \int \sin x \sin 5x \, dx = \frac{1}{25}(\cos x \sin 5x - 5 \sin x \cos 5x)$$

$$+ \frac{24}{25} C$$

$$\int \sin x \sin 5x \, dx = \frac{1}{24}(\cos x \sin 5x - 5 \sin x \cos 5x) + C .$$

41. $u = \sin \theta \qquad v' = \sin \theta$

$u' = \cos \theta \qquad v = -\cos \theta$

$$\int \sin^2\theta \, d\theta = -\sin\theta \cos \theta + \int \cos^2\theta \, d\theta = -\sin \theta \cos \theta$$

$$+ \int (1 - \sin^2\theta) d\theta$$

$$= - \sin \theta \cos \theta + \theta - \int \sin^2\theta \, d\theta$$

$$2 \int \sin^2\theta \, d\theta = \theta - \sin\theta \cos \theta$$

$$\int \sin^2\theta \, d\theta = \frac{1}{2}(\theta - \sin\theta \cos \theta) + C .$$

By the double-angle formula,

$$\int \sin^2\theta \, d\theta = \frac{1}{2}(\theta - \frac{1}{2} \sin 2\theta) + C = \frac{1}{2}(\theta - \sin\theta \cos \theta) + C .$$

45. By Example 2, $\int u \ln u \, du = \frac{u^2}{2} \ln u - \frac{u^2}{4} + C.$

By Example 4, $\int \ln u \, du = u \ln u - u + C$

$3x - x^3 = x(3 - x^2) > 0$ if $0 < x < \sqrt{3}$ or $x < -\sqrt{3}$.

If $0 < x < \sqrt{3}$, then $\int x \ln(3x - x^3) dx$

$$= \int x \ln x + \int x \ln(3 - x^2) dx$$

$$= \int x \ln x - \frac{1}{2}\int \ln(3 - x^2)(-2x)dx$$

$$= \frac{x^2}{2}\ln x - \frac{x^2}{4} - \frac{1}{2}(3 - x^2)\ln(3 - x^2) + \frac{1}{2}(3 - x^2) + C$$

$$= \frac{1}{4}(2x^2\ln x - 2(3 - x^2)\ln(3 - x^2) - 3x^2 + 6) + C$$

If $x < -\sqrt{3}$, then $\int x \ln(3x - x^3)dx$

$$= \int (-x)\ln(-x)(-dx) + \frac{1}{2}\int \ln(x^2 - 3)(2x\,dx)$$

$$= \frac{x^2\ln x}{2} - \frac{x^2}{4} + \frac{1}{2}(x^2 - 3)\ln(x^2 - 3) - \frac{(x^2 - 3)}{2} + C$$

$$= \frac{1}{4}(2x^2\ln x + 2(x^2 - 3)\ln(x^2 - 3) + 6 - 3x^2) + C.$$

49. $\int_0^1 \frac{nx^{n-1}}{1 + x}dx$

$$u = \frac{1}{1 + x} \qquad v' = nx^{n-1}$$

$$u' = \frac{x^n}{(1 + x)^2} \qquad v = x^n$$

$$= \frac{x^n}{1 + x}\Big|_0^1 + \int_0^1 \frac{x^n}{(1 + x)^2}dx = 1 + \int_0^1 \frac{x^n\,dx}{(1 + x)^2}.$$

$\lim_{n\to\infty} x^n = 0$ if $0 \leq x \leq 1$, so

$\lim_{n\to\infty} \int_0^1 \frac{x^n\,dx}{(1 + x)^2} = 0$. Thus

$\lim_{n\to\infty} \int_0^1 \frac{nx^{n-1}}{1 + x}dx = 1.$

CHAPTER 10, SECTION 8 (p. 504)

1. $\frac{1}{x(x+1)} = \frac{A}{x} + \frac{B}{x+1}$,

$1 = A(x+1) + Bx = (A+B)x + A$,

$A + B = 0,\ A = 1,\ B = -1$

$$\int \frac{dx}{x(x+1)} = \int \left(\frac{1}{x} - \frac{1}{x+1}\right)dx = \ln|x| - \ln|x+1| + C$$

$$= \ln\left|\frac{x}{x+1}\right| + C .$$

5. $\int \frac{x^2+3}{x^2+3x}\,dx = \int \left[1 + \frac{-3x+3}{x(x+3)}\right]dx,\ \frac{-3x+3}{x(x+3)} = \frac{A}{x} + \frac{B}{x+3}$,

$-3x + 3 = A(x+3) + Bx = (A+B)x + 3A,\ A + B = -3,\ 3A = 3;$

$A = 1,\ B = -4,$

$$\int \frac{x^2+3}{x^2+3x}\,dx = \int \left(1 + \frac{1}{x} - \frac{4}{x+3}\right)dx = x + \ln|x| - 4\ln|x+3| + C$$

$$= x + \ln\left|\frac{x}{(x+3)^4}\right| + C .$$

9. $\int \frac{(x^3+8)dx}{x^3-3x^2+2x} = \int \left[1 + \frac{3x^2-2x+8}{x(x-1)(x-2)}\right]dx,$

$\frac{3x^2-2x+8}{x(x-1)(x-2)} = \frac{A}{x} + \frac{B}{x-1} + \frac{C}{x-2}$,

$3x^2 - 2x + 8 = A(x-1)(x-2) + Bx(x-2) + Cx(x-1)$

$= (A+B+C)x^2 + (-3A-2B-C)x + 2A$,

$A + B + C = 3,\ -3A - 2B - C = -2,\ 2A = 8;$

$A = 4,\ B = -9,\ C = 8$,

$$\int \frac{(x^3+8)dx}{x^3-3x^2+2x} = \int \left(1 + \frac{4}{x} - \frac{9}{x-1} + \frac{8}{x-2}\right)dx$$

$= x + 4 \ln|x| - 9 \ln|x-1| + 8 \ln|x-2| + C$

$= x + \ln\left|\frac{x^4(x-2)^8}{(x-1)^9}\right| + C .$

13. $\int \frac{x^2 dx}{(x-1)^2} = \int [1 + \frac{2x-1}{(x-1)^2}]dx, \quad \frac{2x-1}{(x-1)^2} = \frac{A}{x-1} + \frac{B}{(x-1)^2} ,$

$2x - 1 = A(x-1) + B = Ax + (-A + B), \; A = 2, \; -A + B = -1; \; B = 1,$

$\int \frac{x^2 dx}{(x-1)^2} = \int [1 + \frac{2}{x-1} + \frac{1}{(x-1)^2}]dx$

$= x + 2 \ln|x-1| - \frac{1}{x-1} + C .$

17. $\frac{x+2}{x^2(x+3)(x-1)} = \frac{A}{x} + \frac{B}{x^2} + \frac{C}{x+3} + \frac{D}{x-1} ,$

$x + 2 = Ax(x+3)(x-1) + B(x+3)(x-1) + Cx^2(x-1) + Dx^2(x+3)$

$= (A + C + D)x^3 + (2A + B - C + 3D)x^2 + (-3A + 2B)x - 3B,$

$A + C + D = 0, \; 2A + B - C + 3D = 0, \; -3A + 2B = 1, \; -3B = 2;$

$A = -\frac{7}{9}, \; B = -\frac{2}{3}, \; C = \frac{1}{36}, \; D = \frac{3}{4},$

$\int \frac{(x+2)dx}{x^4 + 2x^3 - 3x^2} = \int (-\frac{7/9}{x} - \frac{2/3}{x^2} + \frac{1/36}{x+3} + \frac{3/4}{x-1})dx$

$= -\frac{7}{9} \ln|x| + \frac{2}{3x} + \frac{1}{36} \ln|x+3| + \frac{3}{4} \ln|x-1| + C .$

21. $u = \sin x, \quad du = \cos x \, dx,$

$\int \frac{\cos x \, dx}{\sin^3 x + \sin^2 x} = \int \frac{du}{u^3 + u^2} = \int \frac{du}{u^2(u+1)}$

$\frac{1}{u^2(u+1)} = \frac{A}{u} + \frac{B}{u^2} + \frac{C}{u+1} ,$

$1 = Au(u+1) + B(u+1) + Cu^2 = (A + C)u^2 + (A + B)u + B,$

$A + C = 0, \; A + B = 0, \; B = 1; \; A = -1, \; B = 1, \; C = 1,$

$$= \int \left(-\frac{1}{u} + \frac{1}{u^2} + \frac{1}{u+1}\right) du = -\ln|u| - \frac{1}{u} + \ln|u+1| + C$$

$$= \ln\left|\frac{u+1}{u}\right| - \frac{1}{u} + C = \ln\left|\frac{1 + \sin x}{\sin x}\right| - \frac{1}{\sin x} + C$$

$$= \ln\left|\frac{1 + \sin x}{\sin x}\right| - \csc x + C .$$

CHAPTER 10, SECTION 9 (p. 509)

1. $$\frac{1}{x^2(x^2+1)} = \frac{A}{x} + \frac{B}{x^2} + \frac{Cx + D}{x^2 + 1},$$

$$1 = A(x^3 + x) + B(x^2 + 1) + Cx^3 + Dx^2 ,$$

$$A + C = 0, \ B + D = 0, \ A = 0, \ B = 1;$$

$$A = 0, \ B = 1, \ C = 0, \ D = -1,$$

$$\int \frac{dx}{x^2(x^2+1)} = \int \left(\frac{1}{x^2} - \frac{1}{x^2+1}\right) dx = -\frac{1}{x} - \text{Arctan } x + C .$$

5. $$\frac{3}{(x+1)(x^2+1)} = \frac{A}{x+1} + \frac{Bx + C}{x^2 + 1},$$

$$3 = A(x^2 + 1) + (Bx + C)(x + 1) = (A + B)x^2 + (B + C)x + (A + C),$$

$$A + B = 0, \ B + C = 0, \ A + C = 3; \ A = +\frac{3}{2}, \ B = -\frac{3}{2}, \ C = \frac{3}{2},$$

$$\int \frac{3\ dx}{x^3 + x^2 + x + 1} = \frac{3}{2} \int \left(\frac{1}{x+1} + \frac{1 - x}{x^2 + 1}\right) dx$$

$$= \frac{3}{2} \int \left(\frac{1}{x+1} - \frac{1}{2}\frac{2x}{x^2+1} + \frac{1}{x^2+1}\right) dx$$

$$= \frac{3}{2} \left[\ln|x+1| - \frac{1}{2}\ln(x^2+1) + \text{Arctan } x\right] + C$$

$$= \frac{3}{4} \left(\ln \frac{(x+1)^2}{x^2+1} + 2 \text{ Arctan } x\right) + C .$$

9. $\dfrac{2x}{(x+1)(x^2+1)} = \dfrac{A}{x+1} + \dfrac{Bx+C}{x^2+1}$,

$2x = A(x^2+1) + (Bx+C)(x+1) = (A+B)x^2 + (B+C)x + (A+C)$,

$A+B = 0,\ B+C = 2,\ A+C = 0;\ A = -1,\ B = 1,\ C = 1,$

$$\int \frac{2x\,dx}{(x+1)(x^2+1)} = \int\left(-\frac{1}{x+1} + \frac{x+1}{x^2+1}\right)dx$$

$$= \int\left(-\frac{1}{x+1} + \frac{1}{2}\cdot\frac{2x}{x^2+1} + \frac{1}{1+x^2}\right)dx$$

$$= -\ln|x+1| + \frac{1}{2}\ln(x^2+1) + \text{Arctan } x + C$$

$$= \frac{1}{2}\ln\frac{x^2+1}{(x+1)^2} + \text{Arctan } x + C .$$

13. $\dfrac{16}{x(x^2+4)^2} = \dfrac{A}{x} + \dfrac{Bx+C}{x^2+4} + \dfrac{Dx+E}{(x^2+4)^2}$,

$16 = A(x^2+4)^2 + (Bx+C)x(x^2+4) + (Dx+E)x$

$= (A+B)x^4 + Cx^3 + (8A+4B+D)x^2 + (4C+E)x + 16A,$

$A+B = 0,\ C = 0,\ 8A+4B+D = 0,\ 4C+E = 0,\ 16A = 16;$

$A = 1,\ B = -1,\ C = 0,\ D = -4,\ E = 0$

$$\int \frac{16\,dx}{x(x^2+4)^2} = \int\left(\frac{1}{x} - \frac{x}{x^2+4} - \frac{4x}{(x^2+4)^2}\right)dx$$

$$= \ln|x| - \frac{1}{2}\ln(x^2+4) + \frac{2}{x^2+4} + C$$

$$= \frac{1}{2}\ln\frac{x^2}{x^2+4} + \frac{2}{x^2+4} + C .$$

17. $\dfrac{x^2+3x+4}{(x^2+1)(x^2+4)^2} = \dfrac{Ax+B}{x^2+1} + \dfrac{Cx+D}{x^2+4} + \dfrac{Ex+F}{(x^2+4)^2}$,

$x^2+3x+4 = A(x^5+8x^3+16x) + B(x^4+8x^2+16)$

$$+ C(x^5 + 5x^3 + 4x) + D(x^4 + 5x^2 + 4)$$

$$+ E(x^3 + x) + F(x^2 + 1),$$

$A + C = 0,\quad 8A + 5C + E = 0,\quad 16A + 4C + E = 3,$

$B + D = 0,\quad 8B + 5D + F = 1,\quad 16B + 4D + F = 4,$

$A = \frac{1}{3},\ B = \frac{1}{3},\ C = -\frac{1}{3},\ D = -\frac{1}{3},\ E = -1,\ F = 0,$

$$3\int \frac{(x^2 + 3x + 4)dx}{(x^2 + 1)(x^2 + 4)^2} = \int [\frac{x + 1}{x^2 + 1} - \frac{x + 1}{x^2 + 4} - \frac{3x}{(x^2 + 4)^2}]dx$$

$$= \int [\frac{1}{2} \cdot \frac{2x}{x^2 + 1} + \frac{1}{x^2 + 1} - \frac{1}{2} \cdot \frac{2x}{x^2 + 4} - \frac{1}{2} \cdot \frac{1/2}{(x/2)^2 + 1}$$

$$- \frac{3}{2} \cdot \frac{2x}{(x^2 + 4)^2}]dx$$

$$= \frac{1}{2} \ln(x^2 + 1) + \text{Arctan } x - \frac{1}{2} \ln(x^2 + 4) - \frac{1}{2} \text{Arctan } \frac{x}{2}$$

$$+ \frac{3}{2} \cdot \frac{1}{x^2 + 4} + C$$

$$= \frac{1}{2}(\ln \frac{x^2 + 1}{x^2 + 4} + 2 \text{ Arctan } x - \text{Arctan } \frac{x}{2} + \frac{3}{x^2 + 4}) + C .$$

21. $\frac{1}{T^4 - 16} = \frac{A}{T + 2} + \frac{B}{T - 2} + \frac{CT + D}{T^2 + 4}$

$1 = A(T - 2)(T^2 + 4) + B(T + 2)(T^2 + 4) + (CT + D)(T^2 - 4)$

$A + B + C = 0,\ -2A + 2B + D = 0,\ 4A + 4B - 4C = 0,\ -8A + 8B - 4D = 1.$

$A = -\frac{1}{32},\ B = \frac{1}{32},\ C = 0,\ D = -\frac{1}{8}$

$\frac{T'}{T^4 - 16} = -0.01,\ \int \frac{dT}{T^4 - 16} = -0.01 \int dt$

$\frac{1}{32}\int (-\frac{1}{T + 2} + \frac{1}{T - 2} - \frac{4}{T^2 + 4})dT = -0.01t + \frac{C}{32}$

$-\ln|T + 2| + \ln|T - 2| - 2 \text{ Arctan } \frac{T}{2} = -0.32t + C$

$$\ln \frac{T - 2}{T + 2} - 2 \text{ Arctan } \frac{T}{2} = C - 0.32t$$

$t = 0,\ T = 100$: $C = \ln \frac{98}{102} - 2 \text{ Arctan } 50$

$$t = 20:\ \ln \frac{T-2}{T+2} - 2 \text{ Arctan } \frac{T}{2} = \ln \frac{98}{102} - 2 \text{ Arctan } 50 - 6.4$$
$$= -9.542$$

Using a scientific calculator, $T = 2.00138$.

CHAPTER 10, SECTION 10 (pp. 513-514)

1. $u = \sqrt{x+2}$, $x = u^2 - 2$, $dx = 2u\ du$,

$$\int \frac{x\ dx}{x + 2 + \sqrt{x+2}} = \int \frac{(u^2-2)\cdot 2u\ du}{u^2 + u} = 2\int \frac{u^2-2}{u+1}\ du$$

$$= 2\int (u - 1 - \frac{1}{u+1})du = u^2 - 2\ \ln|u+1| + C$$

$$= x + 2 - 2\sqrt{x+2} - 2\ \ln(\sqrt{x+2} + 1) + C.$$

5. $u = \sqrt[4]{x}$, $x = u^4$, $dx = 4u^3 du$,

$$\int \frac{dx}{\sqrt{x} - \sqrt[4]{x}} = \int \frac{4u^3\ du}{u^2 - u} = \int \frac{4u^2\ du}{u-1} = \int (4u + 4 + \frac{4}{u-1})du$$

$$= 2u^2 + 4u + 4\ \ln|u-1| + C = 2\sqrt{x} + 4\sqrt[4]{x} + 4\ \ln|\sqrt[4]{x} - 1| + C.$$

9. $u = \sqrt[3]{x}$, $x = u^3$, $dx = 3u^2\ du$,

$$\int \frac{\sqrt[3]{x}+1}{\sqrt[3]{x}-1}dx = \int \frac{u+1}{u-1}\ 3u^2 du = \int (3u^2 + 6u + 6 + \frac{6}{u-1})du$$

$$= u^3 + 3u^2 + 6u + 6\ \ln|u-1| + C$$

$$= x + 3x^{2/3} + 6\sqrt[3]{x} + 6\ \ln|\sqrt[3]{x} - 1| + C.$$

13. $u = \sqrt{x+2}$, $x = u^2 - 2$, $dx = 2u\ du$,

$$\int \frac{dx}{(x+2)^2 + \sqrt{x+2}} = \int \frac{2u\ du}{u^4 + u} = \int \frac{2\ du}{u^3 + 1}$$

$$\frac{2}{(u+1)(u^2-u+1)} = \frac{A}{u+1} + \frac{Bu+C}{u^2-u+1}$$

$$2 = A(u^2-u+1) + (Bu+C)(u+1)$$

$$= (A+B)u^2 + (-A+B+C)u + (A+C)$$

$A + B = 0$, $-A + B + C = 0$, $A + C = 2$;

$A = \frac{2}{3}$, $B = -\frac{2}{3}$, $C = \frac{4}{3}$,

$$\int \frac{dx}{(x+2)^2 + \sqrt{x+2}} = \frac{1}{3}\int \left(\frac{2}{u+1} - \frac{2u-4}{u^2-u+1}\right) du$$

$$= \frac{1}{3}\int \left(\frac{2}{u+1} - \frac{2u-1-3}{u^2-u+1}\right) du$$

$$= \frac{1}{3}\int \left(\frac{2}{u+1} - \frac{2u-1}{u^2-u+1} + \frac{3}{u^2-u+1/4+3/4}\right) du$$

$$= \frac{2}{3}\ln|u+1| - \frac{1}{3}\ln(u^2-u+1) + \int \frac{4/3}{4(u-1/2)^2/3+1}\, du$$

$$= \frac{1}{3}\ln\frac{(u+1)^2}{u^2-u+1} + \frac{2}{\sqrt{3}}\int \frac{2/\sqrt{3}}{(2u-1)^2/3+1}\, du$$

$$= \frac{1}{3}\ln\frac{(u+1)^2}{u^2-u+1} + \frac{2}{\sqrt{3}}\,\mathrm{Arctan}\,\frac{2u-1}{\sqrt{3}} + C$$

$$= \frac{1}{3}\ln\frac{(1+\sqrt{x+2})^2}{x+3-\sqrt{x+2}} + \frac{2}{\sqrt{3}}\,\mathrm{Arctan}\,\frac{2\sqrt{x+2}-1}{\sqrt{3}} + C.$$

17. $u = \tan\frac{\theta}{2}$, $\sin\theta = \frac{2u}{1+u^2}$, $\cos\theta = \frac{1-u^2}{1+u^2}$, $d\theta = \frac{2\,du}{1+u^2}$,

$$\int \frac{d\theta}{3\cos\theta - 2\sin\theta + 1} = \int \frac{\frac{2\,du}{1+u^2}}{\frac{3-3u^2}{1+u^2} - \frac{4u}{1+u^2} + \frac{1+u^2}{1+u^2}}$$

$$= \int \frac{2\,du}{-2u^2-4u+4} = \int \frac{du}{2-2u-u^2}$$

$$= \int \frac{du}{-(u^2 + 2u + 1) + 1 + 2} = \int \frac{du}{3 - (u+1)^2}$$

$$= \frac{1}{\sqrt{3}} \int \frac{1/\sqrt{3}}{1 - (u+1)^2/3} du = \frac{1}{2\sqrt{3}} \ln\left|\frac{1 + (u+1)/\sqrt{3}}{1 - (u+1)/\sqrt{3}}\right| + C$$

$$= \frac{1}{2\sqrt{3}} \ln\left|\frac{\sqrt{3} + u + 1}{\sqrt{3} - u - 1}\right| + C = \frac{1}{2\sqrt{3}} \ln\left|\frac{\sqrt{3} + \tan(\theta/2) + 1}{\sqrt{3} - \tan(\theta/2) - 1}\right| + C.$$

21. $u = \sqrt{x^2 - 4}$, $u^2 = x^2 - 4$, $x = \sqrt{u^2 + 4}$, $dx = \dfrac{u\,du}{\sqrt{u^2 + 4}}$,

$$\int x^3\sqrt{x^2 - 4}\,dx = \int (u^2 + 4)^{3/2} u \frac{u\,du}{\sqrt{u^2 + 4}} = \int u^2(u^2 + 4)du$$

$$= \int (u^4 + 4u^2)du = \frac{u^5}{5} + \frac{4u^3}{3} + C = \frac{u^3}{15}(3u^2 + 20) + C$$

$$= \frac{1}{15}(x^2 - 4)^{3/2}(3x^2 - 12 + 20) + C$$

$$= \frac{1}{15}(x^2 - 4)^{3/2}(3x^2 + 8) + C.$$

25. $x = \sec\theta$, $dx = \sec\theta\tan\theta\,\phi\theta$

$$\int \frac{dx}{(5x^2 + 4x)\sqrt{x^2 - 1}} = \int \frac{\sec\theta\tan\theta\,d\theta}{(5\sec^2\theta + 4\sec\theta)\tan\theta}$$

$$= \int \frac{d\theta}{5\sec\theta + 4} = \int \frac{\cos\theta\,d\theta}{5 + 4\cos\theta}$$

$$\tan\frac{\theta}{2} = u,\ \cos\theta = \frac{1 - u^2}{1 + u^2},\ d\theta = \frac{2\,du}{1 + u^2}$$

$$= \int \frac{\frac{1 - u^2}{1 + u^2}\cdot\frac{2\,du}{1 + u^2}}{5 + 4\left(\frac{1 - u^2}{1 + u^2}\right)} = 2\int \frac{1 - u^2}{(u^2 + 1)(u^2 + 9)}du$$

$$\frac{1 - u^2}{(u^2 + 1)(u^2 + 9)} = \frac{Au + B}{u^2 + 1} + \frac{Cu + D}{u^2 + 9}$$

$$1 - u^2 = (Au + B)(u^2 + 9) + (Cu + D)(u^2 + 1)$$

$$A + C = 0,\ B + D = -1,\ 9A + C = 0,\ 9B + D = 1$$

$$A = 0,\ C = 0,\ B = \frac{1}{4},\ D = -\frac{5}{4}$$

$$2\int \frac{1 - u^2}{(u^2 + 1)(u^2 + 9)}du = \frac{1}{2}\int\left(\frac{1}{u^2 + 1} - \frac{5}{u^2 + 9}\right)du$$

$$= \frac{1}{2}\text{ Arctan } u - \frac{5}{6}\text{ Arctan }\frac{u}{3} + C$$

$$= \frac{1}{2}\text{ Arctan tan }\frac{\theta}{2} - \frac{5}{6}\text{ Arctan}\left(\frac{1}{3}\tan\frac{\theta}{2}\right) + C$$

$$\tan\frac{\theta}{2} = \frac{1 - \cos\theta}{\sin\theta} = \frac{1 - 1/x}{\sqrt{x^2 - 1}} = \frac{x - 1}{\sqrt{(x + 1)(x - 1)}} = \sqrt{\frac{x - 1}{x + 1}}$$

$$= \frac{1}{4}\theta - \frac{5}{6}\text{ Arctan }\frac{1}{3}\sqrt{\frac{x - 1}{x + 1}} + C$$

$$= \frac{1}{4}\text{ Arcsec } x - \frac{5}{6}\text{ Arctan }\frac{1}{3}\sqrt{\frac{x - 1}{x + 1}} + C.$$

CHAPTER 10, SECTION 11 (p. 517)

1. $\displaystyle\int \frac{u\,du}{\sqrt{a + bu}} = \frac{2}{3b^2}(bu - 2a)\sqrt{a + bu} + C$, by formula 36,

   $a = 5,\ b = 2,\ u = x,\ du = dx,$

   $$\int \frac{x\,dx}{\sqrt{2x + 5}} = \frac{2}{12}(2x - 10)\sqrt{5 + 2x} + C = \frac{1}{3}(x - 5)\sqrt{2x + 5} + C.$$

5. $\displaystyle\int e^{au}\sin bu\,du = \frac{e^{au}(a\sin bu - b\cos bu)}{a^2 + b^2} + C$,

   by formula 119,

   $a = -6,\ b = 8,\ u = x,\ du = dx,$

$$\int e^{-6x}\sin 8x\,dx = \frac{e^{-6x}(-6\sin 8x - 8\cos 8x)}{36+64} + C$$

$$= -\frac{e^{-6x}}{50}(3\sin 8x + 4\cos 8x) + C\,.$$

9. $\int \cos^n u\,du = \frac{1}{n}\cos^{n-1}u\sin u + \frac{n-1}{n}\int \cos^{n-2}u\,du,$

by formula 85,

$u = 3x$, $du = 3\,dx$, $n = 6$,

$$\int \cos^6 3x\,dx = \frac{1}{3}\int \cos^6 3x(3\,dx)$$

$$= \frac{1}{3}[\frac{1}{6}\cos^5 3x\sin 3x + \frac{5}{6}\int \cos^4 3x(3\,dx)]\,.$$

Use formula 85 again, with $n = 4$, and then $n = 2$,

$$\int \cos^4 3x(3\,dx) = \frac{1}{4}\cos^3 3x\sin 3x + \frac{3}{4}\int \cos^2 3x(3\,dx)$$

$$\int \cos^2 3x(3\,dx) = \frac{1}{2}\cos 3x\sin 3x + \frac{1}{2}\int (\cos 3x)^0(3\,dx)$$

$$= \frac{1}{2}\cos 3x\sin 3x + \frac{3}{2}x + C\,.$$

$$\int \cos^6 3x\,dx = \frac{1}{3}\{\frac{1}{6}\cos^5 3x\sin 3x + \frac{5}{6}[\frac{1}{4}\cos^3 3x\sin 3x$$

$$+ \frac{3}{4}(\frac{1}{2}\cos 3x\sin 3x + \frac{3x}{2})]\} + K$$

$$= \frac{1}{144}[8\cos^5 3x\sin 3x + 10\cos^3 3x\sin 3x$$

$$+ 15(\cos 3x\sin 3x + 3x)] + K\,.$$

13. By formula 96,

$$\int x^2 \cos x\,dx = (x^2 - 2)\sin x + 2x\cos x + C.$$

17. $\displaystyle\int \frac{du}{\sqrt{u^2 - a^2}} = \ln|u + \sqrt{u^2 - a^2}| + C$ by formula 49

$$\int \frac{dx}{\sqrt{4x^2 - 8x - 21}} = \frac{1}{2}\,\frac{2\,dx}{\sqrt{4(x^2 - 2x + 1) - 25}} = \int \frac{dx}{\sqrt{(2x - 2)^2 - 25}}$$

$u = 2x - 2$, $du = 2\,dx$, $a = 5$,

$$= \frac{1}{2}\ln|2x - 2 + \sqrt{4x^2 - 8x - 21}| + C.$$

21. $\displaystyle\int (x^2 - 2x + 6)\sin 2x\,dx = (x^2 - 2x + 6)\left(-\frac{\cos 2x}{2}\right)$

$$- (2x - 2)\left(-\frac{\sin 2x}{4}\right) + (2)\left(\frac{\cos 2x}{8}\right) + C$$

$$= \frac{1}{4}(-2x^2 + 4x - 12 + 1)\cos 2x + \frac{1}{4}(2x - 2)\sin 2x + C$$

$$= \frac{1}{4}[(-2x^2 + 4x - 1)\cos 2x + (2x - 2)\sin 2x] + C.$$

CHAPTER 10, REVIEW (pp. 518-519)

1. $\displaystyle\int (1 + \tan 2\theta)^2 d\theta = \int (1 + 2\tan 2\theta + \tan^2 2\theta)\,d\theta$

$$= \int (2\tan 2\theta + \sec^2 2\theta)\,d\theta = -\ln|\cos 2\theta| + \frac{1}{2}\tan 2\theta + C.$$

5. $\int_{\pi/4}^{\pi/2} (\csc x + 1)(\cot x - 1)dx$

$= \int_{\pi/4}^{\pi/2} (\csc x \cot x - \csc x + \cot x - 1)dx$

$= (-\csc x - \ln|\csc x - \cot x| + \ln|\sin x| - x)\Big|_{\pi/4}^{\pi/2}$

$= [-1 - \ln(1 + 0) + \ln 1 - \frac{\pi}{2}] - [-\sqrt{2} - \ln(\sqrt{2} - 1) + \ln\frac{1}{\sqrt{2}} - \frac{\pi}{4}]$

$= \sqrt{2} - 1 - \frac{\pi}{4} + \ln(\sqrt{2} - 1) + \frac{1}{2}\ln 2 .$

9. $u = \text{Arctan } x \qquad v' = x^2$

$u' = \frac{1}{1 + x^2} \qquad v = \frac{x^3}{3}$

$\int x^2 \text{Arctan } x \, dx = \frac{1}{3} x^3 \text{Arctan } x - \frac{1}{3}\int \frac{x^3}{1 + x^2} dx$

$= \frac{1}{3} x^3 \text{Arctan } x - \frac{1}{3}\int (x - \frac{x}{1 + x^2}) dx$

$= \frac{1}{3} x^3 \text{Arctan } x - \frac{x^2}{6} + \frac{1}{6}\ln(1 + x^2) + C .$

13. $\int_1^2 \frac{dx}{2x^2 - 6x + 5} = \frac{1}{2}\int_1^2 \frac{dx}{x^2 - 3x + 9/4 + 1/4}$

$= \frac{1}{2}\int_1^2 \frac{dx}{(x - 3/2)^2 + 1/4}$

$x - \frac{3}{2} = \frac{1}{2}\tan\theta , \quad dx = \frac{1}{2}\sec^2\theta \, d\theta$

$= \frac{1}{2}\int_{-\pi/4}^{\pi/4} \frac{\frac{1}{2}\sec^2\theta \, d\theta}{\frac{1}{4}\tan^2\theta + \frac{1}{4}} = \int_{-\pi/4}^{\pi/4} d\theta = \theta\Big|_{-\pi/4}^{\pi/4} = \pi/2 .$

17. $\int_1^2 x\sqrt{3x-2}\,dx$,

$u = \sqrt{3x-2}$, $x = \frac{u^2+2}{3}$, $dx = \frac{2}{3}u\,du$

$$= \int_1^2 \frac{u^2+2}{3} u \cdot \frac{2}{3} u\,du = \frac{2}{9}\int_1^2 (u^4 + 2u^2)\,du$$

$$= \frac{2}{9}\left(\frac{u^5}{5} + \frac{2u^3}{3}\right)\Big|_1^2 = \frac{2}{9}\left[\left(\frac{32}{5}+\frac{16}{3}\right) - \left(\frac{1}{5}+\frac{2}{3}\right)\right] = \frac{326}{135}.$$

21. $\int \frac{1+\tan^2 x}{1+\tan x}\,dx = \int \frac{\sec^2 x}{1+\tan x}\,dx = \ln|1+\tan x| + C.$

25. $\int \frac{dx}{2\sqrt{x} - 3\sqrt[3]{x}}$, $x = u^6$, $dx = 6u^5\,du$

$$= \int \frac{6u^5\,du}{2u^3 - 3u^2} = \int \frac{6u^3\,du}{2u-3} = \int \left(3u^2 + \frac{9}{2}u + \frac{27}{4} + \frac{81/4}{2u-3}\right)du$$

$$= u^3 + \frac{9}{4}u^2 + \frac{27}{4}u + \frac{81}{8}\ln|2u-3| + C$$

$$= \sqrt{x} + \frac{9}{4}\sqrt[3]{x} + \frac{27}{4}\sqrt[6]{x} + \frac{81}{8}\ln|2\sqrt[6]{x} - 3| + C.$$

29. $\frac{-16}{x^2(x-2)^3} = \frac{A}{x} + \frac{B}{x^2} + \frac{C}{x-2} + \frac{D}{(x-2)^2} + \frac{E}{(x-2)^3}$,

$$-16 = A(x^4 - 6x^3 + 12x^2 - 8x) + B(x^3 - 6x^2 + 12x - 8)$$
$$+ C(x^4 - 4x^3 + 4x^2) + D(x^3 - 2x^2) + Ex^2,$$

$A + C = 0$, $-6A + B - 4C + D = 0$,

$12A - 6B + 4C - 2D + E = 0$, $-8A + 12B = 0$, $-8B = -16$,

$A = 3$, $B = 2$, $C = -3$, $D = 4$, $E = -4$,

$$\int \frac{-16\,dx}{x^2(x-2)^3} = \int \left(\frac{3}{x} + \frac{2}{x^2} - \frac{3}{x-2} + \frac{4}{(x-2)^2} - \frac{4}{(x-2)^3}\right)dx$$

$$= 3\ \ln|x| - \frac{2}{x} - 3\ \ln|x-2| - \frac{4}{x-2} + \frac{2}{(x-2)^2} + C$$

$$= 3\ \ln\left|\frac{x}{x-2}\right| - \frac{2}{x} - \frac{4}{x-2} + \frac{2}{(x-2)^2} + C .$$

33. $$\int \frac{(x+1)^2}{1-4x-x^2}\,dx = \int\left(-1 + \frac{-2x+2}{1-4x-x^2}\right)dx$$

$$= \int\left(-1 + \frac{-2x-4}{1-4x-x^2} + \frac{6}{5-(x+2)^2}\right)dx$$

$$= \int\left(-1 + \frac{-2x-4}{1-4x-x^2} + \frac{6}{\sqrt{5}}\,\frac{1/\sqrt{5}}{1-(x+2)^2/5}\right)dx$$

$$= -x + \ln|1-4x-x^2| + \frac{6}{\sqrt{5}}\int \frac{1/\sqrt{5}}{1-(x+2)^2/5}\,dx$$

$$\frac{x+2}{\sqrt{5}} = \sin\theta\ , \quad \frac{dx}{\sqrt{5}} = \cos\theta\ d\theta\ ,$$

$$\int \frac{1/\sqrt{5}}{1-(x+2)^2/5}\,dx = \int\frac{\cos\theta\ d\theta}{1-\sin^2\theta} = \int \sec\theta\ d\theta$$

$$= \ln|\sec\theta + \tan\theta| + C = \ln\left|\frac{\sqrt{5}}{\sqrt{1-4x-x^2}} + \frac{x+2}{\sqrt{1-4x-x^2}}\right| + C$$

$$= \ln|x+2+\sqrt{5}| - \frac{1}{2}\ln|1-4x-x^2| + C$$

$$\int \frac{(x+1)^2}{1-4x-x^2}\,dx = -x + \ln|1-4x-x^2| + \frac{6}{\sqrt{5}}\ln|x+2+\sqrt{5}|$$

$$- \frac{3}{\sqrt{5}}\ln|1-4x-x^2| + C$$

$$= -x - \frac{3-\sqrt{5}}{\sqrt{5}}\ln|1-4x-x^2| + \frac{6}{\sqrt{5}}\ln|x+2+\sqrt{5}| + C .$$

37. $\dfrac{x^5 + x^4 + 4x^3 + 8x^2 + 12x + 16}{(x^2+4)^3} = \dfrac{Ax + B}{x^2 + 4} + \dfrac{Cx + D}{(x^2+4)^2} + \dfrac{Ex + F}{(x^2+4)^3}$

$x^5 + x^4 + 4x^3 + 8x^2 + 12x + 16 = A(x^5 + 8x^3 + 16x)$

$+ B(x^4 + 8x^2 + 16) + C(x^3 + 4x) + D(x^2 + 4) + Ex + F ,$

$A = 1,\ B = 1,\ 8A + C = 4,\ 8B + D = 8,$

$16A + 4C + E = 12, \quad 16B + 4D + F = 16;$

$A = 1,\ B = 1,\ C = -4,\ D = 0,\ E = 12,\ F = 0,$

$$\int \frac{x^5 + x^4 + 4x^3 + 8x^2 + 12x + 16}{(x^2+4)^3}\,dx$$

$$= \int \left(\frac{x + 1}{x^2 + 4} - \frac{4x}{(x^2+4)^2} + \frac{12x}{(x^2+4)^3}\right) dx$$

$$= \int \left(\frac{1}{2}\,\frac{2x}{x^2 + 4} + \frac{1}{2}\,\frac{1/2}{(x/2)^2 + 1} - 2\,\frac{2x}{(x^2+4)^2} + 6\,\frac{2x}{(x^2+4)^3}\right) dx$$

$$= \frac{1}{2}\ln(x^2 + 4) + \frac{1}{2}\operatorname{Arctan}\frac{x}{2} + \frac{2}{x^2 + 4} - \frac{3}{(x^2+4)^2} + C$$

$$= \frac{1}{2}\ln(x^2 + 4) + \frac{1}{2}\operatorname{Arctan}\frac{x}{2} + \frac{2x^2 + 5}{(x^2+4)^2} + C .$$

41. $\displaystyle\int_1^2 x^4 e^{2x}dx$

u	v'
x^4	e^{2x}
$4x^3$	$\frac{1}{2}e^{2x}$
$12x^2$	$\frac{1}{4}e^{2x}$
$24x$	$\frac{1}{8}e^{2x}$
24	$\frac{1}{16}e^{2x}$
0	$\frac{1}{32}e^{2x}$

$$= e^{2x}\left(\frac{x^4}{2} - x^3 + \frac{3x^2}{2} - \frac{3x}{2} + \frac{3}{4}\right)\Bigg|_1^2$$

$$= e^4\left(8 - 8 + 6 - 3 + \frac{3}{4}\right) - e^2\left(\frac{1}{2} - 1 + \frac{3}{2} - \frac{3}{2} + \frac{3}{4}\right)$$

$$= \frac{15}{4}e^4 - \frac{e^2}{4} = \frac{e^2}{4}(15e^2 - 1) .$$

45. $u = \sqrt{4 - x}$, $x = 4 - u^2$, $dx = -2u\,du$

$$\int x \cos\sqrt{4 - x}\,dx = \int (4 - u^2)\cos u(-2u\,du)$$

$$= -8 \int u \cos u\,du + 2 \int u^3 \cos u\,du$$

By formula 94, $\int u \cos u\,du = \cos u + u \sin u + k_1$

By formula 99, $\int u^3 \cos u\,du = u^3 \sin u - 3 \int u^2 \sin u\,du$

By formula 95, $\int u^2 \sin u\,du = 2u \sin u + (2 - u^2)\cos u + k_2$

$$-8 \int u \cos u\,du + 2 \int u^3 \cos u\,du$$

$$= -8 \cos u - 8u \sin u - 8k_1 + 2u^3 \sin u$$
$$-6[2u \sin u + (2 - u^2) \cos u + k_2]$$
$$= (2u^3 - 20u)\sin u + (6u^2 - 20)\cos u + C$$
$$= -2\sqrt{4 - x}(6 + x)\sin\sqrt{4 - x} + (4 - 6x)\cos\sqrt{4 - x} + C.$$

Tabular integration can also be used to evaluate this integral.

49. $\int_a^b (\int_a^x f(t)dt)dx$

$u = \int_a^x f(t)dt \qquad v' = 1$

$u' = f(x) \qquad v = x$

$$= x \int_a^x f(t)dt - 0 - \int_a^b x\,f(x)dx$$

$$= b \int_a^b f(t)dt - 0 - \int_a^b x\,f(x)dx$$

$$= \int_a^b (b - x)f(x)dx.$$

Chapter 11
Further Applications of the Integral

CHAPTER 11, SECTION 1 (pp. 529-530)

1. $\int_2^{+\infty} \frac{dx}{x^3} = \lim_{k\to+\infty} \int_2^k \frac{dx}{x^3} = \lim_{k\to+\infty} \left(-\frac{1}{2x^2}\right)\Big|_2^k = \lim_{k\to+\infty} \left(\frac{1}{8} - \frac{1}{2k^2}\right)$

5. $\int_1^{+\infty} \frac{dx}{3x+1} = \lim_{k\to+\infty} \int_1^k \frac{dx}{3x+1} = \lim_{k\to+\infty} \frac{1}{3}\ln(3x+1)\Big|_1^k$

$= \lim_{k\to+\infty} \frac{1}{3}[\ln(3k+1) - \ln 4] = +\infty.$

9. $\int_2^3 \frac{dx}{\sqrt{x-2}} = \lim_{\varepsilon\to0^+} \int_{2+\varepsilon}^3 \frac{dx}{\sqrt{x-2}} = \lim_{\varepsilon\to0^+} 2\sqrt{x-2}\Big|_{2+\varepsilon}^3$

$= \lim_{\varepsilon\to0^+} 2(1 - \sqrt{\varepsilon}) = 2.$

13. $\int_{-1}^{27} \frac{dx}{\sqrt[3]{x}} = \int_{-1}^{0} \frac{dx}{\sqrt[3]{x}} + \int_0^{27} \frac{dx}{\sqrt[3]{x}} = \lim_{\varepsilon\to0^+} \int_{-1}^{-\varepsilon} \frac{dx}{\sqrt[3]{x}} + \lim_{\delta\to0^+} \int_\delta^{27} \frac{dx}{\sqrt[3]{x}}$

$= \lim_{\varepsilon\to0^+} \frac{3}{2}x^{2/3}\Big|_{-1}^{-\varepsilon} + \lim_{\delta\to0^+} \frac{3}{2}x^{2/3}\Big|_\delta^{27} = \lim_{\varepsilon\to0^+} \frac{3}{2}(\varepsilon^{2/3} - 1)$

$+ \lim_{\delta\to0^+} \frac{3}{2}(9 - \delta^{2/3}) = -\frac{3}{2} + \frac{27}{2} = 12.$

17. $\int_{-3}^{6} \frac{dx}{(x+1)^{4/3}} = \int_{-3}^{-1} \frac{dx}{(x+1)^{4/3}} + \int_{-1}^{6} \frac{dx}{(x+1)^{4/3}}$

$$= \lim_{\varepsilon\to 0^+} \int_{-3}^{-1-\varepsilon} \frac{dx}{(x+1)^{4/3}} + \lim_{\delta\to 0^+} \int_{-1+\delta}^{6} \frac{dx}{(x+1)^{4/3}}$$

$$= \lim_{\varepsilon\to 0^+} [-3(\frac{1}{\sqrt[3]{-\varepsilon}} - \frac{1}{\sqrt[3]{-2}})] + \lim_{\delta\to 0^+} [-3(\frac{1}{\sqrt[3]{7}} - \frac{1}{\sqrt[3]{\delta}})]$$

$$= -3 \lim_{\substack{\varepsilon\to 0^+ \\ \delta\to 0^+}} (-\frac{1}{\sqrt[3]{\varepsilon}} + \frac{1}{\sqrt[3]{2}} + \frac{1}{\sqrt[3]{7}} - \frac{1}{\sqrt[3]{\delta}}) = +\infty.$$

21. $A = \int_2^{\infty} \frac{dx}{x^2} = \lim_{k\to+\infty} \int_2^k \frac{dx}{x^2} = \lim_{k\to+\infty} (-\frac{1}{x})\Big|_2^k = \lim_{k\to+\infty} (\frac{1}{2} - \frac{1}{k}) = \frac{1}{2}.$

25. $\int_{-1}^{0} \frac{dx}{x} = \lim_{\varepsilon\to 0^-} \int_{-1}^{\varepsilon} \frac{dx}{x} = \lim_{\varepsilon\to 0^-} \ln|x|\Big|_{-1}^{\varepsilon} = \lim_{\varepsilon\to 0^-} \ln|\varepsilon| = -\infty.$

29. $\int_1^{+\infty} \frac{dx}{x^n} = \lim_{k\to+\infty} \int_1^k x^{-n}\, dx = \lim_{k\to+\infty} \frac{x^{1-n}}{1-n}\Big|_1^k \qquad (n \neq 1)$

$$= \lim_{k\to+\infty} \frac{k^{1-n} - 1}{1-n} = \begin{cases} 1/(n-1) & \text{if } n > 1 \\ +\infty & \text{if } n < 1 \end{cases}$$

If $n = 1$, $\int_1^{+\infty} \frac{dx}{x^n} = \lim_{k\to+\infty} \int_1^k \frac{dx}{x} = \lim_{k\to+\infty} \ln x\Big|_1^k$

$$= \lim_{k\to+\infty} \ln k = +\infty.$$

33. Let f be the function whose graph is

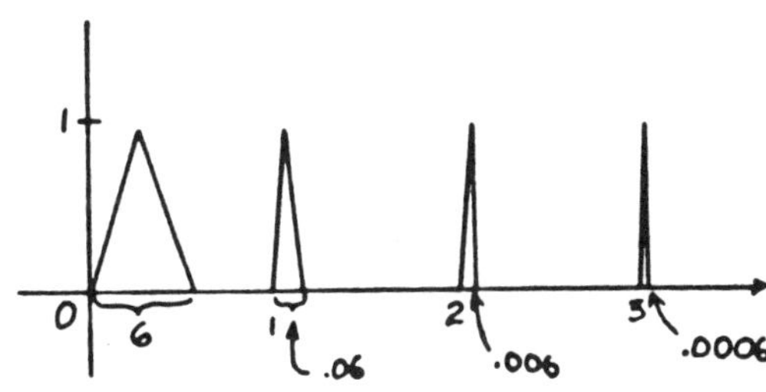

$\int_0^{+\infty} f(x)dx$ is the area under the curve, which is

$0.3 + 0.03 + 0.003 + 0.0003 + \cdots = 0.333 \cdots = 1/3.$

Clearly $\lim_{x\to+\infty} f(x)$ does not exist.

CHAPTER 11, SECTION 2 (pp. 536-538)

1. $V = \int_0^1 \pi y^2 dx = \int_0^1 \pi x^6 dx = \left.\frac{\pi x^7}{7}\right|_0^1 = \frac{\pi}{7}.$

5. $V = \int_0^{\pi} \pi y^2 dx = \pi \int_0^{\pi} \sin^2 x\, dx = \frac{\pi}{2}\int_0^{\pi} (1 - \cos 2x)dx$

$= \left.\frac{\pi}{2}(x - \frac{1}{2}\sin 2x)\right|_0^{\pi} = \frac{\pi^2}{2}.$

9. $V = \int_0^1 \pi(y_1^2 - y_2^2)dx$

$= \pi\int_0^1 (x^2 - x^6)dx$

$= \left.\pi(\frac{x^3}{3} - \frac{x^7}{7})\right|_0^1 = \pi(\frac{1}{3} - \frac{1}{7})$

$= \frac{4\pi}{21}.$

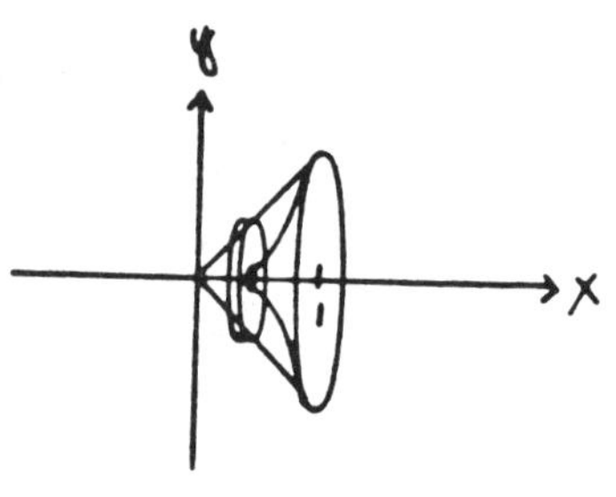

13. $V = \int_0^1 \pi(x_1^2 - x_2^2)dy$

$= \pi \int_0^1 (1 - y^{2/3})dy$

$$= \pi\left(y - \frac{3}{5}y^{5/3}\right)\Big|_0^1$$

$$= \pi\left(1 - \frac{3}{5}\right) = \frac{2\pi}{5}.$$

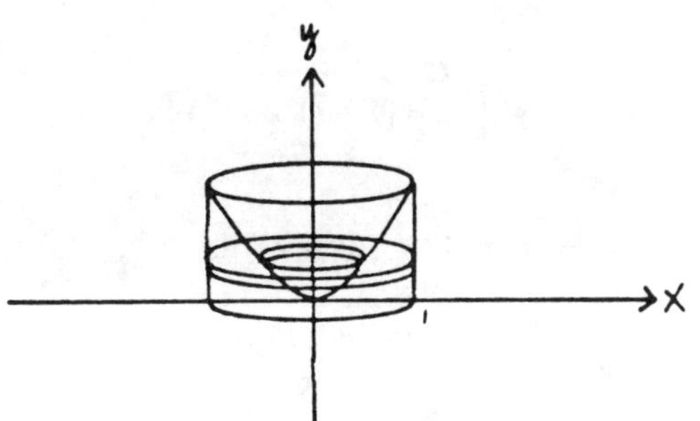

17. $V = \int_0^1 \pi x^2 dy = \pi \int_0^1 \text{Arccos}^2 y dy$

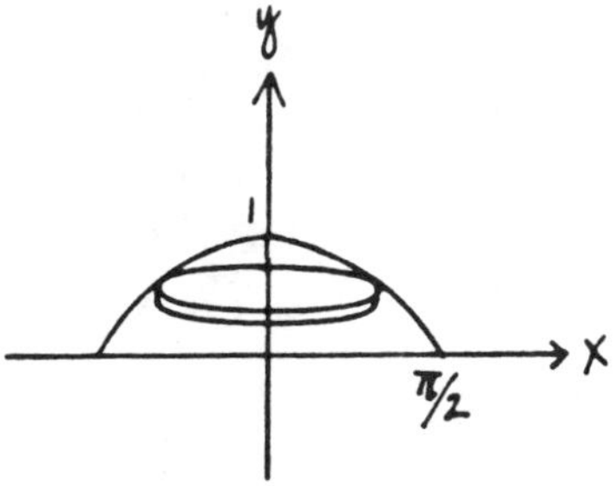

$u = \text{Arccos}^2 y \qquad v' = 1$

$u' = \dfrac{-2 \text{ Arccos } y}{\sqrt{1 - y^2}} \qquad v = y$

$$= \pi\left[y \text{ Arccos}^2 y\Big|_0^1 + \int_0^1 \frac{2y \text{ Arccos } y}{\sqrt{1 - y^2}} dy\right]$$

$u = \text{Arccos } y \qquad v' = \dfrac{2y}{\sqrt{1 - y^2}}$

$u' = \dfrac{-1}{\sqrt{1 - y^2}} \qquad v = -2\sqrt{1 - y^2}$

$$= \pi\left[\left(y \text{ Arccos}^2 y - 2\sqrt{1 - y^2} \text{ Arccos } y\right)\Big|_0^1 - \int_0^1 2 \, dy\right]$$

$$= \pi\left(y \text{ Arccos}^2 y - 2\sqrt{1 - y^2} \text{ Arccos } y - 2y\right)\Big|_0^1$$

$$= \pi(-2 + \pi) = \pi(\pi - 2).$$

21. $V = \int_0^{\pi/2} \pi[1 - (1 - y)^2]dx$

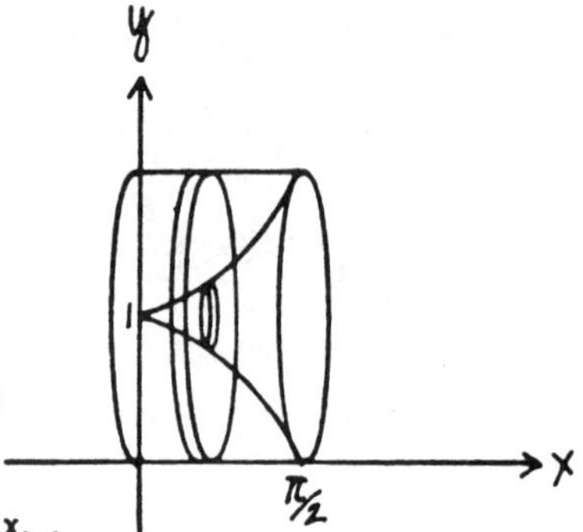

$= \pi \int_0^{\pi/2} (2y - y^2)dx$

$= \pi \int_0^{\pi/2} (2 \cos x - \cos^2 x)dx$

$= \pi \int_0^{\pi/2} (2 \cos x - \frac{1 + \cos 2x}{2})dx$

$= \pi(2 \sin x - \frac{x}{2} - \frac{1}{4} \sin 2x)\Big|_0^{\pi/2}$

$= \pi(2 - \frac{\pi}{4}) = \frac{\pi(8 - \pi)}{4}$.

25. $y = x^2 - x$

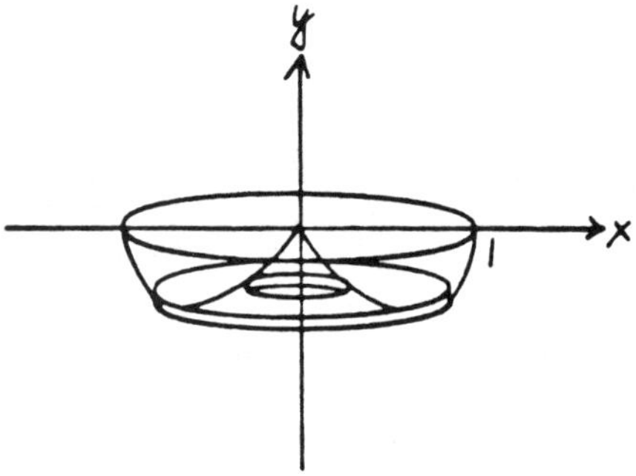

$x^2 - x - y = 0$

$x = \frac{1 \pm \sqrt{1 + 4y}}{2}$

$x_1 = \frac{1 - \sqrt{1 + 4y}}{2}$

$x_2 = \frac{1 + \sqrt{1 + 4y}}{2}$

$y' = 2x - 1 = 0,\ x = \frac{1}{2}$,

$y = -\frac{1}{4}$

$V = \int_{-1/4}^{0} \pi(x_2^2 - x_1^2)dy$

$= \pi \int_{-1/4}^{0} [\frac{(1 + \sqrt{1 + 4y})^2}{4} - \frac{(1 - \sqrt{1 + 4y})^2}{4}]\, dy$

$$= \pi \int_{-1/4}^{0} \sqrt{1 + 4y}\, dy = \frac{\pi}{4}\,\frac{2}{3}(1 + 4y)^{3/2}\Big|_{-1/4}^{0}$$

$$= \frac{\pi}{6}(1 + 4y)^{3/2}\Big|_{-1/4}^{0} = \frac{\pi}{6}\,.$$

29. $V = 2\int_0^1 \pi(x_1^2 - x_2^2)dy$

$(x - 2)^2 + y^2 = 1$

$x = 2 \pm \sqrt{1 - y^2}$

$$V = 2\pi \int_0^1 [(2 + \sqrt{1 - y^2})^2 - (2 - \sqrt{1 - y^2})^2]dy$$

$$= 2\pi \int_0^1 8\sqrt{1 - y^2}\, dy$$

$y = \sin\theta, \quad dy = \cos\theta\, d\theta$

$$= 16\pi \int_0^{\pi/2} \sqrt{1 - \sin^2\theta}\, \cos\theta\, d\theta$$

$$= 16\pi \int_0^{\pi/2} \cos^2\theta\, d\theta$$

$$= 8\pi \int_0^{\pi/2} (1 + \cos 2\theta)d\theta$$

$$= 8\pi(\theta + \frac{1}{2}\sin 2\theta)\Big|_0^{\pi/2} = 4\pi^2.$$

33. $A = 2\int_1^{+\infty} y dx$

$$= 2 \lim_{k\to+\infty} \int_1^k \frac{1}{x}\, dx$$

$$= 2 \lim_{k\to+\infty} \ln x\Big|_1^k$$

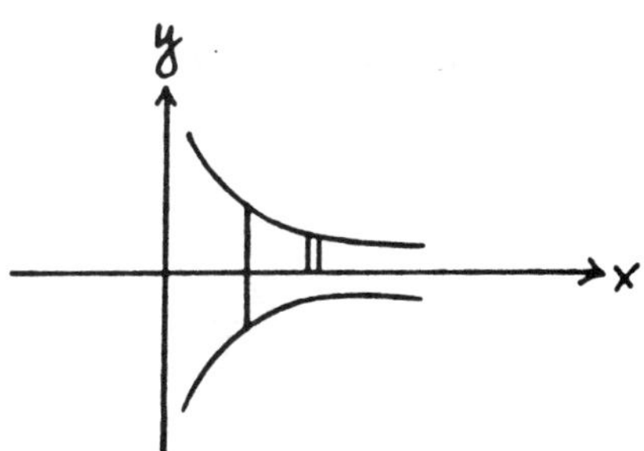

$$= 2 \lim_{k \to +\infty} \ln k = +\infty,$$

$$V = \int_1^{+\infty} \pi y^2 dx$$

$$= \pi \lim_{k \to +\infty} \int_1^k \frac{1}{x^2}\, dx = \pi \lim_{k \to +\infty} -\frac{1}{x}\Big|_1^k$$

$$= \lim_{k \to +\infty} (1 - \frac{1}{k}) = \pi.$$

Thus we have a solid with finite volume but having a cross-section with infinite area. This appears to be a contradiction, but area and volume are not comparable.

CHAPTER 11, SECTION 3 (pp. 543-544)

1. $$V = \int_0^1 2\pi yx\, dy$$

$$= 2\pi \int_0^1 y \cdot y^{1/2} dy$$

$$= \frac{4\pi\, y^{5/2}}{5}\Big|_0^1$$

$$= \frac{4\pi}{5}.$$

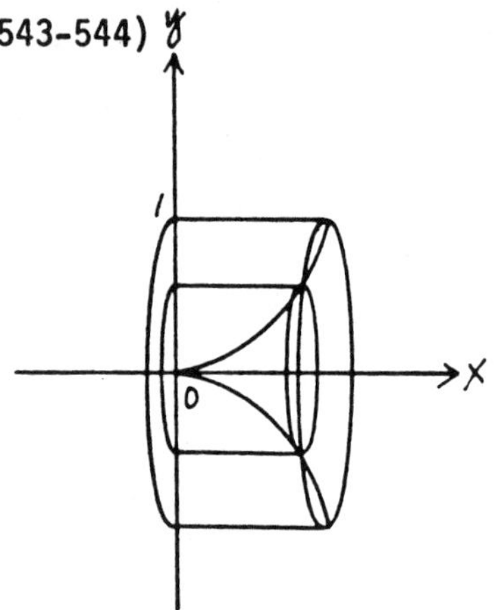

5. $$V = \int_0^1 2\pi(1 - x)y\, dx$$

$$= 2\pi \int_0^1 (x^2 - x^3) dx$$

$$= 2\pi (\frac{x^3}{3} - \frac{x^4}{4})\Big|_0^1$$

$$= 2\pi(\frac{1}{3} - \frac{1}{4}) = \frac{\pi}{6}.$$

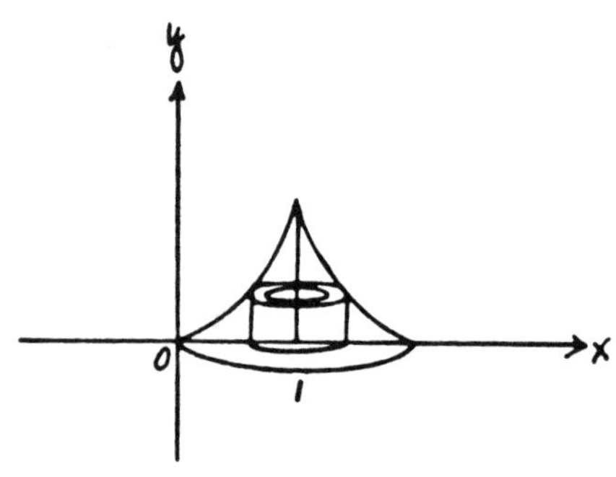

9. $V = \int_0^3 2\pi\ y(-x)dy$

$= 2\pi \int_0^3 y(3y - y^2)dy$

$= 2\pi \int_0^3 (3y^2 - y^3)dy$

$= 2\pi(y^3 - \frac{y^4}{4})\Big|_0^3 = \frac{27\pi}{2}.$

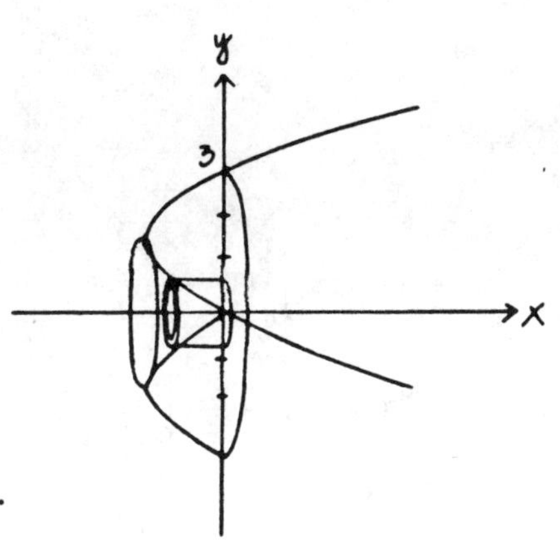

13. $V = \int_0^2 \pi y^2 dx = \pi \int_0^2 x^2(x - 2)^4 dx$

$= \pi \int_0^2 (x^6 - 8x^5 + 24x^4 - 32x^3 + 16x^2)dx$

$= \pi(\frac{x^7}{7} - \frac{4x^6}{3} + \frac{24x^5}{5} - 8x^4 + \frac{16x^3}{3})\Big|_0^2$

$= \pi(\frac{128}{7} - \frac{256}{3} + \frac{768}{5} - 128 + \frac{128}{3}) = \frac{128\pi}{105}.$

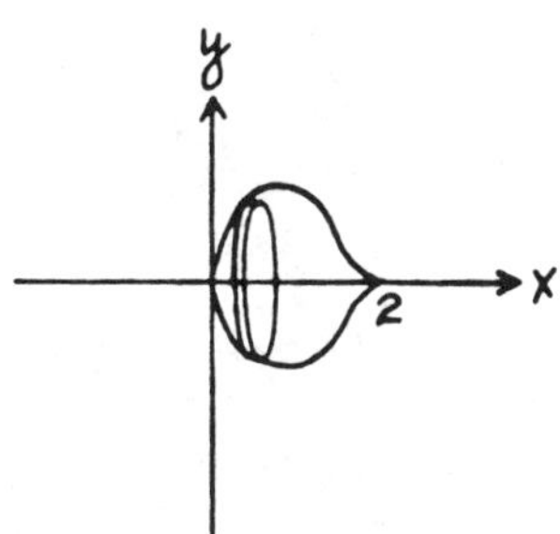

17. $V = \int_0^1 2\pi y(x - 1)dy$

$= 2\pi \int_0^1 y(\frac{1}{y} - 1)dy$

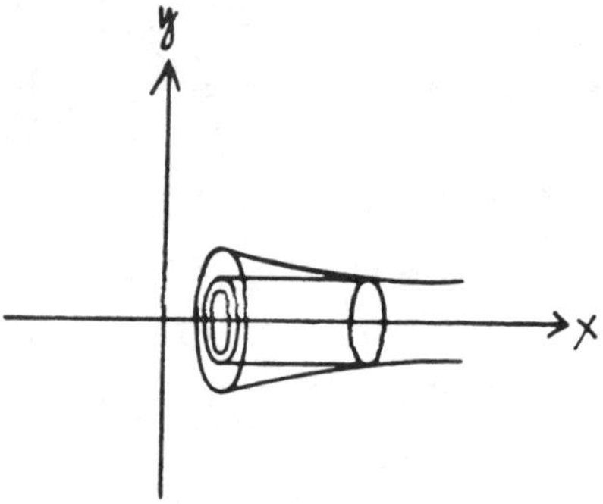

$$= 2\pi \lim_{\varepsilon \to 0} \int_{\varepsilon}^{1} (1 - y)dy$$

$$= 2\pi \lim_{\varepsilon \to 0} \left(y - \frac{y^2}{2}\right)\Big|_{\varepsilon}^{1}$$

$$= 2\pi \lim_{\varepsilon \to 0} \left[\frac{1}{2} - \left(\varepsilon - \frac{\varepsilon^2}{2}\right)\right] = \pi \lim_{\varepsilon \to 0} [1 - \varepsilon(2 - \varepsilon)] = \pi .$$

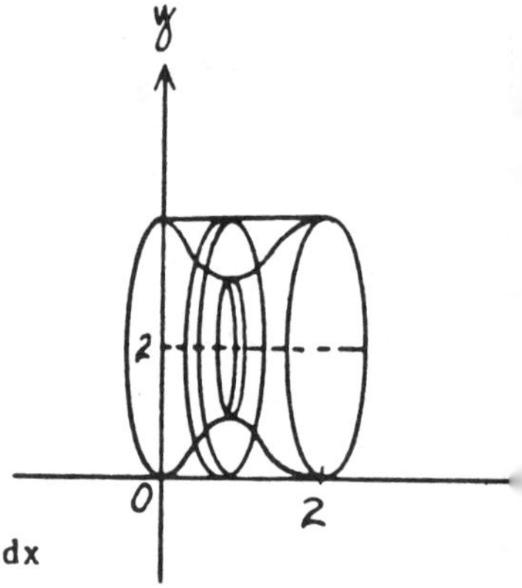

21. $V = \int_0^2 \pi[4 - (2 - y)^2]dx$

$$= \pi \int_0^2 (4y - y^2)dx$$

$$= \pi \int_0^2 [4x(x - 2)^2 - x^2(x - 2)^4]dx$$

$$= \pi \int_0^2 (4x^3 - 16x^2 + 16x - x^6 + 8x^5 - 24x^4 + 32x^3 - 16x^2)dx$$

$$= \pi \int_0^2 (-x^6 + 8x^5 - 24x^4 + 36x^3 - 32x^2 + 16x)dx$$

$$= \pi\left(-\frac{x^7}{7} + \frac{4x^6}{3} - \frac{24x^5}{5} + 9x^4 - \frac{32x^3}{3} + 8x^2\right)\Big|_0^2$$

$$= \pi\left(-\frac{128}{7} + \frac{256}{3} - \frac{768}{5} + 144 - \frac{256}{3} + 32\right) = \frac{144\pi}{35} .$$

25. $V = \int_0^1 2\pi\ xy\ dy$

$= \pi \int_0^1 2y(y^2 - 1)^2 dy$

$= \frac{\pi}{3}(y^2 - 1)^3 \Big|_0^1 = \frac{\pi}{3}$.

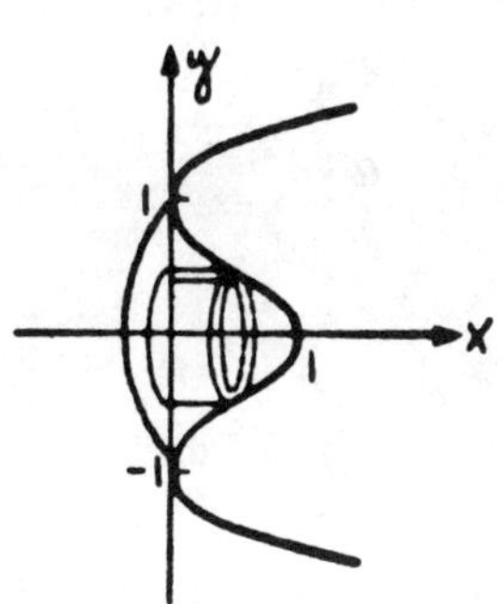

CHAPTER 11, SECTION 4 (pp. 548-549)

1. $V = 4 \int_0^1 y\ dx$

$= 4 \int_0^1 \sqrt{1 - x^2}\ dx$

$x = \sin\theta,\ dx = \cos\theta\, d\theta$

$= 4 \int_0^{\pi/2} \cos^2\theta\ d\theta$

$= 2 \int_0^{\pi/2} (1 + \cos 2\theta) d\theta = 2(\theta + \frac{1}{2} \sin 2\theta) \Big|_0^{\pi/2} = \pi$.

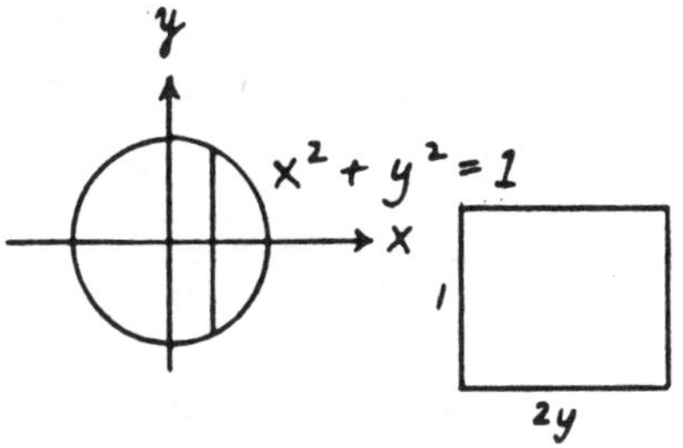

5. $V = 2 \int_0^1 \frac{1}{2} \pi y^2 dx = \frac{2\pi}{3}$.

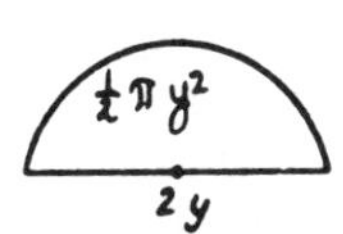

9. $V = \int_0^4 4x\ dy$

$= 4 \int_0^4 \sqrt{4 - y}\ dy = -\frac{8}{3}(4 - y)^{3/2} \Big|_0^4$

$= \frac{8}{3} \cdot 8 = \frac{64}{3}$.

2

2x

13. $V = \int_0^4 \frac{1}{2}\pi x^2 dy = 4\pi$.

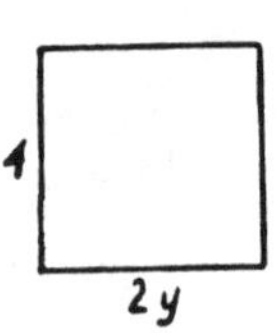

17. $V = 2\int_0^4 8y\ dx$

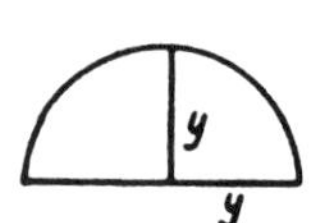

$= 16\int_0^4 \frac{3}{4}\sqrt{16 - x^2}\ dx$

$x = 4\sin\theta\ ,\quad dx = 4\cos\theta\, d\theta$

$= 12\int_0^{\pi/2} 4\cos\theta\, 4\cos\theta\, d\theta$

$= 192\int_0^{\pi/2} \cos^2\theta\, d\theta = 96\int_0^{\pi/2} (1 + \cos 2\theta)d\theta$

$= 96(\theta + \frac{1}{2}\sin 2\theta)\Big|_0^{\pi/2} = 48\pi$.

21. $V = 2\int_0^4 \frac{\pi}{2} y^2 dx = 24\pi$.

25. $V = 2\int_0^2 4y^2 dx$

$= 8\int_0^2 (4 - x^2)dx$

$= 8(4x - \frac{x^3}{3})\Big|_0^2$

$= 8 \cdot \frac{2\cdot 8}{3} = \frac{128}{3}$.

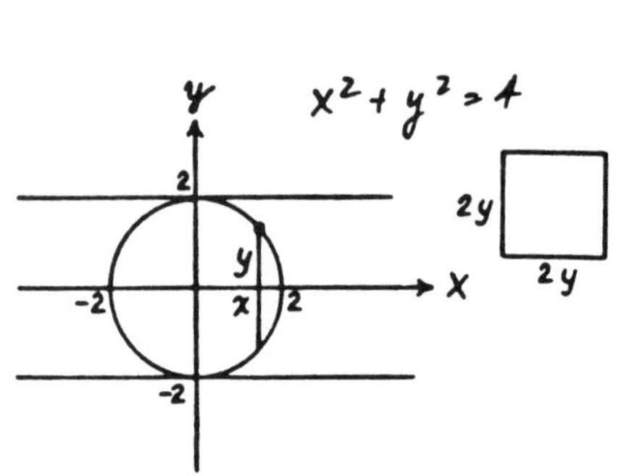

CHAPTER 11, SECTION 5 (pp. 555-556)

1. $M_y = 2\cdot 4 + 10\cdot 2 + 4(-6) = 4,$

   $m = 2 + 10 + 4 = 16,\ \bar{x} = \dfrac{4}{16} = \dfrac{1}{4},\ \bar{y} = 0.$

5. $M_y = 10\cdot 2 + 3\cdot 6 + 7\cdot 0 = 38,\ M_x = 10\cdot 2 + 3(-2) + 7(-4) = -14,$

   $m = 10 + 3 + 7 = 20,\ \bar{x} = \dfrac{38}{20} = \dfrac{19}{10},\ \bar{y} = \dfrac{-14}{20} = \dfrac{-7}{10}.$

9. $M_y = 2\cdot 5 + 4(-2) + 4x = 2 + 4x,$

   $M_x = 2\cdot 2 + 4\cdot 4 + 4y = 20 + 4y,$

   $m = 2 + 4 + 4 = 10,\quad \dfrac{2 + 4x}{10} = 0,\quad x = -\dfrac{1}{2};$

   $\dfrac{20 + 4y}{10} = 0,\quad y = -5.$

13. $M_x = 20\cdot 2 + 2\cdot\dfrac{1}{2} + 8\cdot 2 = 57,\ M_y = 20\cdot\dfrac{5}{2} + 2\cdot 6 + 8\cdot 8 = 126,$

    $A = 20 + 2 + 8 = 30,$

    $\bar{x} = \dfrac{126}{30} = \dfrac{21}{5},$

    $\bar{y} = \dfrac{57}{30} = \dfrac{19}{10}.$

17. $M_x = 4\cdot 1 + 2\sqrt{3}\,(2 + \dfrac{\sqrt{3}}{3}) = 6 + 4\sqrt{3},$

    $m = 4 + 2\sqrt{3},$

    $\bar{y} = \dfrac{6 + 4\sqrt{3}}{4 + 2\sqrt{3}}$

    $= \dfrac{3 + 2\sqrt{3}}{2 + \sqrt{3}}$

    $= \dfrac{6 + 4\sqrt{3} - 3\sqrt{3} - 6}{4 - 3}$

    $= \sqrt{3}.$

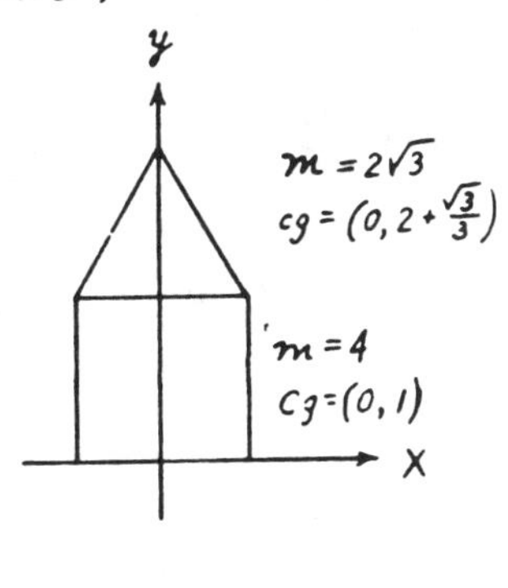

CHAPTER 11, SECTION 6 (pp. 560-561)

1. $A = \int_0^2 x^2 dx = \frac{x^3}{3}\Big|_0^2 = \frac{8}{3}$,

$M_x = \int_0^2 \frac{y}{2} y \, dx = \int_0^2 \frac{x^4}{2} dx = \frac{x^5}{10}\Big|_0^2 = \frac{16}{5}$,

$M_y = \int_0^2 xy \, dx = \int_0^2 x^3 dx = \frac{x^4}{4}\Big|_0^2 = 4$,

$\bar{x} = \frac{4}{8/3} = \frac{3}{2}$,

$\bar{y} = \frac{16/5}{8/3} = \frac{6}{5}$.

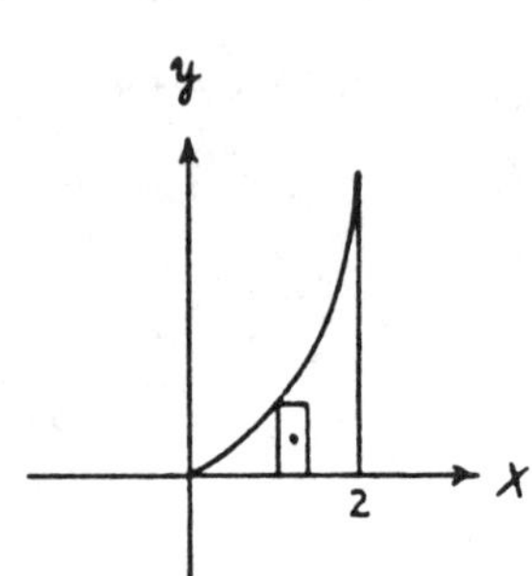

5. $A = \int_0^1 \sqrt{y} \, dy = \frac{2}{3} y^{3/2}\Big|_0^1 = \frac{2}{3}$,

$M_x = \int_0^1 yx \, dy = \int_0^1 y^{3/2} dy = \frac{2}{5} y^{5/2}\Big|_0^1 = \frac{2}{5}$,

$M_y = \int_0^1 \frac{x}{2} \cdot x \, dy = \int_0^1 \frac{y}{2} dy = \frac{y^2}{4}\Big|_0^1 = \frac{1}{4}$,

$\bar{x} = \frac{1/4}{2/3} = \frac{3}{8}$, $\bar{y} = \frac{2/5}{2/3} = \frac{3}{5}$.

9. $A = \int_0^1 (y^4 + 1) dy = \left(\frac{y^5}{5} + y\right)\Big|_0^1 = \frac{6}{5}$,

$M_x = \int_0^1 (y^5 + y) dy = \left(\frac{y^6}{6} + \frac{y^2}{2}\right)\Big|_0^1 = \frac{2}{3}$,

$$M_y = \int_0^1 \frac{(y^4+1)^2}{2}\,dy$$

$$= \frac{1}{2}\int_0^1 (y^8 + 2y^4 + 1)\,dy$$

$$= \frac{1}{2}\left(\frac{y^9}{9} + \frac{2y^5}{5} + y\right)\Big|_0^1$$

$$= \frac{1}{2}\left(\frac{1}{9} + \frac{2}{5} + 1\right) = \frac{34}{45},$$

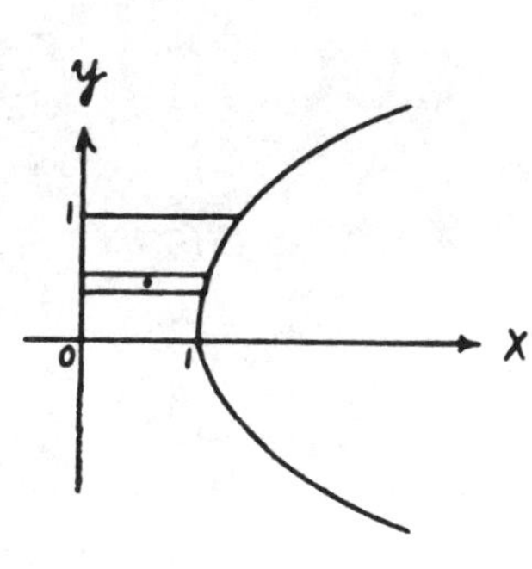

$$\bar{x} = \frac{34/45}{6/5} = \frac{17}{27}, \quad \bar{y} = \frac{2/3}{6/5} = \frac{5}{9}.$$

13. $A = \int_0^1 (y_1 - y_2)\,dx = \int_0^1 (x - x^4)\,dx$

$$= \left(\frac{x^2}{2} - \frac{x^5}{5}\right)\Big|_0^1 = \frac{1}{2} - \frac{1}{5} = \frac{3}{10},$$

$$M_x = \int_0^1 \frac{y_1 + y_2}{2}(y_1 - y_2)\,dx$$

$$= \frac{1}{2}\int_0^1 (y_1^2 - y_2^2)\,dx$$

$$= \frac{1}{2}\int_0^1 (x^2 - x^8)\,dx = \frac{1}{2}\left(\frac{x^3}{3} - \frac{x^9}{9}\right)\Big|_0^1 = \frac{1}{2}\left(\frac{1}{3} - \frac{1}{9}\right) = \frac{1}{9},$$

$$M_y = \int_0^1 x(y_1 - y_2)\,dx = \int_0^1 (x^2 - x^5)\,dx = \left(\frac{x^3}{3} - \frac{x^6}{6}\right)\Big|_0^1$$

$$= \frac{1}{3} - \frac{1}{6} = \frac{1}{6},$$

$$\bar{x} = \frac{1/6}{3/10} = \frac{5}{9}, \quad \bar{y} = \frac{1/9}{3/10} = \frac{10}{27}.$$

17. $9 - x^2 = \frac{x^2}{4} - 1$, $36 - 4x^2 = x^2 - 4$,

$5x^2 = 40$, $x = \pm 2\sqrt{2}$,

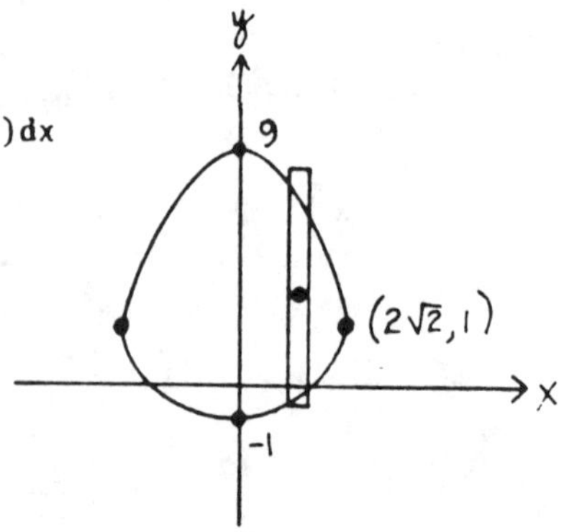

$$A = 2\int_0^{2\sqrt{2}} (9 - x^2 - \frac{x^2}{4} + 1)dx$$

$$= 2\int_0^{2\sqrt{2}} (10 - \frac{5x^2}{4})dx$$

$$= 20x - \frac{5x^3}{6}\Big|_0^{2\sqrt{2}}$$

$$= 40\sqrt{2} - \frac{40\sqrt{2}}{3} = \frac{80\sqrt{2}}{3} ,$$

$$M_x = 2 \cdot \frac{1}{2}\int_0^{2\sqrt{2}} [(9 - x^2)^2 - (\frac{x^2}{4} - 1)^2]dx$$

$$= \int_0^{2\sqrt{2}} (81 - 18x^2 + x^4 - \frac{x^4}{16} + \frac{x^2}{2} - 1)dx$$

$$= \int_0^{2\sqrt{2}} (\frac{15}{16}x^4 - \frac{35}{2}x^2 + 80)dx$$

$$= \frac{3x^5}{16} - \frac{35x^3}{6} + 80x\Big|_0^{2\sqrt{2}} = 2\sqrt{2}\ [64(\frac{3}{16}) - 8(\frac{35}{6}) + 80]$$

$$= \frac{272}{3}\sqrt{2} ,$$

$\bar{x} = 0$ by symmetry, $\bar{y} = \frac{272\sqrt{2}/3}{80\sqrt{2}/3} = \frac{17}{5}$.

21. $\sqrt{x} + \sqrt{y} = \sqrt{a}$, $y = (\sqrt{a} - \sqrt{x})^2 = a - 2\sqrt{ax} + x$,

$$A = \int_0^a (a - 2\sqrt{ax} + x)dx = (ax - \frac{4}{3}\sqrt{a}\ x^{3/2} + \frac{x^2}{2})\Big|_0^a$$

$$= a^2 - \frac{4}{3} a^2 + \frac{1}{2} a^2 = \frac{a^2}{6},$$

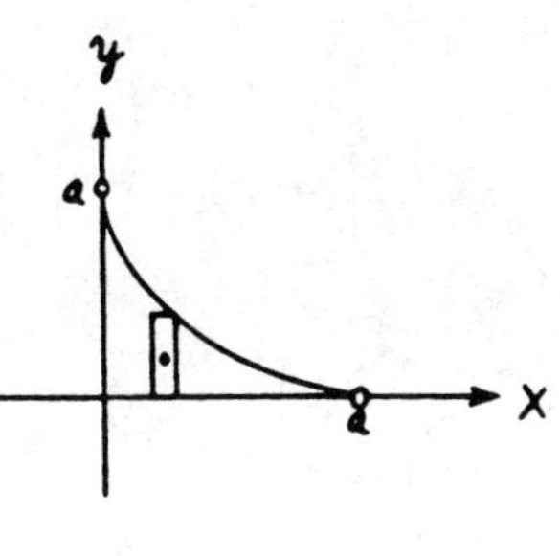

$$M_y = \int_0^a (ax - 2\sqrt{a}\, x^{3/2} + x^2)dx$$

$$= (\frac{ax^2}{2} - \frac{4\sqrt{a}\, x^{5/2}}{5} + \frac{x^3}{3})\Big|_0^a$$

$$= \frac{a^3}{2} - \frac{4a^3}{5} + \frac{a^3}{3} = \frac{15 - 24 + 10}{30} a^3 = \frac{a^3}{30},$$

$$\bar{x} = \bar{y} = \frac{a^3/30}{a^2/6} = \frac{a}{5}.$$

25. $A = \int_0^1 x\, dy = \int_0^1 \sqrt{y}\, dy = \frac{2}{3} y^{3/2}\Big|_0^1 = \frac{2}{3},$

$$M_x = \int_0^1 xy\, dy = \int_0^1 y^{3/2} dy = \frac{2}{5} y^{5/2}\Big|_0^1 = \frac{2}{5},$$

$$\bar{y} = \frac{2/5}{2/3} = \frac{3}{5},$$

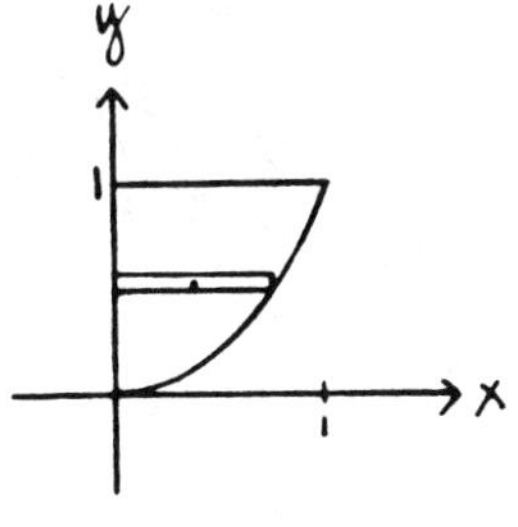

$$V = \frac{2}{3} \cdot 2\pi \cdot \frac{3}{5} = \frac{4\pi}{5}$$

$$V = \int_0^1 2\pi\, xy\, dy$$

$$= 2\pi \int_0^1 y^{3/2} dy$$

$$= 2\pi \cdot \frac{2}{5} y^{5/2}\Big|_0^1 = \frac{4\pi}{5}.$$

CHAPTER 11, SECTION 7 (p. 567)

1. $V = \int_0^1 \pi\, y^2 dx = \pi \int_0^1 x^6 dx$

$= \frac{\pi x^7}{7}\Big|_0^1 = \frac{\pi}{7}$,

$M_{yz} = \int_0^1 \pi\, y^2 x\, dx = \pi \int_0^1 x^7 dx$

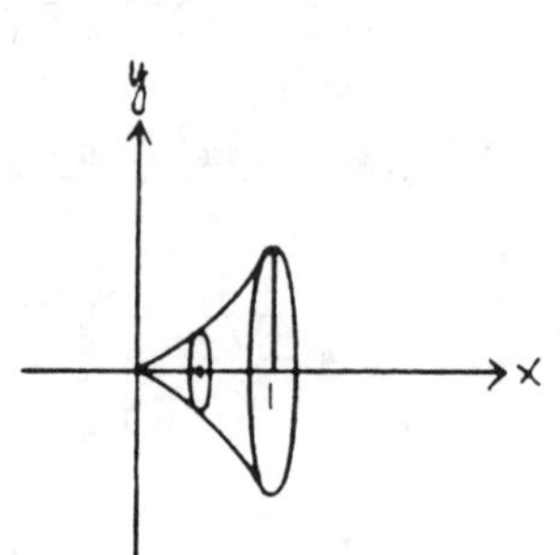

$= \frac{\pi x^8}{8}\Big|_0^1 = \frac{\pi}{8}$,

$\bar{x} = \frac{\pi/8}{\pi/7} = \frac{7}{8}$.

5. $V = \int_0^2 \pi\, x^2 dy = \pi \int_0^2 (1 - \frac{y}{2})^2 dy$

$= \pi \int_0^2 (1 - y + \frac{y^2}{4}) dy$

$= \pi (y - \frac{y^2}{2} + \frac{y^3}{12}) \Big|_0^2$

$= \pi (2 - 2 + \frac{2}{3}) = \frac{2\pi}{3}$,

$M_{xz} = \int_0^2 \pi\, x^2 y\, dy$

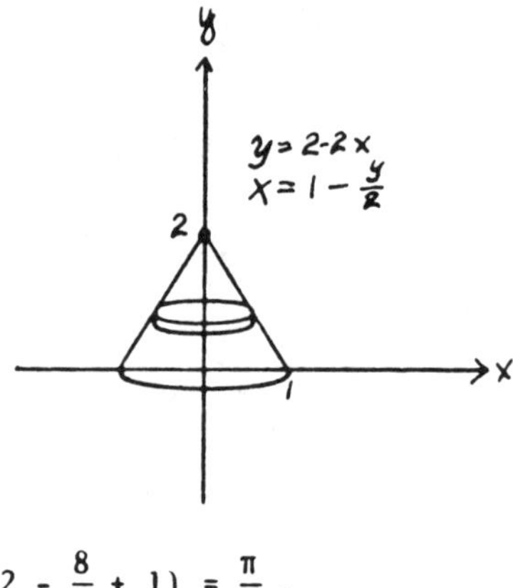

$= \pi \int_0^2 (y - y^2 + \frac{y^3}{4}) dy$

$= \pi (\frac{y^2}{2} - \frac{y^3}{3} + \frac{y^4}{16}) \Big|_0^2 = \pi (2 - \frac{8}{3} + 1) = \frac{\pi}{3}$,

$\bar{y} = \frac{\pi/3}{2\pi/3} = \frac{1}{2}$.

9. $\dfrac{x^2}{a^2} + \dfrac{y^2}{b^2} = 1$,

$$V = \int_0^a \pi y^2 dx = \pi \int_0^a \frac{b^2}{a^2}(a^2 - x^2)dx$$

$$= \frac{\pi b^2}{a^2}(a^2 x - \frac{x^3}{3})\Big|_0^a = \frac{\pi b^2}{a^2}\frac{2a^3}{3} = \frac{2\pi ab^2}{3} ,$$

$$M_{yz} = \int_0^a \pi y^2 x dx = \frac{\pi b^2}{a^2}\int_0^a (a^2 x - x^3)dx$$

$$= \frac{\pi b^2}{a^2}(\frac{a^2 x^2}{2} - \frac{x^4}{4})\Big|_0^a$$

$$= \frac{\pi b^2}{a^2}\frac{a^4}{4} = \frac{\pi a^2 b^2}{4} ,$$

$$\bar{x} = \frac{\pi a^2 b^2/4}{2\pi ab^2/3} = \frac{3a}{8} .$$

13. $V = \displaystyle\int_0^1 2\pi\ x(-y)dx = 2\pi \int_0^1 (x^2 - x^3)dx$

$$= 2\pi(\frac{x^3}{3} - \frac{x^4}{4})\Big|_0^1$$

$$= 2\pi(\frac{1}{3} - \frac{1}{4}) = \frac{\pi}{6} ,$$

$$M_{xz} = \int_0^1 \frac{y}{2}\, 2\pi x(-y)dx$$

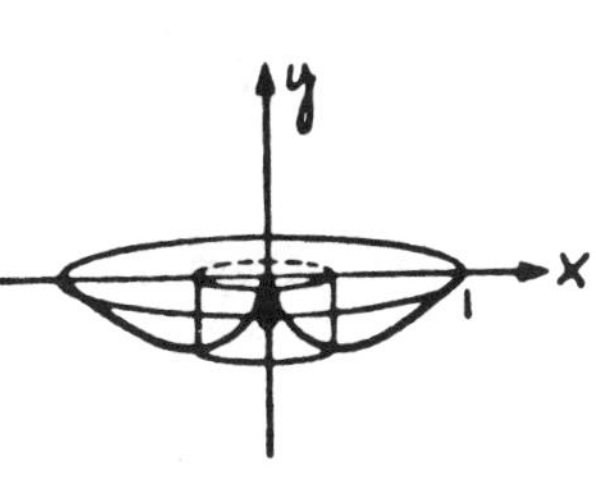

$$= \pi \int_0^1 - xy^2 dx$$

$$= \pi \int_0^1 (-x^5 + 2x^4 - x^3)dx$$

$$= \pi\left(-\frac{x^6}{6}+\frac{2x^5}{5}-\frac{x^4}{4}\right)\Big|_0^1 = \pi\left(-\frac{1}{6}+\frac{2}{5}-\frac{1}{4}\right) = -\frac{\pi}{60},$$

$$\bar{y} = \frac{-\pi/60}{\pi/6} = -\frac{1}{10}.$$

17. $$V = \int_0^{+\infty} \pi y^2 dx = \pi \lim_{k\to+\infty} \int_0^k e^{-2x} dx$$

$$= -\frac{\pi}{2} \lim_{k\to+\infty} e^{-2x}\Big|_0^k$$

$$= -\frac{\pi}{2} \lim_{k\to+\infty} (e^{-2k} - 1) = \frac{\pi}{2},$$

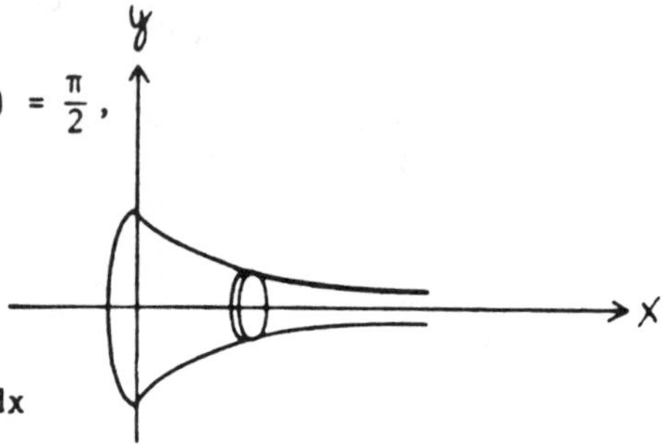

$$M_{yz} = \int_0^{+\infty} x\pi y^2 dx$$

$$= \pi \lim_{k\to+\infty} \int_0^k xe^{-2x} dx$$

$$u = x \quad v' = e^{-2x}$$

$$u' = 1 \quad v = -\frac{1}{2} e^{-2x}$$

$$= \pi \lim_{k\to+\infty} \left(-\frac{x}{2} e^{-2x} - \frac{1}{4} e^{-2x}\right)\Big|_0^k$$

$$= \pi \lim_{k\to+\infty} \left[e^{-2k}\left(-\frac{k}{2} - \frac{1}{4}\right) + \frac{1}{4}\right] = \frac{\pi}{4},$$

$$\bar{x} = \frac{\pi/4}{\pi/2} = \frac{1}{2}.$$

CHAPTER 11, SECTION 8 (pp. 575-576)

1. $I_y = 4\cdot 2^2 + 2(-3)^2 = 34,\ k = \sqrt{\frac{I}{m}} = \sqrt{\frac{34}{6}} = \sqrt{\frac{17}{3}}$.

5. $m = \int_0^1 6(1 - y)dx = 6 \int_0^1 (1 - x^2)dx$

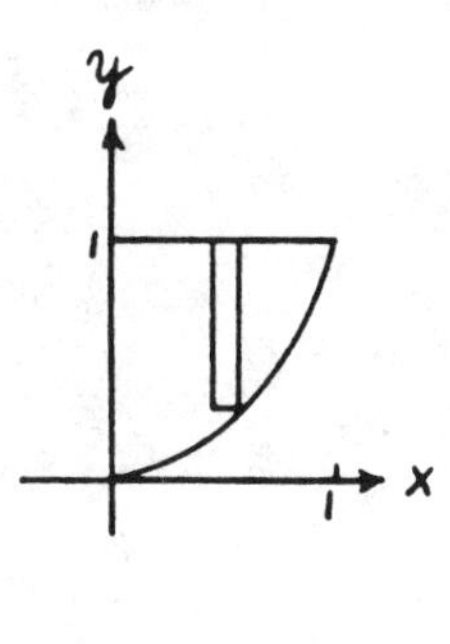

$= 6(x - \frac{x^3}{3})\Big|_0^1 = 6(1 - \frac{1}{3}) = 4$,

$I_y = \int_0^1 6x^2(1 - y)dx = 6 \int_0^1 (x^2 - x^4)dx$

$= 6(\frac{x^3}{3} - \frac{x^5}{5})\Big|_0^1 = 6(\frac{1}{3} - \frac{1}{5}) = \frac{4}{5}$,

$k = \sqrt{\frac{I}{m}} = \sqrt{\frac{4/5}{4}} = \frac{1}{\sqrt{5}}$.

9. $m = 4 + 6 + 2 = 12$,

$I_y = 4(\frac{1}{2})^2 + 6(\frac{1}{2})^2 + 2 \int_0^{1/2} 2x^2dx$

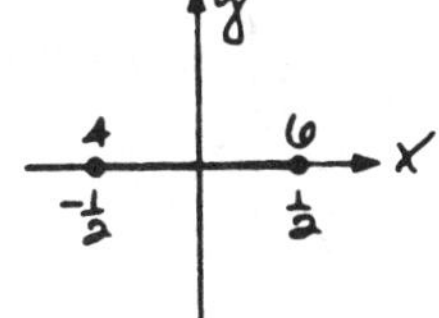

$= \frac{5}{2} + \frac{4}{3}x^3\Big|_0^{1/2} = \frac{5}{2} + \frac{1}{6} = \frac{8}{3}$.

$k = \sqrt{\frac{I}{m}} = \sqrt{\frac{8/3}{12}} = \frac{\sqrt{2}}{3}m$.

13. $m = \int_{-1}^0 7y\ dx = \int_{-1}^0 7\ e^x dx = 7e^x\Big|_{-1}^0 = 7(1 - \frac{1}{e}) = \frac{7}{e}(e - 1)$,

$I_y = \int_{-1}^0 7x^2y\ dx = 7 \int_{-1}^0 x^2e^x dx$

u	v'
x^2	e^x
$2x$	e^x
2	e^x
0	e^x

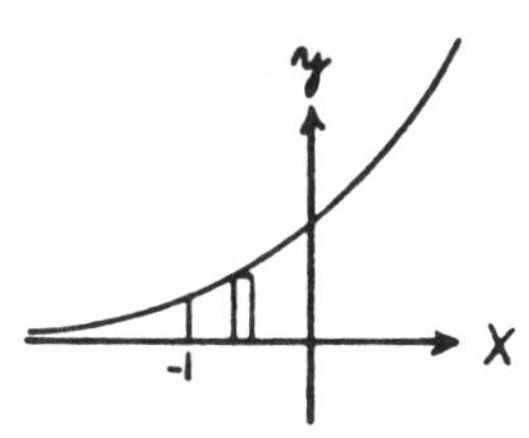

$$= 7(x^2e^x - 2xe^x + 2e^x)\Big|_{-1}^{0} = 7(2 - e^{-1} - 2e^{-1} - 2e^{-1})$$

$$= 7(2 - \frac{5}{e}) = \frac{7(2e - 5)}{e},$$

$$k = \sqrt{\frac{I}{m}} = \sqrt{\frac{7(2e-5)/e}{7(e-1)/e}} = \sqrt{\frac{2e-5}{e-1}}.$$

17. $m = 4 \cdot \pi(3)^2 = 36\pi$ by geometry.

$$x^2 + (y - 4)^2 = 9, \; x = \sqrt{9 - (y - 4)^2}$$

$$I_x = 2\int_1^7 4y^2x\,dy = 8\int_1^7 y^2\sqrt{9 - (y - 4)^2}\,dy$$

$$y - 4 = 3\sin\theta, \; dy = 3\cos\theta\,d\theta$$

$$= 8\int_{-\pi/2}^{\pi/2} (4 + 3\sin\theta)^2 \cdot 9\cos^2\theta\,d\theta$$

$$= 72\int_{-\pi/2}^{\pi/2} (16\cos^2\theta + 24\cos^2\theta\sin\theta + 9\sin^2\theta\cos^2\theta)d\theta$$

$$= 72\int_{-\pi/2}^{\pi/2} (8 + 8\cos 2\theta + 24\cos^2\theta\sin\theta + \frac{9}{4}\sin^2 2\theta)d\theta$$

$$= 72\int_{-\pi/2}^{\pi/2} (8 + 8\cos 2\theta + 24\cos^2\theta\sin\theta + \frac{9}{8} - \frac{9}{8}\cos 4\theta)$$

$$= 72(\frac{73\theta}{8} + 4\sin 2\theta - 8\cos^3\theta - \frac{9}{32}\sin 4\theta\Big|_{-\pi/2}^{\pi/2}$$

$$= 72(\frac{73\pi}{16} + \frac{73\pi}{16}) = 657\pi,$$

$$k = \sqrt{\frac{I}{m}} = \sqrt{\frac{657\pi}{36\pi}} = \frac{1}{2}\sqrt{73}.$$

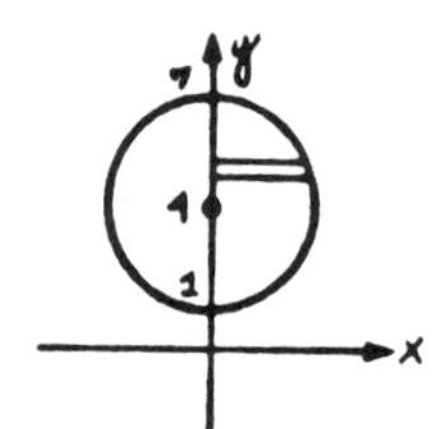

21. $m = 12$, 6 faces each of area $\frac{1}{4}$,

$$\delta = \frac{12}{6(\frac{1}{4})} = 8$$

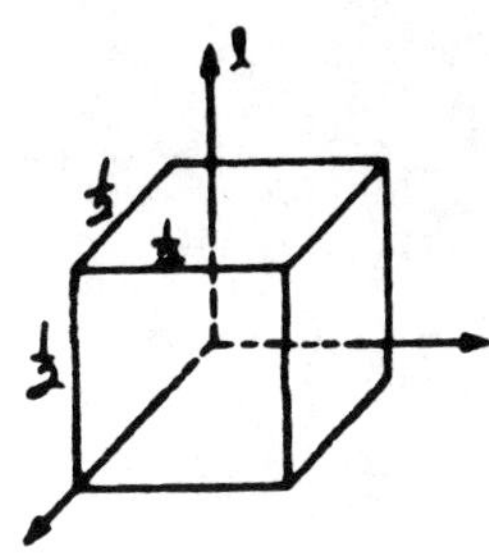

Moment of inertia of upper and lower faces about ℓ:

$$I_x = 8\int_0^{1/4} 2y^2 \cdot \frac{1}{2}dy = \frac{8y^3}{3}\Big|_0^{1/4} = \frac{1}{24}$$

$$I_\ell = I_x + I_y = \frac{1}{24} + \frac{1}{24} = \frac{1}{12}$$

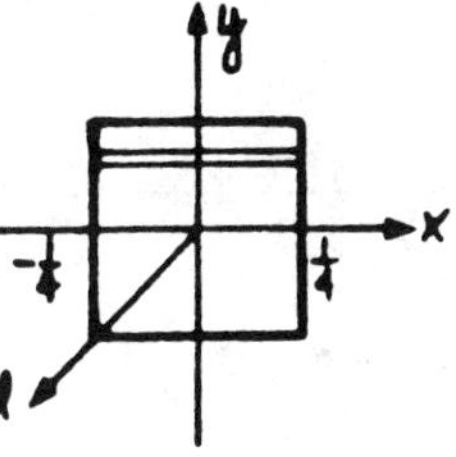

Moment of inertia of vertical faces:

$$I_\ell = 8\int_0^{1/4} 2(x^2 + \frac{1}{16})\frac{1}{2}dx$$

$$= \frac{8x^3}{3} + \frac{x}{2}\Big|_0^{1/4} = \frac{1}{24} + \frac{1}{8} = \frac{1}{6}$$

$$I_\ell = \frac{1}{12} + \frac{1}{12} + 4(\frac{1}{6}) = \frac{5}{6}$$

$$k = \sqrt{\frac{I}{m}} = \sqrt{\frac{5/6}{12}} = \frac{1}{6}\sqrt{\frac{5}{2}}$$

$$r^2 = x^2 + \frac{1}{16}$$

CHAPTER 11, SECTION 9 (pp. 581-582)

1. $F = \int_3^5 9800y \cdot 3dy$

$= 29{,}400\frac{y^2}{2}\Big|_3^5$

$= 29{,}400(\frac{25}{2} - \frac{9}{2})$

$= 2.352 \times 10^5 N.$

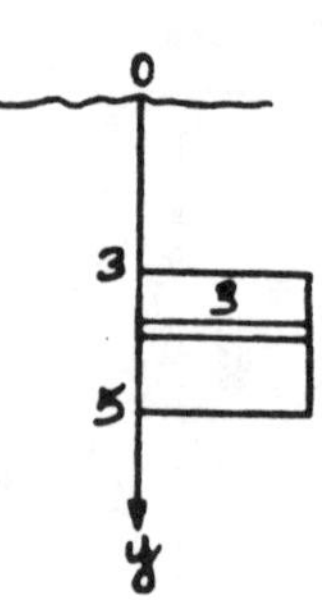

5. $F = \int_{-1}^{0} 9800(-y)\,2x\,dy$

$= -\frac{19{,}600}{\sqrt{3}}\int_{-1}^{0} y(1 + y)dy$

$= -\frac{19{,}600}{\sqrt{3}}(\frac{y^2}{2} + \frac{y^3}{3})\Big|_{-1}^{0}$

$= 1.866 \times 10^3 N.$

$F = \int_{-1}^{0} 9800(-y) \cdot 6 \cdot \frac{2}{\sqrt{3}}\,dy$

$= -\frac{58{,}800}{\sqrt{3}}y^2\Big|_{-1}^{0} = 3.395 \times 10^4\ N.$

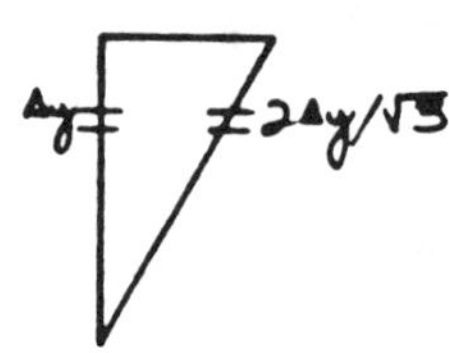

9. $F = \int_{-4}^{-3} 9800(-y)2x\,dy$

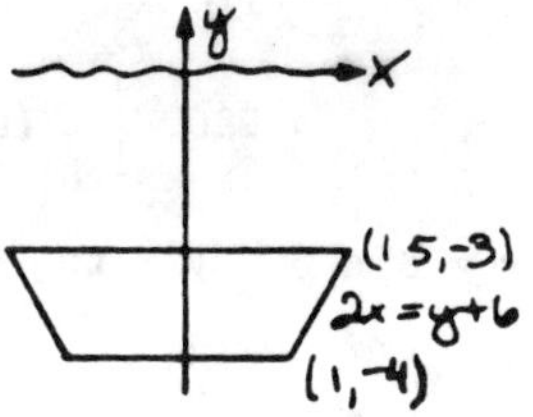

$= -9800 \int_{-4}^{-3} y(y + 6)\,dy$

$= -9800(\frac{y^3}{3} + 3y^2)\Big|_{-4}^{-3}$

$= -9800(-9 + 27 + \frac{64}{3} - 48)$

$= 8.493 \times 10^4$ N.

(1.5, -3)
2x = y + 6
(1, -4)

13. $F = \int_{-1/2}^{h - 1/2} 9800(h - \frac{1}{2} - y) \cdot 2x\,dy$

$= 19{,}600 \int_{-1/2}^{h-1/2} (h - \frac{1}{2})\sqrt{\frac{1}{4} - y^2}\,dy$

$- 9800 \int_{-1/2}^{h-1/2} 2y\sqrt{\frac{1}{4} - y^2}\,dy$

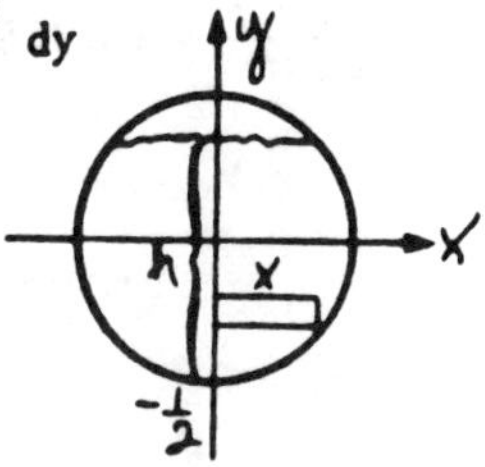

$y = \frac{1}{2} \sin \theta$, $dy = \frac{1}{2} \cos \theta\, d\theta$

$= \frac{19{,}600}{8} \int_{-\pi/2}^{\text{Arcsin}(2h-1)} (2h - 1)\cos^2\theta\, d\theta$

$+ 9800 \cdot \frac{2}{3}(\frac{1}{4} - y^2)^{3/2}\Big|_{-1/2}^{h-1/2}$

$= 1225(2h - 1)(\theta + \sin\theta\cos\theta)\Big|_{-\pi/2}^{\text{Arcsin}(2h-1)}$

$+ \frac{19{,}600}{3}(h - h^2)^{3/2}$

$= 1225(2h - 1)[\text{Arcsin}(2h - 1) + (2h - 1)\sqrt{h - h^2} + \frac{\pi}{2}]$
$+ \frac{19{,}600}{3}(h - h^2)^{3/2}$.

17. $600{,}000 = \int_{-1}^{1} 9800(h - y)2x\,dy$

$$= 9800 \int_{-1}^{1} (h - y)2\sqrt{1 - y^2}\,dy$$

$y = \sin\theta,\ dy = \cos\theta\,d\theta$

$$= 9800h \int_{-\pi/2}^{\pi/2} 2\cos^2\theta\,d\theta + 9800 \int_{-1}^{1} -2y\sqrt{1 - y^2}\,dy$$

$$= 9800h \int_{-\pi/2}^{\pi/2} (1 + \cos 2\theta)d\theta + 9800 \cdot \frac{2}{3}(1 - y^2)^{3/2}\Big|_{-1}^{1}$$

$$= 9800h(\theta + \frac{1}{2}\sin 2\theta)\Big|_{-\pi/2}^{\pi/2} + 0 = 9800\pi h.$$

$h = \dfrac{600{,}000}{9800\pi} = 19.488$ m above the center.

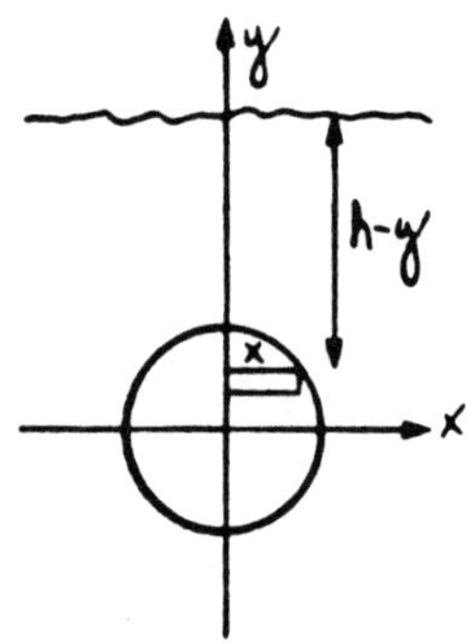

21. $F = \int_{-8}^{0} 9800(-y)x\,\frac{5}{4}dy$

$$= -12{,}250 \int_{-8}^{0} -y(\frac{3}{2}y)dy$$

$$= -18{,}375(-\frac{y^3}{3})\Big|_{-8}^{0}$$

$$= 3.136 \times 10^6 \text{ N}.$$

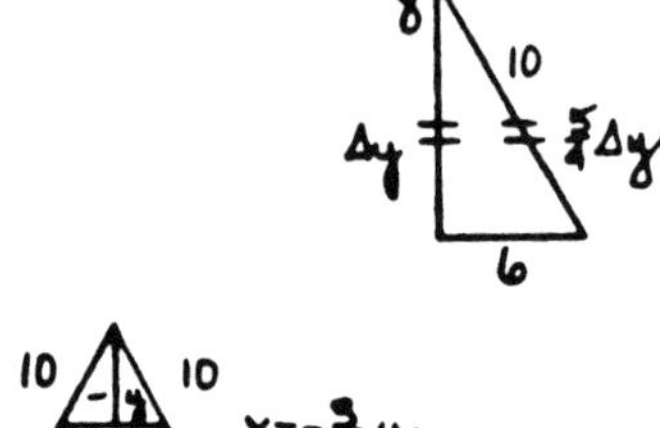

CHAPTER 11, REVIEW (pp. 582-584)

1. $\int_{-\infty}^{+\infty} \frac{x\,dx}{x^4+1} = 0$ provided $\int_{0}^{+\infty} \frac{x\,dx}{x^4+1}$ converges since $\frac{x}{x^4+1}$ is an odd function.

$$\int_{0}^{+\infty} \frac{x\,dx}{x^4+1} = \lim_{k\to+\infty} \frac{1}{2} \int_{0}^{k} \frac{2x\,dx}{1+x^4} = \lim_{k\to+\infty} \frac{1}{2} \text{Arctan } x^2 \Big|_0^k$$

$$= \lim_{k\to+\infty} \frac{1}{2} \text{Arctan } k^2 = \frac{\pi}{4}. \quad \text{Hence } \int_{-\infty}^{+\infty} \frac{x\,dx}{x^4+1} = 0.$$

5. $V = \int_0^1 \pi(e^2 - y^2)dx$

$$= \int_0^1 \pi(e^2 - e^{2x})dx$$

$$= \pi(e^2x - \frac{1}{2} e^{2x})\Big|_0^1$$

$$= \pi(e^2 - \frac{1}{2} e^2 + \frac{1}{2})$$

$$= \frac{(e^2+1)\pi}{2}.$$

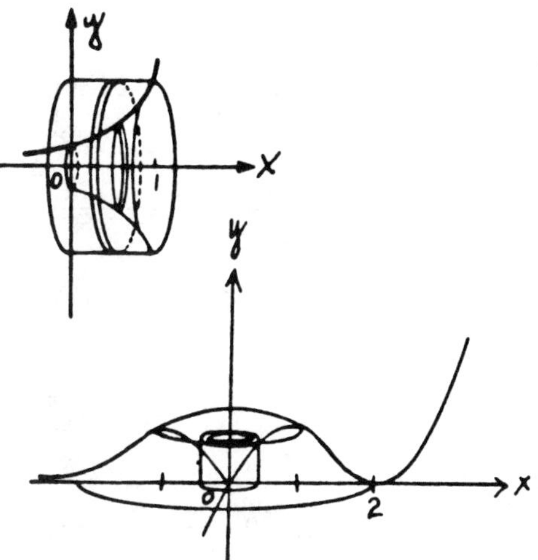

9. $V = 2\pi \int_0^2 xy\,dx = 2\pi \int_0^2 x^2(x-2)^2 dx$

$$= 2\pi \int_0^2 (x^4 - 4x^3 + 4x^2)dx = 2\pi(\frac{x^5}{5} - x^4 + \frac{4x^3}{3})\Big|_0^2$$

$$= 2\pi(\frac{32}{5} - 16 + \frac{32}{3}) = \frac{32\pi}{15}.$$

13. $m = 12 + 16 + 12 = 40$,

$M_x = 12 \cdot 2 + 16 \cdot 2 + 12 \cdot \frac{3}{2} = 74$,

$M_y = 12(-1) + 16 \cdot 2 + 12 \cdot 5 = 80$,

$\bar{x} = \frac{M_y}{m} = \frac{80}{40} = 2$, $\bar{y} = \frac{M_x}{m} = \frac{74}{40} = \frac{37}{20}$.

17. $V = \pi \int_0^2 y^2 dx = \pi \int_0^2 x^4(x - 2)^2 dx$

$= \pi \int_0^2 (x^6 - 4x^5 + 4x^4) dx$

$= \pi(\frac{x^7}{7} - \frac{2x^6}{3} + \frac{4x^5}{5}) \Big|_0^2$

$= \pi(\frac{128}{7} - \frac{128}{3} + \frac{128}{5}) = \frac{128\pi}{105}$,

$M_{yz} = \pi \int_0^2 xy^2 dx$

$= \pi \int_0^2 (x^7 - 4x^6 + 4x^5) dx = \pi(\frac{x^8}{8} - \frac{4x^7}{7} + \frac{2x^6}{3}) \Big|_0^2$

$= \pi(\frac{256}{8} - \frac{512}{7} + \frac{128}{3}) = \frac{128\pi}{84}$,

$\bar{x} = \frac{M_{yz}}{V} = \frac{128\pi/84}{128\pi/105} = \frac{5}{4}$.

21. $m = \int_0^1 4(x_1 - x_2) dy = \int_0^1 4(\sqrt{y} - y) dy$

$= 4(\frac{2}{3} y^{3/2} - \frac{y^2}{2}) \Big|_0^1 = 4(\frac{2}{3} - \frac{1}{2}) = \frac{2}{3}$,

$I_x = \int_0^1 4y^2(x_1 - x_2) dy$

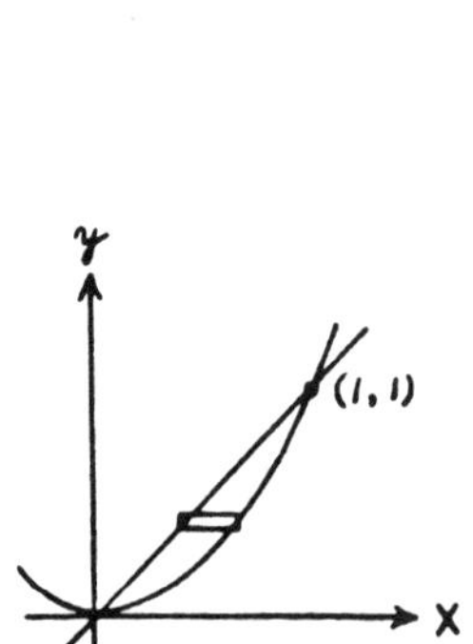

$$= 4\int_0^1 (y^{5/2} - y^3)\,dy$$

$$= 4\left(\frac{2}{7}y^{7/2} - \frac{y^4}{4}\right)\Big|_0^1 = 4\left(\frac{2}{7} - \frac{1}{4}\right) = \frac{1}{7},$$

$$k = \sqrt{\frac{I}{m}} = \sqrt{\frac{1/7}{2/3}} = \sqrt{\frac{3}{14}}.$$

25. $A = \frac{1}{2}\pi(4x) = 2\pi x$

$$= 2\pi\left(2 - \frac{y}{\sqrt{3}}\right),$$

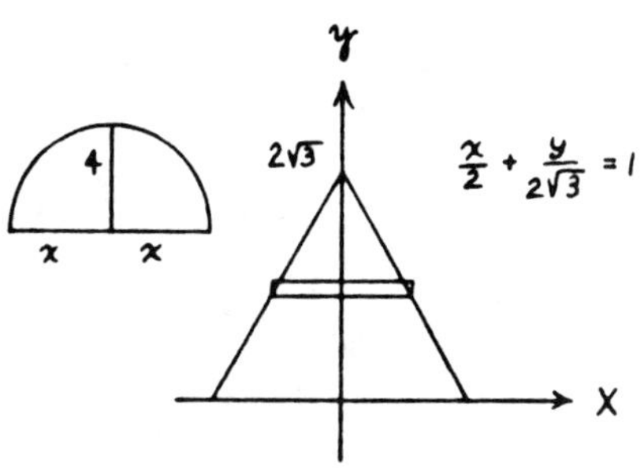

$$V = \int_0^{2\sqrt{3}} 2\pi\left(2 - \frac{y}{\sqrt{3}}\right) dy$$

$$= 2\pi\left(2y - \frac{y^2}{2\sqrt{3}}\Big|_0^{2\sqrt{3}}\right)$$

$$= 2\pi\left(4\sqrt{3} - \frac{12}{2\sqrt{3}}\right) = 4\pi\sqrt{3}.$$

29. $y = \frac{x^2}{10}$,

$$F = \int_0^{40} 9800(40 - y)2x\,dy$$

$$= 19{,}600\int_0^{40} (40 - y)\sqrt{10y}\,dy$$

$$= 19{,}600\sqrt{10}\int_0^{40} (40\sqrt{y} - y^{3/2})\,dy$$

$$= 19{,}600\sqrt{10}\left(\frac{80}{3}y^{3/2} - \frac{2}{5}y^{5/2}\right)\Big|_0^{40}$$

$$= \frac{2}{15}\cdot 19{,}600\sqrt{10}\,y^{3/2}(200 - 3y)\Big|_0^{40}$$

$$= \frac{2}{15}19{,}600\sqrt{10}\cdot 80\sqrt{10}\cdot 80 = 1.673\times 10^8\text{ N}$$

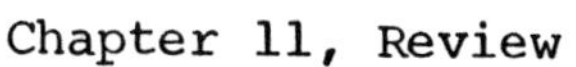

$$F = \int_0^{30} 19,600(30 - y)\sqrt{10y}\,dy$$

$$= 19,600\sqrt{10}\int_0^{30} (30\sqrt{y} - y^{3/2})dy$$

$$= 19,600\sqrt{10}\,(20y^{3/2} - \frac{2}{5}y^{5/2})\Big|_0^{30}$$

$$= 19,600\sqrt{10}\cdot\frac{2}{5}y^{3/2}(50 - y)\Big|_0^{30}$$

$$= 19,600\sqrt{10}\cdot\frac{2}{5}\cdot 30\sqrt{30}\cdot 20 = 8.1476 \times 10^7 \text{ N.}$$

33. $F = \int_0^{10} 9.8\frac{600 + 1000}{2}y(10)dy = 98\int_0^{10} 800\,y\,dy$

$$= 78,400\,\frac{y^2}{2}\Big|_0^{10} = 78,400\cdot 50 = 3.92 \times 10^6\text{N.}$$

$$F = \int_0^{5} 9.8[600\cdot 5 + 1000(5 - y)]10\,dy$$

$$+ \int_5^{10} 9.8\cdot 600(10 - y)10\,dy$$

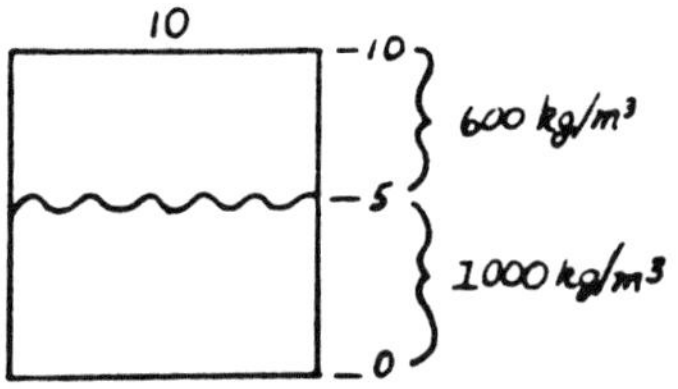

$$= 9800[\int_0^{5} (80 - 10y)dy$$

$$+ \int_5^{10} (60 - 6y)dy]$$

$$= 9800[(80y - 5y^2)\Big|_0^{5} + 60y - 3y^2\Big|_5^{10}]$$

$$= 9800(400 - 125 + 600 - 300 - 300 + 75)$$

$$= 9800\cdot 350 = 3.43 \times 10^6 \text{ N.}$$

Chapter 12
Conic Sections

CHAPTER 12, SECTION 2 (pp. 590-592)

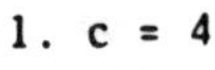

1. $c = 4$

   Axis: x axis

   $V(0, 0)$

   $F(4, 0)$

   D: $x = -4$

   $\ell r = 4 \cdot 4 = 16.$

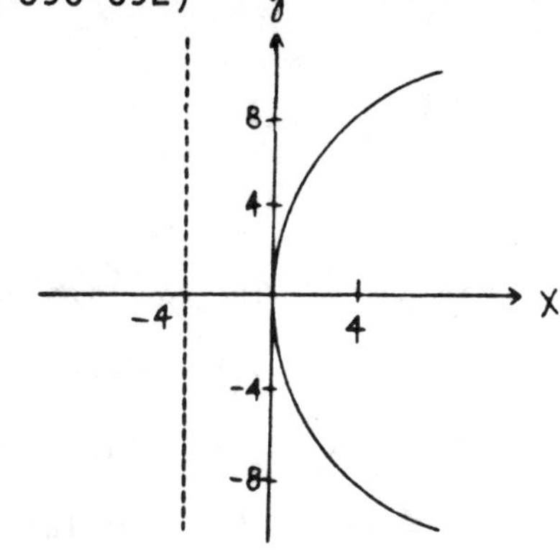

5. $c = \frac{5}{2}$

   Axis: x axis

   $V(0, 0)$

   $F(\frac{5}{2}, 0)$

   D: $x = -\frac{5}{2}$

   $\ell r = 4 \cdot \frac{5}{2} = 10.$

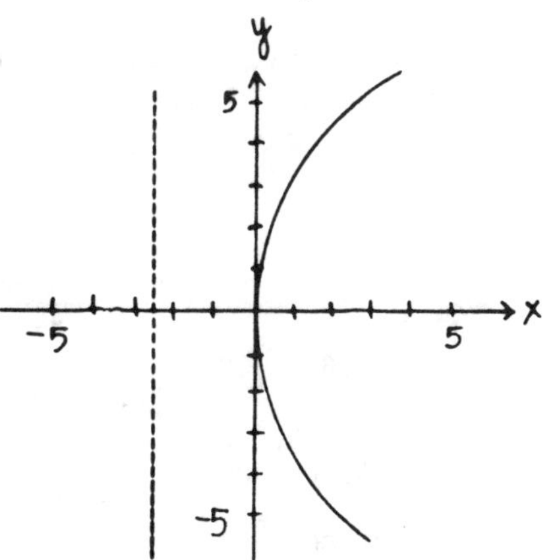

9. $c = -\frac{1}{2}$

   Axis: y axis

   $V(0, 0)$

   $F(0, -\frac{1}{2})$

   D: $y = \frac{1}{2}$

   $\ell r = 4 \cdot \frac{1}{2} = 2.$

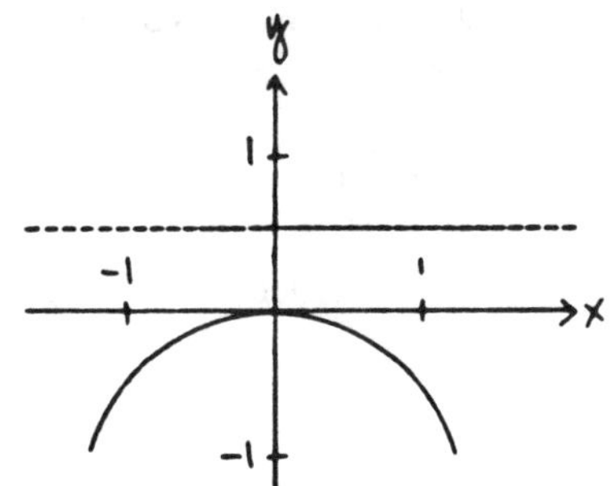

13. $y^2 = 4cx$, $\ell r = 4|c| = 5$, $|c| = \frac{5}{4}$, $c = \pm\frac{5}{4}$, $y^2 = \pm 5x$.

17. $x^2 = 4cy$, $4 = 4c \cdot 3$, $4c = \frac{4}{3}$, $x^2 = \frac{4}{3}y$.

21. $2x = -5y'$, $y' = -\frac{2x}{5} = -2$, $y + 5 = -2(x - 5)$,

$2x + y - 5 = 0$.

25. $2x = -8y'$, $y' = -\frac{x}{4} = -\frac{x_0}{4}$, where (x_0, y_0) is the point of tangency. The equation of the tangent line is

$y = -\frac{x_0}{4}(x - 4)$. Since (x_0, y_0) is on the tangent line,

$4y_0 = -x_0^2 + 4x_0$. Since (x_0, y_0) is on the parabola,

$x_0^2 = -8y_0$. From the last two equations,

$x_0^2 = 2x_0^2 - 8x_0$ $\qquad x_0(x_0 - 8) = 0$

$x_0 = 0$ $\qquad x_0 = 8$

$y = 0.$ $\qquad y = -2(x - 4)$

$\qquad 2x + y - 8 = 0$.

29. $$A_1 = \int_{25,000,000}^{75,000,000} (y_1 - y_2)dx$$

$$= \int_{25,000,000}^{75,000,000} [10,000\sqrt{x} - \sqrt{3}(x - 25,000,000)]dx$$

$$= \frac{20,000}{3}x^{3/2} - \frac{\sqrt{3}}{2}x^2 + 25,000,000\sqrt{3}x \Big|_{25,000,000}^{75,000,000}$$

$$= \frac{20,000}{3}(75,000,000^{3/2} - 25,000,000^{3/2})$$

$$- \frac{\sqrt{3}}{2}(75,000,000^2 - 25,000,000^2)$$

$$+ \; 25{,}000{,}000\sqrt{3}(50{,}000{,}000) = \frac{375\sqrt{3} - 250}{3} \cdot 10^{13}$$

$$A_2 = \int_0^{25{,}000{,}000} 10{,}000\sqrt{x}\, dx = \frac{20{,}000}{3} x^{3/2}\Big|_0^{25{,}000{,}000}$$

$$= \frac{20{,}000}{3} 25{,}000{,}000^{3/2} = \frac{250}{3} \cdot 10^{13}$$

$$\frac{(375\sqrt{3} - 250)/3}{36} = \frac{250/3}{t}$$

$$t = \frac{250 \cdot 36}{375\sqrt{3} - 250} = \frac{72}{3\sqrt{3} - 2}$$

$$= \frac{72(3\sqrt{3} + 2)}{23} \text{ hr}$$

$\approx$ *22* hr 31 min.

33. P_1Q: $yy_1 = 2c(x + x_1)$

P_2Q: $yy_2 = 2c(x + x_2)$

Solving simultaneously, we have

$$y(y_1 - y_2) = 2c(x_1 - x_2)$$

$$y = \frac{2c(x_1 - x_2)}{y_1 - y_2} = \frac{4cx_1 - 4cx_2}{2(y_1 - y_2)} = \frac{y_1^2 - y_2^2}{2(y_1 - y_2)} = \frac{y_1 + y_2}{2}$$

$$\frac{(y_1 + y_2)y_1}{2} = 2c(x + x_1)$$

$$x = \frac{y_1^2 + y_1y_2}{4c} - \frac{4cx_1}{4c} = \frac{y_1y_2}{4c}$$

$$Q: \left(\frac{y_1y_2}{4c}, \frac{y_1 + y_2}{2}\right)$$

$$m_{P_1F} = \frac{y_1}{x_1 - c} = \frac{4cy_1}{4cx_1 - 4c^2}$$

$$= \frac{4cy_1}{y_1^2 - 4c^2}$$

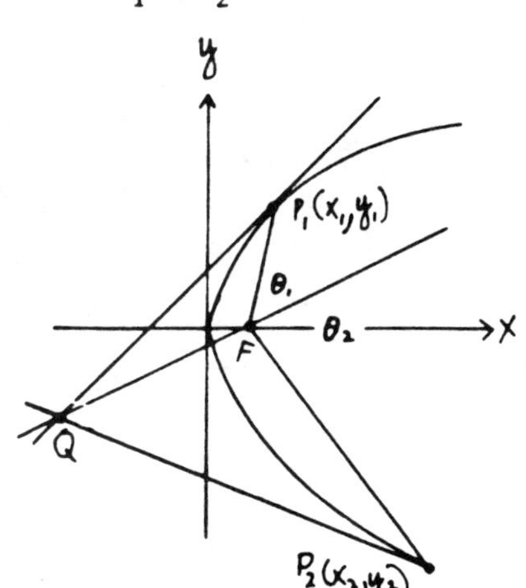

$$m_{P_2F} = \frac{y_2}{x_2 - c} = \frac{4cy_2}{y_2^2 - 4c^2}$$

$$m_{FQ} = \frac{\frac{y_1 + y_2}{2}}{\frac{y_1y_2}{4c} - c} = \frac{2c(y_1 + y_2)}{y_1y_2 - 4c^2}$$

$$\tan \theta_1 = \frac{m_{P_1F} - m_{FQ}}{1 + m_{P_1F}m_{FQ}} = \frac{\frac{4cy_1}{y_1^2 - 4c^2} - \frac{2c(y_1 + y_2)}{y_1y_2 - 4c^2}}{1 + \frac{8c^2y_1(y_1 + y_2)}{(y_1^2 - 4c^2)(y_1y_2 - 4c^2)}}$$

$$= \frac{4cy_1^2y_2 - 16c^3y_1 - 2cy_1^3 - 2cy_1^2y_2 + 8c^3y_1 + 8c^3y_2}{y_1^3y_2 - 4c^2y_1^2 - 4c^2y_1y_2 + 16c^4 + 8c^2y_1^2 + 8c^2y_1y_2}$$

$$= \frac{2cy_1^2y_2 - 8c^3y_1 - 2cy_1^3 + 8c^3y_2}{y_1^3y_2 + 4c^2y_1^2 + 4c^2y_1y_2 + 16c^4}$$

$$= \frac{2c(y_1^2 + 4c^2)(y_2 - y_1)}{(y_1^2 + 4c^2)(y_1y_2 + 4c^2)} = \frac{2c(y_2 - y_1)}{y_1y_2 + 4c^2}$$

$$\tan \theta_2 = \frac{m_{FQ} - m_{P_2F}}{1 + m_{FQ}m_{P_2F}} = \frac{\frac{2c(y_1 + y_2)}{y_1y_2 - 4c^2} - \frac{4cy_2}{y_2^2 - 4c^2}}{1 + \frac{8c^2y_2(y_1 - y_2)}{(y_2^2 - 4c^2)(y_1y_2 - 4c^2)}}$$

$$= \frac{2cy_1y_2^2 - 8c^3y_1 + 2cy_2^3 - 8c^3y_2 - 4cy_1y_2^2 + 16c^3y_2}{y_1y_2^3 - 4c^2y_2^2 - 4c^2y_1y_2 + 16c^4 + 8c^2y_1y_2 + 8c^2y_2^2}$$

$$= \frac{-8c^3y_1 + 2cy_2^3 - 2cy_1y_2^2 + 8c^3y_2}{y_1y_2^3 + 4c^2y_1y_2 + 4c^2y_2^2 + 16c^4}$$

$$= \frac{2c(y_2^2 + 4c^2)(y_2 - y_1)}{(y_2^2 + 4c^2)(y_1y_2 + 4c^2)} = \frac{2c(y_2 - y_1)}{y_1y_2 + 4c^2}.$$

CHAPTER 12, SECTION 3 (pp. 597-599)

1. C(0, 0), V(±13, 0), CV(0, ±5)

   $c^2 = a^2 - b^2 = 169 - 25 = 144$

   F(±12, 0)

   $\ell r = \frac{2b^2}{a} = \frac{2 \cdot 25}{13} = \frac{50}{13}.$

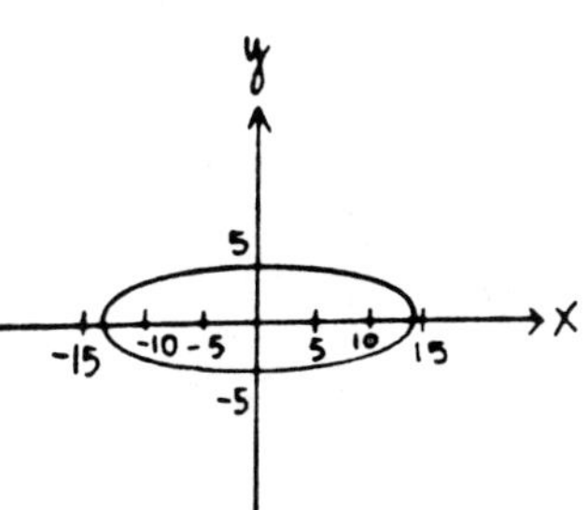

5. C(0, 0), V(0, ±7), CV(±5, 0)

   $c^2 = a^2 - b^2 = 49 - 25 = 24$

   F(0, $\pm 2\sqrt{6}$)

   $\ell r = \frac{2b^2}{a} = \frac{2 \cdot 25}{7} = \frac{50}{7}.$

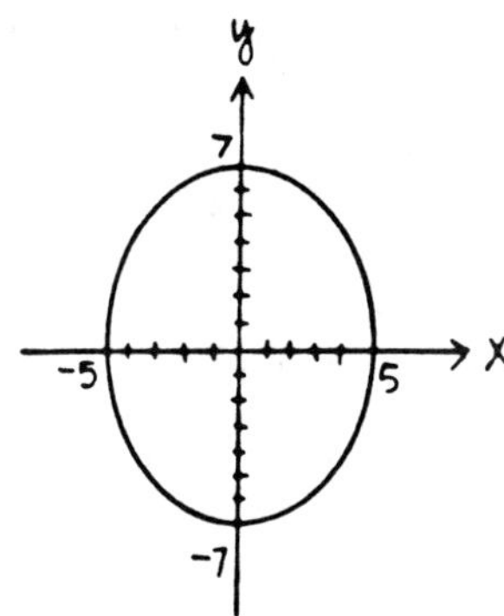

9. $\frac{x^2}{9} + \frac{y^2}{16} = 1$

$C(0, 0)$, $V(0, \pm 4)$, $CV(\pm 3, 0)$

$c^2 = a^2 - b^2 = 16 - 9 = 7$

$F(0, \pm\sqrt{7})$

$\ell r = \frac{2b^2}{a} = \frac{2 \cdot 9}{4} = \frac{9}{2}$.

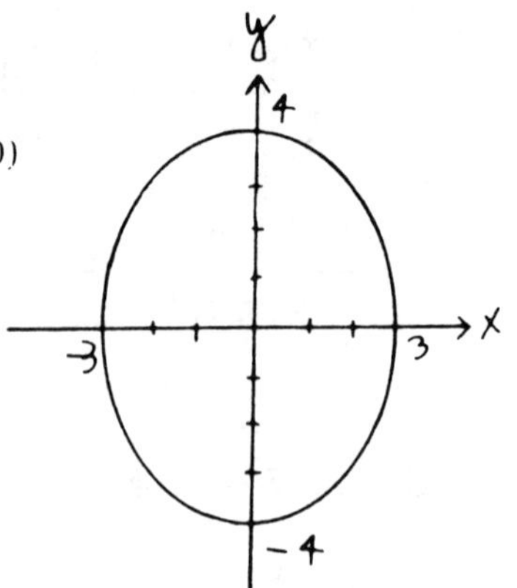

13. $\frac{x^2}{a^2} + \frac{y^2}{b^2} = 1$, $a = 5$, $\frac{x^2}{25} + \frac{y^2}{b^2} = 1$, $\frac{15}{25} + \frac{4}{b^2} = 1$, $\frac{4}{b^2} = \frac{2}{5}$,

$b^2 = 10$, $\frac{x^2}{25} + \frac{y^2}{10} = 1$.

17. $\frac{x^2}{a^2} + \frac{y^2}{b^2} = 1$, $c = 6$, $e = \frac{c}{a}$, $\frac{3}{5} = \frac{6}{a}$, $a = 10$,

$b^2 = a^2 - c^2 = 100 - 36 = 64$, $\frac{x^2}{100} + \frac{y^2}{64} = 1$.

21. $4x + 6yy' = 0$, $y' = -\frac{2x}{3y} = -\frac{4}{3}$, $y - 1 = -\frac{4}{3}(x - 2)$,

$4x + 3y - 11 = 0$.

25. $\dfrac{x^2}{a^2} + \dfrac{y^2}{b^2} = 1.$

By Problem 24, the tangent at (x_0, y_0) has slope

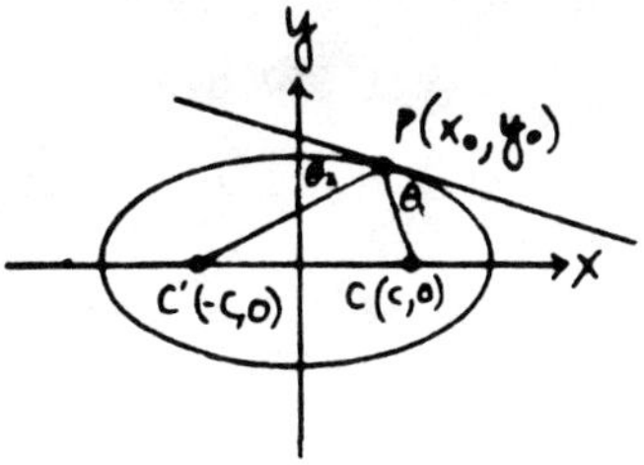

$$m = -\frac{b^2x_0}{a^2y_0}$$

$$m_{PC} = \frac{y_0}{x_0 - c}, \quad m_{PC'} = \frac{y_0}{x_0 + c}$$

$$\tan\theta_1 = \frac{-\dfrac{b^2x_0}{a^2y_0} - \dfrac{y_0}{x_0 - c}}{1 - \dfrac{b^2x_0}{a^2(x_0 - c)}} = \frac{-b^2x_0(x_0 - c) - a^2y_0^2}{a^2y_0(x_0 - c) - b^2x_0y_0}$$

$$= \frac{b^2x_0^2 - b^2cx_0 + a^2y_0^2}{-a^2x_0y_0 + a^2cy_0 + b^2c_0y_0}$$

$$= \frac{a^2b^2 - b^2cx_0}{a^2cy_0 + (b^2 - a^2)x_0y_0} = \frac{b^2(a^2 - cx_0)}{cy_0(a^2 - cx_0)} = \frac{b^2}{cy_0},$$

$$\tan\theta_2 = \frac{\dfrac{y_0}{x_0 + c} + \dfrac{b^2x_0}{a^2y_0}}{1 - \dfrac{b^2x_0}{a^2(x_0 + c)}} = \frac{a^2y_0^2 + b^2x_0(x_0 + c)}{a^2y_0(x_0 + c) - b^2x_0y_0}$$

$$= \frac{a^2y_0^2 + b^2x_0^2 + b^2cx_0}{a^2x_0y_0 + a^2cy_0 - b^2x_0y_0} = \frac{a^2b^2 + b^2cx_0}{a^2cy_0 - (b^2 - a^2)x_0y_0}$$

$$= \frac{b^2(a^2 + cx_0)}{cy_0(a^2 + cx_0)} = \frac{b^2}{cy_0}.$$

29. $a - c = 91{,}446{,}000, \quad a + c = 94{,}560{,}000$

$2a = 186{,}006{,}000, \quad a = 93{,}003{,}000$

$2c = 3{,}114{,}000$

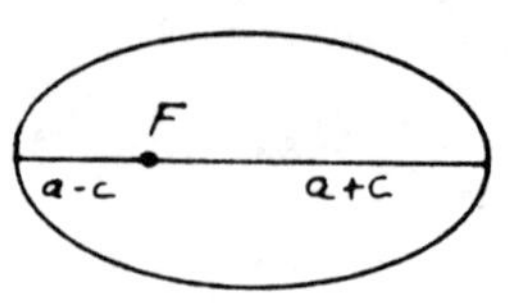

$c = 1{,}557{,}000$

$e = \frac{c}{a} = \frac{1{,}557{,}000}{93{,}003{,}000} = 0.0167.$

CHAPTER 12, SECTION 4 (pp. 605-607)

1. $C(0, 0), \quad V(\pm 4, 0),$

$c^2 = a^2 + b^2 = 16 + 9 = 25$

$F(\pm 5, 0), \quad A: y = \pm \frac{3x}{4}$

$\ell r = \frac{2b^2}{a} = \frac{2 \cdot 9}{4} = \frac{9}{2}.$

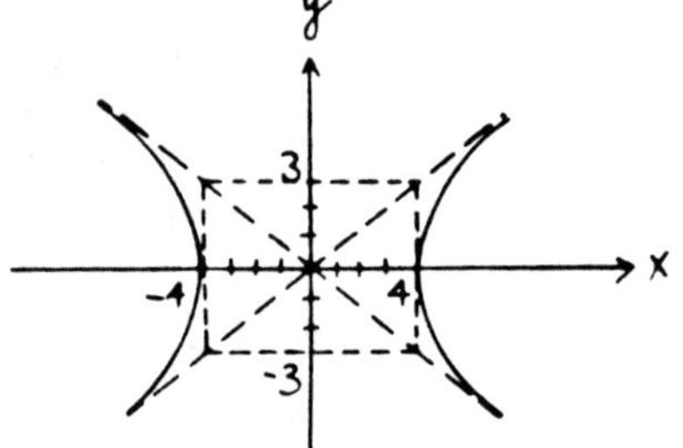

5. $C(0, 0), \quad V(\pm 12, 0)$

$c^2 = a^2 + b^2$

$= 144 + 25 = 169$

$F(\pm 13, 0), \quad A: y = \pm \frac{5x}{12}$

$\ell r = \frac{2b^2}{a} = \frac{2 \cdot 25}{12} = \frac{25}{6}.$

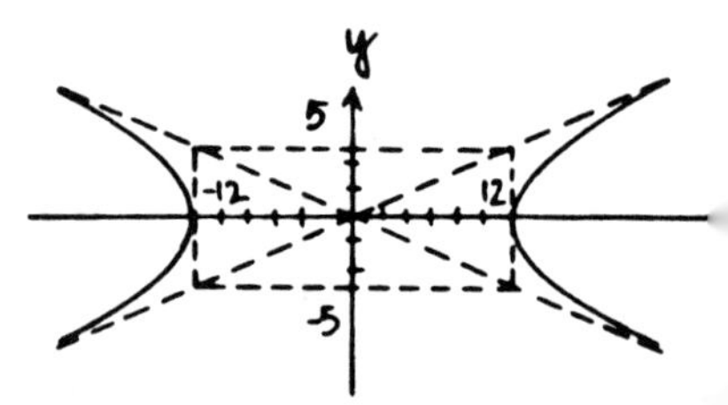

9. $\frac{x^2}{1} - \frac{y^2}{4} = 1$, C(0, 0), V(±1, 0),

$c^2 = a^2 + b^2 = 1 + 4 = 5$

F($\pm\sqrt{5}$, 0), A: $y = \pm 2x$

$\ell r = \frac{2b^2}{a} = \frac{2 \cdot 4}{1} = 8.$

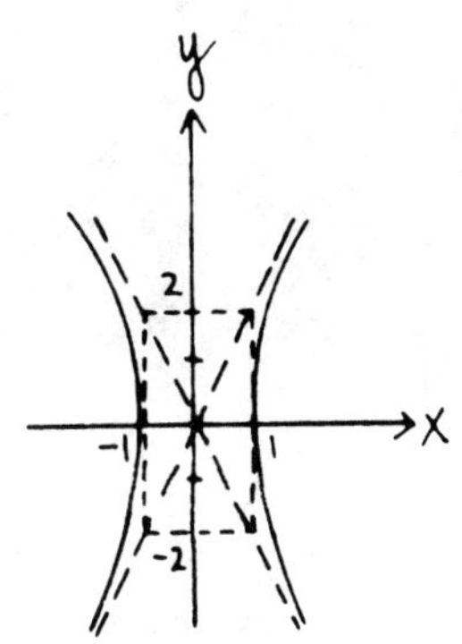

13. $\frac{y^2}{25/4} - \frac{x^2}{9/4} = 1$

C(0, 0), V(0, $\pm\frac{5}{2}$)

$c^2 = a^2 + b^2 = \frac{25}{4} + \frac{9}{4} = \frac{34}{4}$

F(0, $\pm\frac{\sqrt{34}}{2}$), A: $y = \pm\frac{5x}{3}$

$\ell r = \frac{2b^2}{a} = \frac{2 \cdot 9/4}{5/2} = \frac{9}{5}.$

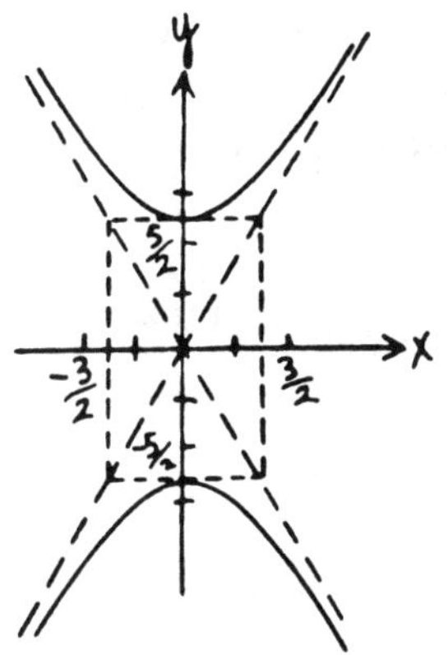

17. $\frac{x^2}{a^2} - \frac{y^2}{b^2} = 1$, $\frac{b}{a} = \frac{2}{3}$, a = 6, b = 4, $\frac{x^2}{36} - \frac{y^2}{16} = 1.$

21. $\frac{x^2}{a^2} - \frac{y^2}{b^2} = 1$, $a = 5$, $\frac{x^2}{25} - \frac{y^2}{b^2} = 1$, $\frac{81/25}{25} - \frac{16}{b^2} = 1$,

$\frac{81b^2}{25} - 400 = 25b^2$, $(81 - 625)b^2 = 400$, $b^2 = -\frac{625}{34} < 0$,

no solution.

25. $\frac{x^2}{a^2} - \frac{y^2}{b^2} = 1$, $\frac{a^2}{c} = \frac{9}{5}$, $e = \frac{c}{a} = \frac{5}{3}$, $c = 5$, $a = 3$, $b^2 = c^2 - a^2$

$= 25 - 9 = 16$, $\frac{x^2}{9} - \frac{y^2}{16} = 1$.

29. $2x - 2yy' = 0$

$y' = \frac{x}{y} = \frac{x_0}{y_0}$, where (x_0, y_0) is the point of tangency. The equation of the tangent line is $y - 9 = \frac{x_0}{y_0}(x - 9)$.

Since (x_0, y_0) is on the tangent line,

$y_0 - 9 = \frac{x_0}{y_0}(x_0 - 9)$, $x_0^2 - y_0^2 = 9x_0 - 9y_0$.

Since (x_0, y_0) is on the hyperbola, $x_0^2 - y_0^2 = 9$.

Solving simultaneously, we have

$9x_0 - 9y_0 = 9$, $y_0 = x_0 - 1$, $x_0^2 - (x_0 - 1)^2 = 9$

$2x_0 = 10$, $x_0 = 5$, $y_0 = 4$

$y - 9 = \frac{5}{4}(x - 9)$, $5x - 4y - 9 = 0$.

There is only one tangent line because the given point is on an asymptote.

33. $\lim_{x\to+\infty} \left[\left(\frac{b}{a}\sqrt{x^2 - a^2}\right) - \left(\frac{b}{a}x\right)\right] = \lim_{x\to+\infty} \frac{b}{a} \frac{(\sqrt{x^2-a^2}-x)(\sqrt{x^2-a^2}+x)}{\sqrt{x^2 - a^2} + x}$

$= \lim_{x\to+\infty} \frac{b}{a} \frac{(x^2 - a^2) - x^2}{\sqrt{x^2 - a^2} + x} = \lim_{x\to+\infty} \frac{b}{a} \frac{-a^2}{\sqrt{x^2 - a^2} + x} = 0.$

37. $\dfrac{\sqrt{(x - 1000)^2 + y^2}}{1100} = 1 + \dfrac{\sqrt{(x + 1000)^2 + y^2}}{1100}$

$$\sqrt{(x - 1000)^2 + y^2} = 1100 + \sqrt{(x + 1000)^2 + y^2}$$

$$x^2 - 2000x + 1,000,000 + y^2$$
$$= 1,210,000 + 2200\sqrt{(x + 1000)^2 + y^2} + x^2 + 2000x + 1,000,000 + y^2$$

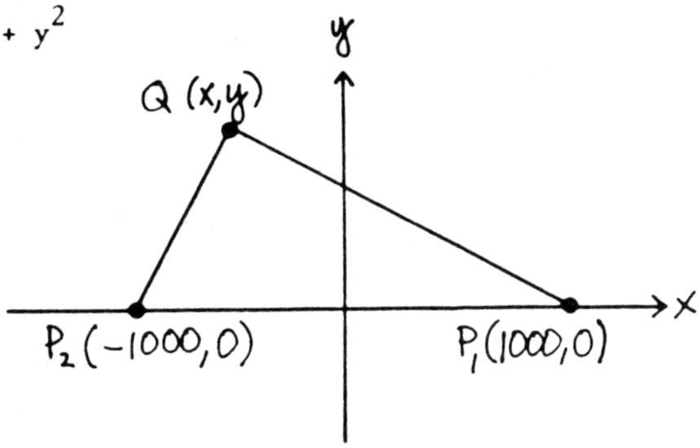

$$-4000x - 1,210,000 = 2200\sqrt{(x + 1000)^2 + y^2}$$

$$-20x - 6050 = 11\sqrt{(x + 1000)^2 + y^2}$$

$$400x^2 + 242,000x + 36,602,500$$
$$= 121(x^2 + 2000x + 1,000,000 + y^2)$$

$$279x^2 - 121y^2 = 84,397,500.$$

CHAPTER 12, SECTION 5 (pp. 616-617)

1. $y^2 - 2y + 1 = 4x - 9 + 1$

$$(y - 1)^2 = 4(x - 2)$$

$$x' = x - 2, \quad y' = y - 1$$

$$y'^2 = 4x'.$$

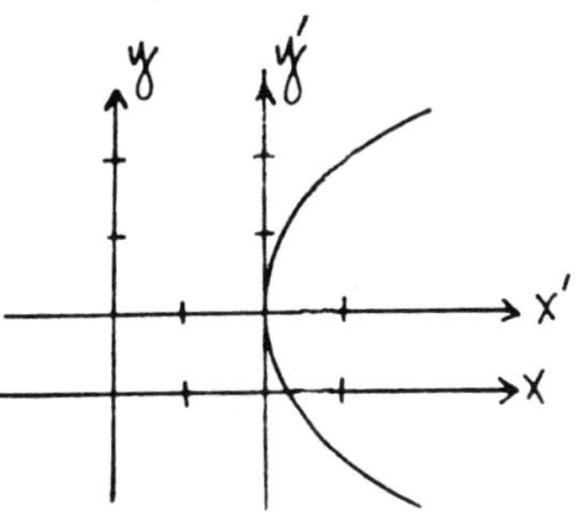

5. $9(x^2 + 10x + 25) - 4(y^2 - 8y + 16)$

$= -125 + 225 - 64$

$9(x + 5)^2 - 4(y - 4)^2 = 36$

$x' = x + 5, \quad y' = y - 4$

$\frac{x'^2}{4} - \frac{y'^2}{9} = 1.$

9. $4(x^2 - x + \frac{1}{4})$

$= 4y + 5 + 1$

$4(x - \frac{1}{2})^2 = 4(y + \frac{3}{2})$

$x' = x - \frac{1}{2}, \quad y' = y + \frac{3}{2}$

$x'^2 = y'.$

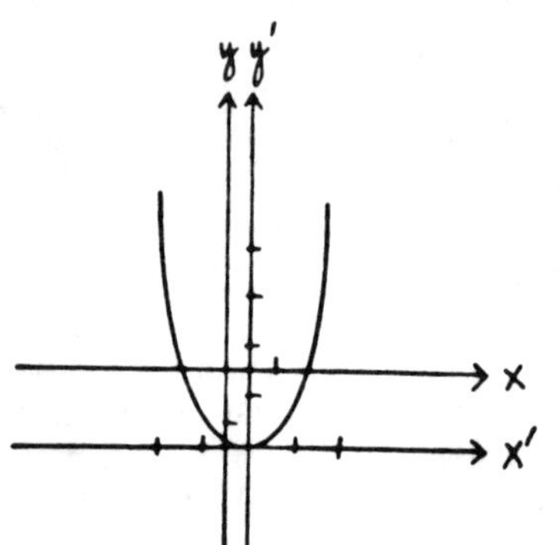

13. $16x^2 - 9(y^2 - 6y + 9) = 225 - 81$

$16x^2 - 9(y - 3)^2 = 144$

$\frac{x^2}{9} - \frac{(y - 3)^2}{16} = 1$

$C(0, 3), \quad V(\pm 3, 3)$

$c^2 = a^2 + b^2 = 9 + 16 = 25$

$F(\pm 5, 3), \quad A: y = 3 \pm \frac{4x}{3}$

$\ell r = \frac{2b^2}{a} = \frac{2 \cdot 16}{3} = \frac{32}{3}.$

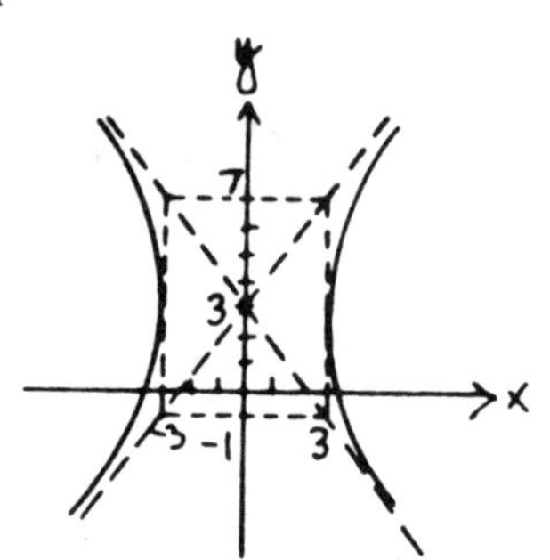

17. $3(x^2 - \frac{2}{3}x + \frac{1}{9})$

$-3(y^2 + \frac{4}{3}y + \frac{4}{9}) = 13 + \frac{1}{3} - \frac{4}{3}$

$3(x - \frac{1}{3})^2 - 3(y + \frac{2}{3})^2 = 12$

$\frac{(x - \frac{1}{3})^2}{4} - \frac{(y + \frac{2}{3})^2}{4} = 1$

$C(\frac{1}{3}, -\frac{2}{3})$, $V(\frac{1}{3} \pm 2, -\frac{2}{3})$

$c^2 = a^2 + b^2 = 4 + 4 = 8$

$F(\frac{1}{3} \pm 2\sqrt{2}, -\frac{2}{3})$

A: $y = -\frac{2}{3} \pm (x - \frac{1}{3})$

$\ell r = \frac{2b^2}{a} = \frac{2 \cdot 4}{2} = 4.$

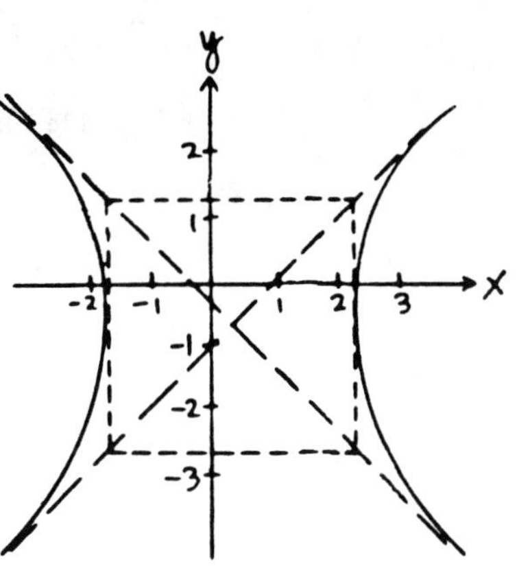

21. $C(-2, 1)$, $c = 4$, $a = 1$; $b^2 = c^2 - a^2 = 16 - 1 = 15$

$\frac{(y - 1)^2}{1} - \frac{(x + 2)^2}{15} = 1$

$x^2 - 15y^2 + 4x + 30y + 4 = 0.$

25. $x = x' + h$, $y = y' + k$

$(x' + h)^2 - 2(x' + h)(y' + k) + 4(y' + k)^2 + 8(x' + h)$

$- 26(y' + k) + 38 = 0$

$x'^2 + 2x'h + h^2 - 2x'y' - 2kx' - 2hy' - 2hk$

$+ h^2 - 2hk + 4k^2 + 8h - 26k + 38 = 0$

$h - k = -4$, $h - 4k = -13$,

$h = -1$, $k = 3$, $x'^2 - 2x'y' + 4y'^2 - 5 = 0.$

29. $x = x' + h, \quad y = y' + k$

$$y' + k = x'^3 + 3hx'^2 + 3h^2x' + h^3 - 3x' - 3h + 6$$

$$y' = x'^3 + 3hx'^2 + (3h^2 - 3)x' + (h^3 - 3h + 6 - k)$$

$$3h^2 - 3 = 0, \quad h = \pm 1, \quad h^3 - 3h + 6 - k = 0$$

$h = 1, \quad k = 4 \qquad\qquad h = -1, \quad k = 8$

$y' = x'^3 + 3x'^2. \qquad\qquad y' = x'^3 - 3x'^2.$

33. $Ax^2 + Bxy + Cy^2 + Dx + Ey + F = 0$

$x = x' + h, \quad y = y' + k$

$$Ax'^2 + 2Ahx' + Ah^2 + Bx'y' + Bkx' + Bhy' + Bhk + Cy'^2 + 2Cky' + Ck^2 + Dx' + Dh + Ey' + Ek + F = 0$$

$$Ax'^2 + Bx'y' + Cy'^2 + (2Ah + Bk + D)x' + (Bh + 2Ck + E)y' + (Ah^2 + Bhk + Ck^2 + Dh + Ek + F) = 0$$

$A' = A, \quad B' = B, \quad C' = C$, no matter what values of h and k are used.

CHAPTER 12, SECTION 6 (pp. 625-627)

1. $\tan\theta = \frac{3}{2}, \quad \sin\theta = \frac{3}{\sqrt{13}}, \quad \cos\theta = \frac{2}{\sqrt{13}},$

$$x = \frac{2x' - 3y'}{\sqrt{13}}, \quad y = \frac{3x' + 2y'}{\sqrt{13}},$$

$$\frac{4x' - 6y' + 9x' + 6y'}{\sqrt{13}} = 6$$

$$\frac{13x'}{\sqrt{13}} = 6, \quad \sqrt{13}\,x' = 6.$$

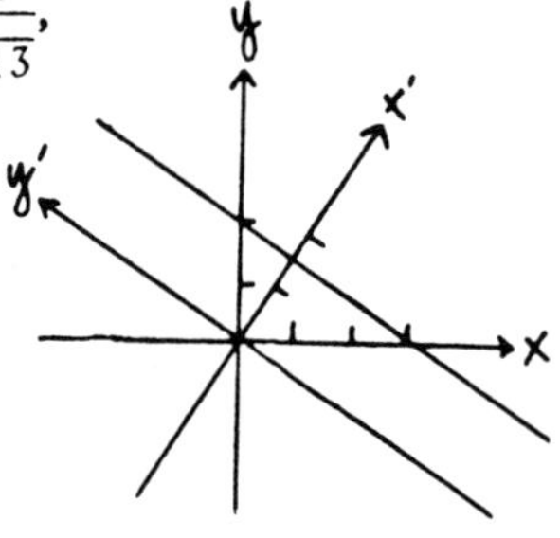

5. $\sin\theta = \frac{1}{2}$, $\cos\theta = \frac{\sqrt{3}}{2}$,

$x = \frac{\sqrt{3}x' - y'}{2}$, $y = \frac{x' + \sqrt{3}y'}{2}$,

$$\frac{93x'^2 - 62\sqrt{3}x'y' + 31y'^2 + 30x'^2 + 20\sqrt{3}x'y' - 30y'^2 + 21x'^2 + 42\sqrt{3}x'y' + 63y'^2}{4} = 144$$

$$\frac{144x'^2 + 64y'^2}{4} = 144$$

$$\frac{x'^2}{4} + \frac{y'^2}{9} = 1.$$

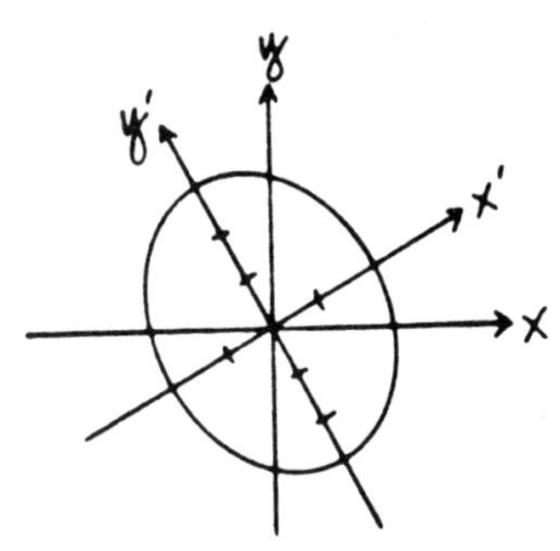

9. $A = C$, $\theta = 45°$, $\sin\theta = \cos\theta = \frac{1}{\sqrt{2}}$,

$x = \frac{x' - y'}{\sqrt{2}}$, $y = \frac{x' + y'}{\sqrt{2}}$,

$$\frac{x'^2 + 2x'y' + y'^2 + x'^2 - y'^2 + x'^2 + 2x'y' + y'^2}{2}$$

$$+ \frac{4\sqrt{2}(x' - y') - 4\sqrt{2}(x' + y')}{\sqrt{2}} = 0$$

$$\frac{3x'^2 + y'^2}{2} - 8y' = 0$$

$$3x'^2 + y'^2 - 16y'^2 + 64 = 64$$

$$\frac{x'^2}{64/3} + \frac{(y' - 8)^2}{64} = 1.$$

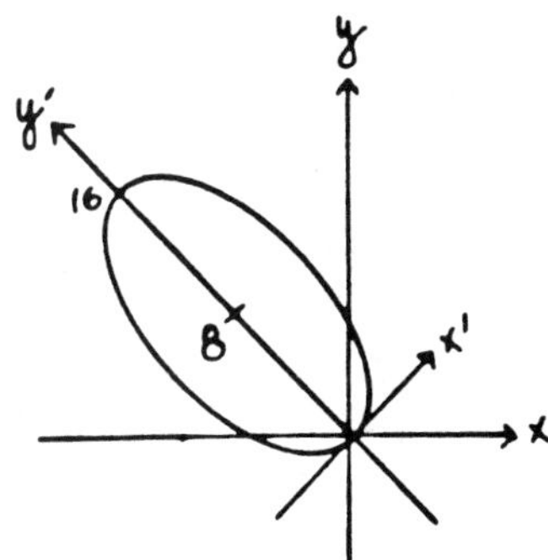

13. $\tan 2\theta = \dfrac{B}{A - C} = \dfrac{-4}{5 - 8} = \dfrac{4}{3}$, $\cos 2\theta = \dfrac{1}{\sqrt{1 + 16/9}} = \dfrac{3}{5}$,

$\sin\theta = \sqrt{\dfrac{1 - 3/5}{2}} = \dfrac{1}{\sqrt{5}}$, $\cos\theta = \sqrt{\dfrac{1 + 3/5}{2}} = \dfrac{2}{\sqrt{5}}$,

$\tan\theta = \dfrac{1}{2}$, $x = \dfrac{2x' - y'}{\sqrt{5}}$, $y = \dfrac{x' + 2y'}{\sqrt{5}}$,

$$\frac{5(4x'^2-4x'y'+y'^2) - 4(2x'^2+3x'y'-2y'^2) + 8(x'^2+4x'y'+4y'^2)}{5}$$

$$- 36 = 0$$

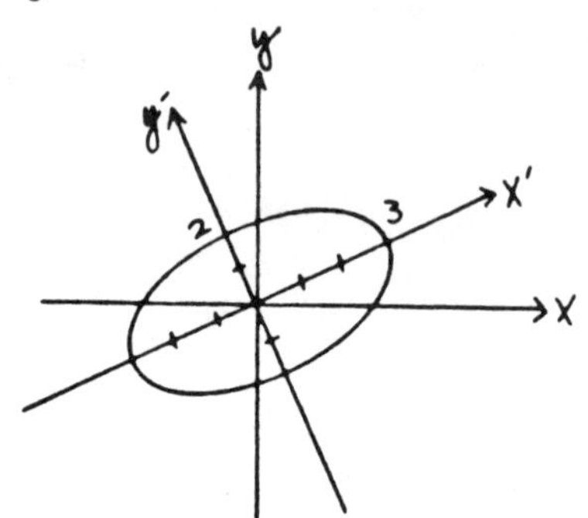

$$4x'^2 + 9y'^2 = 36$$

$$\frac{x'^2}{9} + \frac{y'^2}{4} = 1.$$

17. $\tan 2\theta = \dfrac{B}{A - C} = \dfrac{-6}{9 - 1} = -\dfrac{3}{4}$, $\cos 2\theta = \dfrac{-1}{\sqrt{1 + 9/16}} = -\dfrac{4}{5}$,

$\sin\theta = \sqrt{\dfrac{1 + 4/5}{2}} = \dfrac{3}{\sqrt{10}}$, $\cos\theta = \sqrt{\dfrac{1 - 4/5}{2}} = \dfrac{1}{\sqrt{10}}$,

$\tan\theta = 3$, $x = \dfrac{x' - 3y'}{\sqrt{10}}$, $y = \dfrac{3x' + y'}{\sqrt{10}}$,

$$\frac{9(x'^2-6x'y'+9y'^2) - 6(3x'^2-8x'y'-3y'^2) + (9x'^2+6x'y'+y'^2)}{10}$$

$$+ \frac{-12\sqrt{10}(x' - 3y') - 36\sqrt{10}(3x' + y')}{\sqrt{10}} = 0$$

$$10y'^2 - 120x' = 0$$

$$y'^2 = 12x'.$$

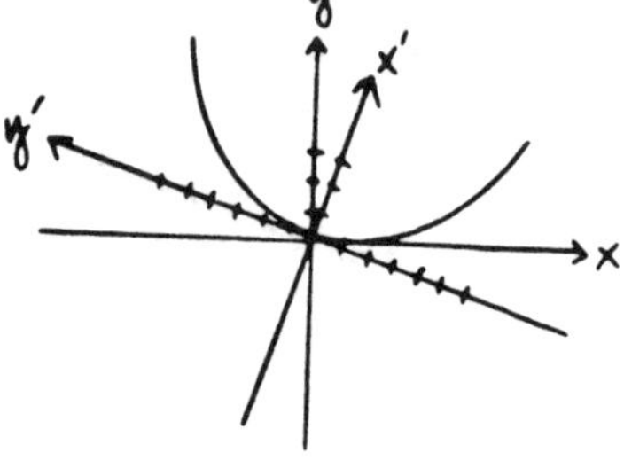

21. $x^2 - xy - y - 4 = 0.$

$$y(x + 1) = x^2 - 4$$

$$y = \frac{x^2 - 4}{x + 1} = x - 1 - \frac{3}{x + 1}.$$

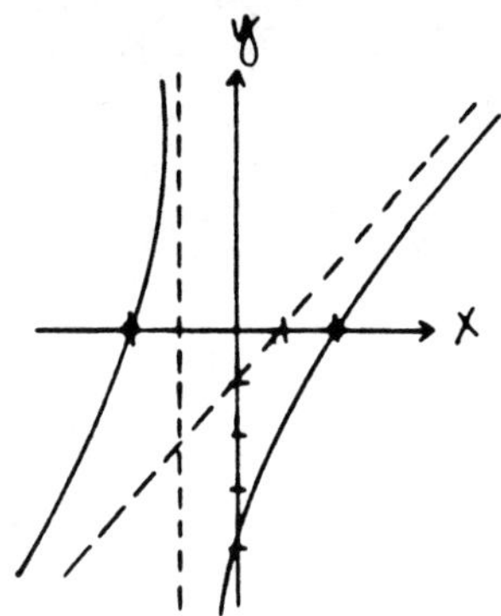

25. $x = x' \cos \theta - y' \sin \theta, \quad y = x' \sin \theta + y' \cos \theta$

$$Ax'^2\cos^2\theta - 2Ax'y' \sin \theta \cos \theta + Ay'^2\sin^2\theta + Bx'^2\sin \theta \cos \theta$$

$$+ B(\cos^2\theta - \sin^2\theta)x'y' - By'^2\sin \theta \cos \theta + Cx'^2\sin^2\theta$$

$$+ 2Cx'y' \sin \theta \cos \theta + Cy'^2\cos^2\theta + Dx' \cos \theta - Dy' \sin \theta$$

$$+ Ex' \sin \theta + Ey' \cos \theta + F = 0$$

$$(A \cos^2\theta + B \sin \theta \cos \theta + C \sin^2\theta)x'^2 + [-2A \sin \theta \cos \theta$$

$$+ B(\cos^2\theta - \sin^2\theta) + 2C \sin \theta \cos \theta]x'y' + (A \sin^2\theta$$

$$- B \sin \theta \cos \theta + C \cos^2\theta)y'^2 + (D \cos \theta + E \sin \theta)x'$$

$$+ (-D \sin \theta + E \cos \theta)y' + F = 0$$

$$A' + C' = (A \cos^2\theta + B \sin \theta \cos \theta + C \sin^2\theta)$$

$$+ (A \sin^2\theta - B \sin \theta \cos \theta + C \cos^2\theta)$$

$$= A(\cos^2\theta + \sin^2\theta) + C(\sin^2\theta + \cos^2\theta) = A + C.$$

CHAPTER 12, REVIEW (pp. 627-628)

1. $c = 4$

   Axis: x axis

   $V(0, 0)$, $F(4, 0)$,

   D: $x = -4$, $\ell r = 4 \cdot 4 = 16$.

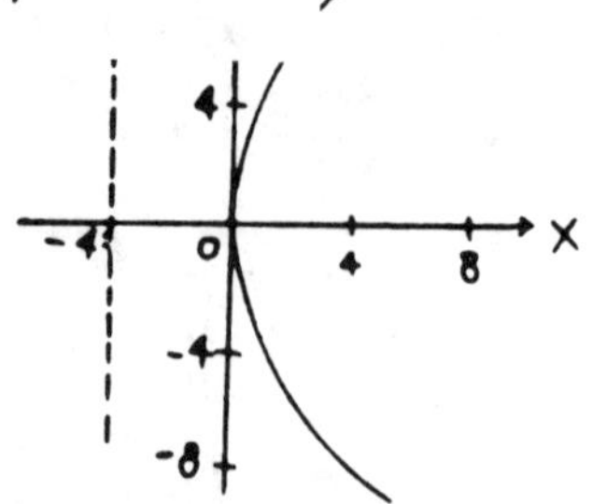

5. $y^2 + 2y + 1 = x - 3$

   $(y + 1)^2 = x - 3$

   Axis: $y = -1$

   $V(3, -1)$, $F(\frac{13}{4}, -1)$, D: $x = \frac{11}{4}$,

   $\ell r = 1$.

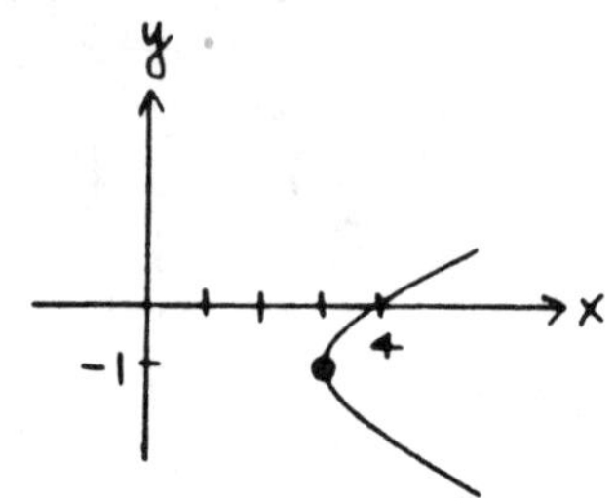

9. $9(x^2 + 2x + 1) - 16(y^2 + y + \frac{1}{4}) = 139 + 9 - 4$

   $9(x + 1)^2 - 16(y + \frac{1}{2})^2 = 144$

   $\frac{(x + 1)^2}{16} - \frac{(y + 1/2)^2}{9} = 1$

   $C(-1, -\frac{1}{2})$, $V(-1 \pm 4, -\frac{1}{2})$

   $c^2 = a^2 + b^2 = 16 + 9 = 25$

   $F(-1 \pm 5, -\frac{1}{2})$,

   A: $y = -\frac{1}{2} \pm \frac{3}{4}(x + 1)$,

   $\ell r = \frac{2b^2}{a} = \frac{2 \cdot 9}{4} = \frac{9}{2}$.

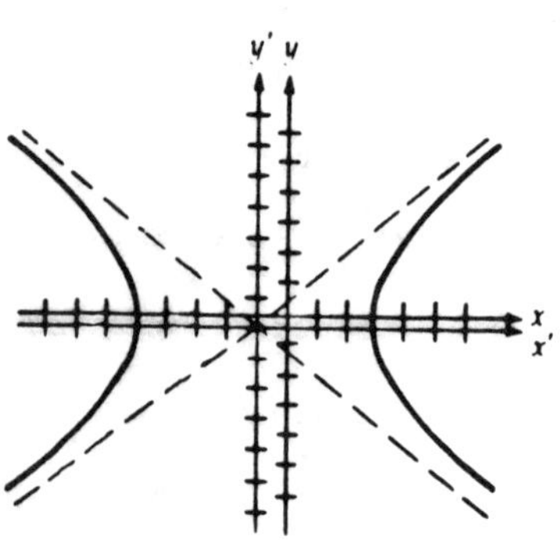

13. $y^2 = 4cx$

$(2, 4)$: $16 = 8c$, $c = 2$

$(8, 8)$: $64 = 32c$, $c = 2$

$y^2 = 8x$.

$x^2 = 4cy$

$(2, 4)$: $4 = 16c$, $c = \frac{1}{4}$

$(8, 8)$: $64 = 8c$, $c = 8$

Impossible.

17. $(y - k)^2 = 4c(x - h)$, $V = (1, 5)$, $c = 2$

$(y - 5)^2 = 8(x - 1)$, $y^2 - 8x - 10y + 33 = 0$.

21. $y = x^3 + 6x^2 + 3x - 14$, $x = x' + h$, $y = y' + k$

$$y' + k = (x' + h)^3 + 6(x' + h)^2 + 3(x' + h) - 14$$

$$= x'^3 + 3hx'^2 + 3h^2x' + h^3 + 6x'^2 + 12hx' + 6h^2 + 3x' + 3h - 14$$

$$y' = x'^3 + (3h + 6)x'^2 + (3h^2 + 12h + 3)x' + (h^3 + 6h^2 + 3h - 14 - k)$$

$3h + 6 = 0$, $h = -2$,

$h^3 + 6h^2 + 3h - 14 - k = 0$,

$k = -8 + 24 - 6 - 14 = -4$

$x = x' - 2$, $y = y' - 4$,

$y' = x'^3 - 9x'$

$= x'(x' + 3)(x' - 3)$.

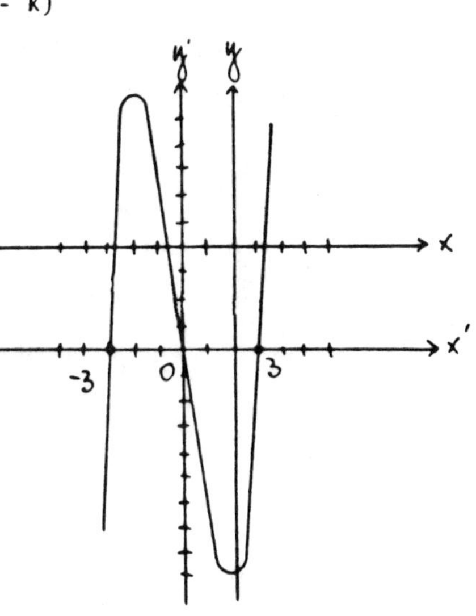

25. $\tan 2\theta = \dfrac{-\sqrt{3}}{2 - 1} = -\sqrt{3}$,

$2\theta = 120°, \quad \theta = 60°$

$\sin\theta = \dfrac{\sqrt{3}}{2}, \quad \cos\theta = \dfrac{1}{2}$

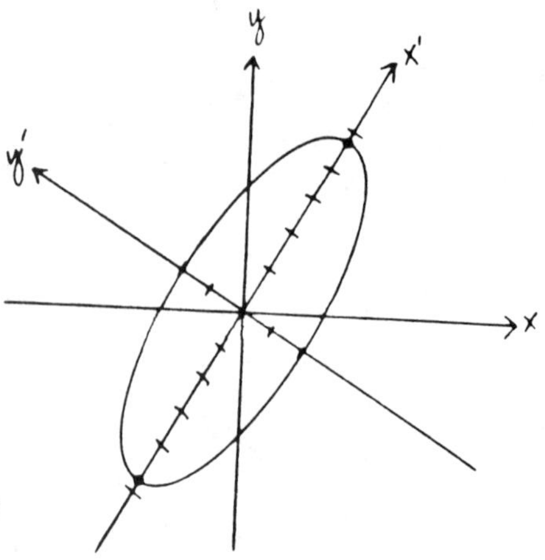

$x = \dfrac{x' - \sqrt{3}y'}{2}$,

$y = \dfrac{\sqrt{3}x' + y'}{2}$

$\frac{1}{4}(2x'^2 - 4\sqrt{3}x'y' + 6y'^2$

$- 3x'^2 + 2\sqrt{3}x'y' + 3y'^2$

$+ 3x'^2 + 2\sqrt{3}x'y' + y'^2) - 10 = 0$

$x'^2 + 5y'^2 - 20 = 0$

$\dfrac{x'^2}{20} + \dfrac{y'^2}{4} = 1.$

29. $P = (c, \dfrac{b^2}{a})$. $\quad y' = -\dfrac{b^2x}{a^2y} = -\dfrac{b^2c}{a^2(b^2/a)} = -\dfrac{c}{a}$

Tangent line: $y - \dfrac{b^2}{a} = -\dfrac{c}{a}(x - c)$

$y = 0$: $-\dfrac{b^2}{a} = -\dfrac{c}{a}(x - c)$, $x = c + \dfrac{b^2}{c} = \dfrac{b^2 + c^2}{c} = \dfrac{a^2}{c}$

At $(c, -\dfrac{b^2}{a})$, $y' = \dfrac{c}{a}$ and the tangent line is

$y + \dfrac{b^2}{a} = \dfrac{c}{a}(x - c)$, $x = \dfrac{a^2}{c}$ when $y = 0$.

Tangent lines intersect at $(\dfrac{a^2}{c}, 0)$ on the directrix $x = \dfrac{a^2}{c}$.

33. $\frac{x^2}{a^2} + \frac{y^2}{a^2 - 1} = 1, \frac{x^2}{b^2} - \frac{y^2}{1 - b^2} = 1$

$$x^2 + \frac{a^2}{a^2 - 1}y^2 = a^2$$

$$x^2 - \frac{b^2}{1 - b^2}y^2 = b^2$$

$$\left(\frac{a^2}{a^2 - 1} + \frac{b^2}{1 - b^2}\right)y^2 = a^2 - b^2$$

$$y^2 = (a^2 - 1)(1 - b^2)$$

$$x^2 = b^2 + \frac{b^2}{1 - b^2}y^2 = b^2 + b^2(a^2 - 1) = a^2b^2$$

$$\frac{y}{x} = \pm \frac{\sqrt{(a^2 - 1)(1 - b^2)}}{ab}$$

$$\frac{x^2}{a^2} + \frac{y^2}{a^2 - 1} = 1, \ y' = -\frac{a^2 - 1}{a^2}\frac{x}{y}$$

$$= -\frac{a^2 - 1}{a^2}\frac{ab}{\sqrt{(a^2 - 1)(1 - b^2)}}$$

$$= -\frac{b\sqrt{a^2 - 1}}{a\sqrt{1 - b^2}}$$

$$\frac{x^2}{b^2} - \frac{y^2}{1 - b^2} = 1, \ y' = \frac{1 - b^2}{b^2}\frac{x}{y}$$

$$= \frac{1 - b^2}{b^2}\frac{ab}{\sqrt{(a^2 - 1)(1 - b^2)}}$$

$$= \frac{a\sqrt{1 - b^2}}{b\sqrt{a^2 - 1}}$$

$-\frac{b\sqrt{a^2 - 1}}{a\sqrt{1 - b^2}} \cdot \frac{a\sqrt{1 - b^2}}{b\sqrt{a^2 - 1}} = 1$, so tangent lines at point of intersection are perpendicular.

Chapter 13
Parametric Equations

CHAPTER 13, SECTION 1 (pp. 633-634)

1. $t = y - 1,$

   $x = (y - 1)^2 - 2,$

   $x + 2 = (y - 1)^2.$

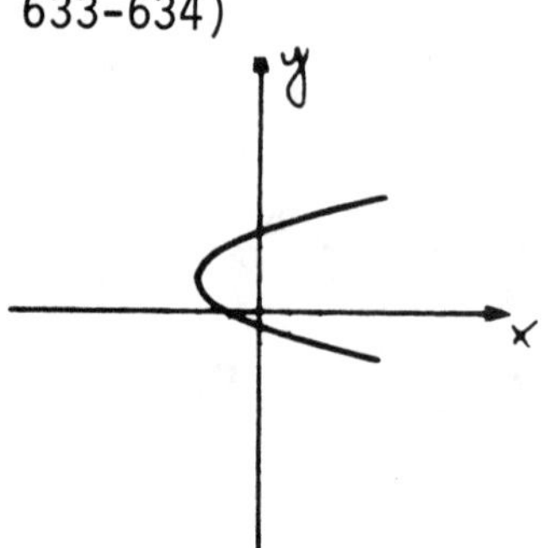

5. $x - y = 2t, \quad t = \frac{x - y}{2},$

   $x = \frac{x^2 - 2xy + y^2}{4} + \frac{x - y}{2}$

   $4x = x^2 - 2xy + y^2 + 2x - 2y$

   $x^2 - 2xy + y^2 - 2x - 2y = 0.$

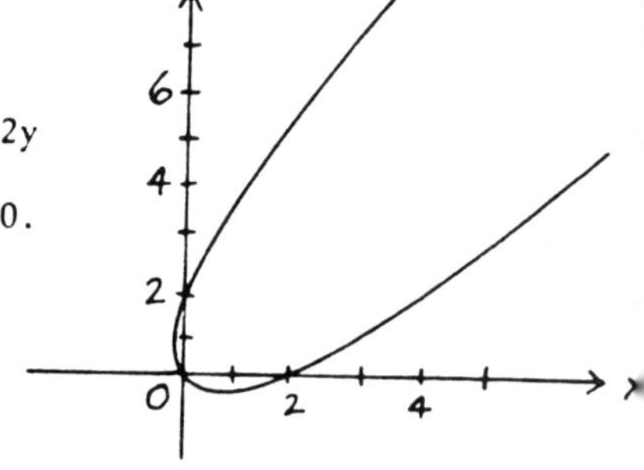

9. $\cos\theta = x - 1$

   $\sin\theta = y + 2$

   $(x - 1)^2 + (y + 2)^2 = 1.$

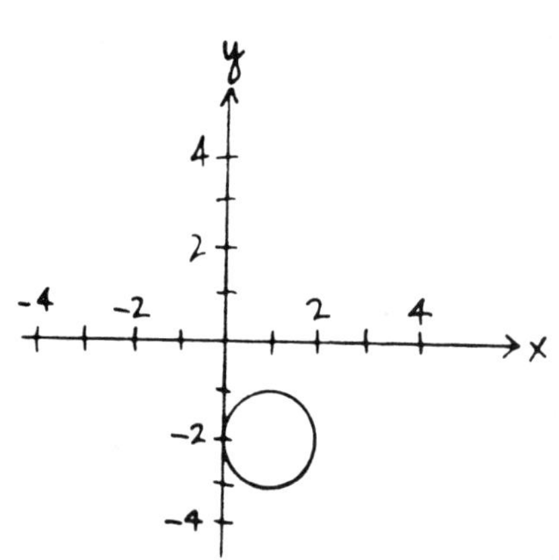

13. $x = t + 2, \quad y = t^2,$

$t = x - 2,$

$y = (x - 2)^2.$

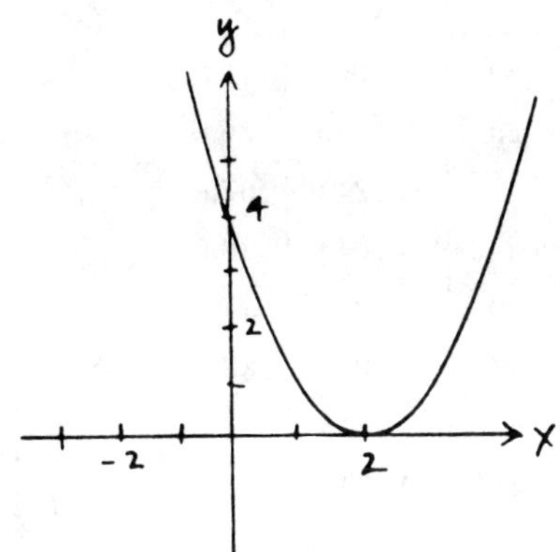

17. $x = x_1 + r(x_2 - x_1) = 1 + 6r, \; y = y_1 + r(y_2 - y_1) = 5 - 2r.$

21. $x = x_1 + r(x_2 - x_1) = -1 + 3r, \; y = y_1 + r(y_2 - y_1) = 3.$

25. $\cos\theta = 1 - y, \; \cos^2\theta = 1 - 2y + y^2, \; 2y - y^2 = 1 - \cos^2\theta = \sin^2\theta$

$\sin\theta = \pm\sqrt{2y - y^2}, \; \theta = \arccos(1 - y),$

$x = \arccos(1 - y) \pm \sqrt{2y - y^2}$

This is sketched more easily by plotting point by point from the parametric equations.

θ	x	y
0	0	0
$\pi/2$	$\pi/2 - 1$	1
π	π	2
$3\pi/2$	$3\pi/2 + 1$	1
2π	2π	0

29.

t	x	y
-3a	-2.00a	.10a
-2a	-1.04a	.26a
-a	-.24a	.67a
-a/2	-.04a	.88a
0	0	a
a/2	.04a	.88a
a	.24a	.67a
2a	1.04a	.26a
3a	2.00a	.10a

33. (a) $y^2 = x^2 - 1$

$x^2 - y^2 = 1,\ y \geq 0.$

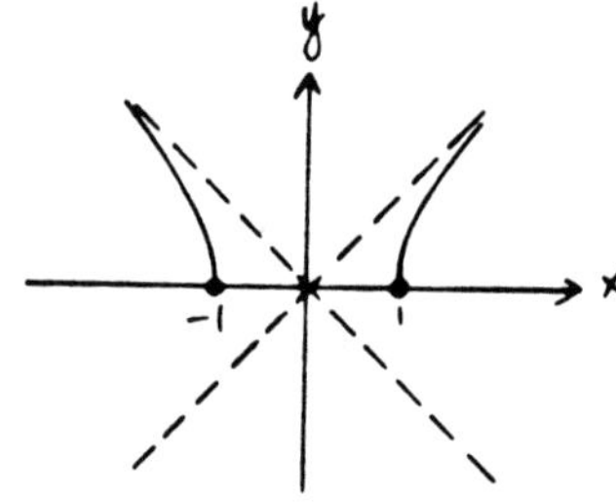

(b) $y^2 = x^2 - 1$

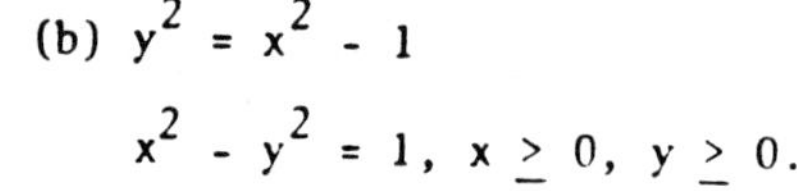
$x^2 - y^2 = 1,\ x \geq 0,\ y \geq 0.$

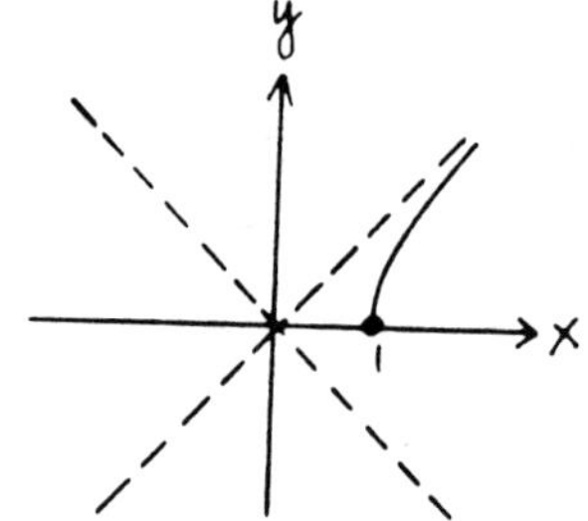

(c) $\sec^2 t = 1 + \tan^2 t$

$x^2 = 1 + y^2$,

$x^2 - y^2 = 1$.

(d) $\cosh^2 t - \sinh^2 t = 1$

$x^2 - y^2 = 1,\ x \geq 1$.

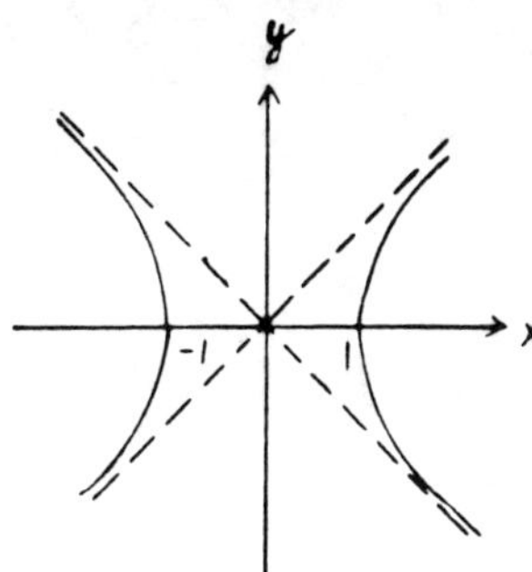

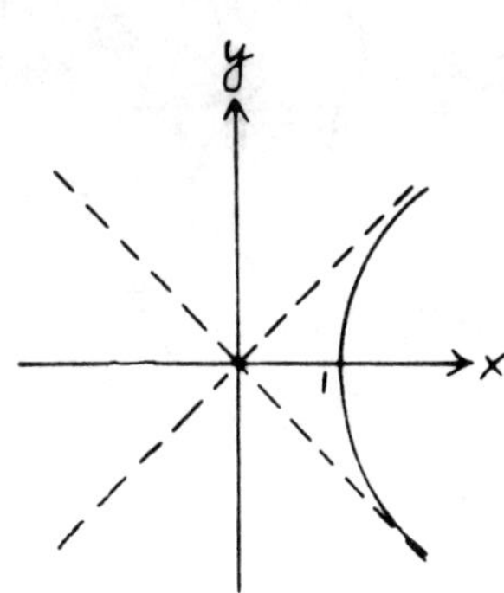

CHAPTER 13, SECTION 2 (pp. 637-639)

1. $v_x^0 = 250 \cos 60° = 125$

$v_x = v_x^0 = 125$

$x = 125t$

$v_y^0 = 250 \sin 60° = 125\sqrt{3}$

$a_y = -9.8$

$v_y = -9.8t + v_y^0 = -9.8t + 125\sqrt{3}$

$y = -4.9t^2 + 125\sqrt{3}t$.

5. $x_0 = y_0 = 0$

$v_x^0 = v_0 \cos \theta,\ v_y^0 = v_0 \sin \theta$

$v_x = v_0 \cos \theta$

$x = (v_0 \cos \theta)t + x_0$

$= (v_0 \cos \theta)t$

$a_y = -9.8$

$v_y = -9.8t + v_y^0$

$= -9.8t + v_0 \sin \theta$

$y = -4.9t^2 + (v_0 \sin \theta)t + y_0$

$= -4.9t^2 + (v_0 \sin \theta)t$.

9. $\overline{OT} = P'T = a\theta$

$\sin\theta = \dfrac{\overline{PQ}}{\overline{PC}}$

$\overline{PQ} = \overline{PC}\sin\theta = b\sin\theta$

$\cos\theta = \dfrac{\overline{CQ}}{\overline{PC}}$

$\overline{CQ} = \overline{PC}\cos\theta = b\cos\theta$

$x = \overline{OR} = \overline{OT} - \overline{RT}$

$\quad = \overline{OT} - \overline{PQ} = a\theta - b\sin\theta$

$y = \overline{PR} = \overline{QT} = \overline{CT} - \overline{CQ} = a - b\cos\theta.$

13.

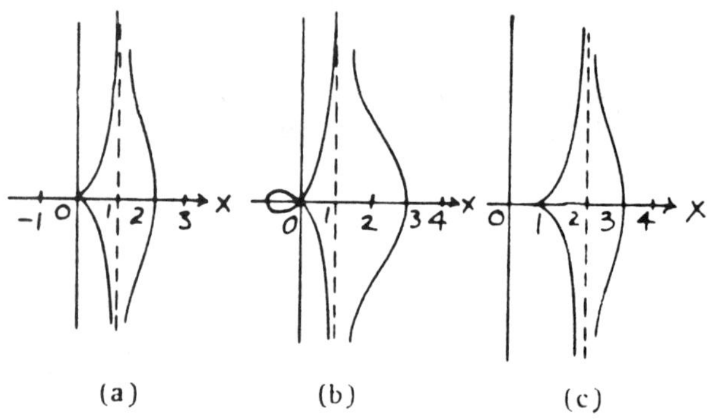

(a) (b) (c)

17. $C = (2a\cos\theta,\ 2a\sin\theta)$

$p = a\cos 2\theta, \quad q = a\sin 2\theta$

$x = 2a\cos\theta - a\cos 2\theta, \quad y = 2a\sin\theta - a\sin 2\theta.$

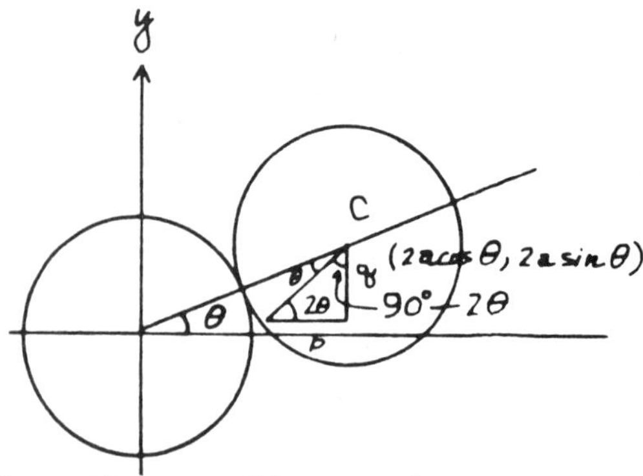

CHAPTER 13, SECTION 3 (pp. 642-643)

1. $\frac{dy}{dx} = \frac{dy/dt}{dx/dt} = \frac{1}{2t + 1}$

$\frac{d^2y}{dx^2} = \frac{dy'/dt}{dx/dt} = \frac{-2/(2t + 1)^2}{2t + 1} = \frac{-2}{(2t + 1)^3}.$

5. $\frac{dy}{dx} = \frac{dy/dt}{dx/dt} = \frac{2t - 1}{2t + 1}$

$\frac{d^2y}{dx^2} = \frac{dy'/dt}{dx/dt} = \frac{[(2t + 1)2 - (2t - 1)2]/(2t + 1)^2}{2t + 1} = \frac{4}{(2t + 1)^3}.$

9. $\frac{dy}{dx} = \frac{dy/dt}{dx/dt} = \frac{\cosh t}{\sinh t} = \coth t$

$\frac{d^2y}{dx^2} = \frac{dy'/dt}{dx/dt} = \frac{-\operatorname{csch}^2 t}{\sinh t} = -\operatorname{csch}^3 t.$

13. $\frac{dy}{dx} = \frac{dy/dt}{dx/dt} = \frac{2t}{3t^2} = \frac{2}{3t}.\quad \frac{dy}{dx} = -\frac{2}{9}$ at $t = -3$.

$\frac{d^2y}{dx^2} = \frac{dy'/dt}{dx/dt} = \frac{-2/3t^2}{3t^2} = -\frac{2}{9t^4}.\quad \frac{d^2y}{dx^2} = -\frac{2}{729}$ at $t = -3$.

17. $\frac{dy}{dx} = \frac{dy/dt}{dx/dt} = \frac{8t - 2}{2} = 4t - 1 = 3$

$x = 1,\quad y = 2$

$y - 2 = 3(x - 1),\quad 3x - y - 1 = 0.$

21. $\frac{dy}{dx} = \frac{dy/dt}{dx/dt} = \frac{3t^2 - 6t}{1} = 3t(t - 2)$

$t = 0,\quad t = 2$

$(3, 0)\quad (5, -4).$

25. $\frac{dy}{dx} = \frac{dy/d\theta}{dx/d\theta} = \frac{\sin\theta}{1 - \cos\theta}$

$\sin\theta = 0,\qquad 1 - \cos\theta = 0$

$\theta = n\pi\qquad \cos\theta = 1$

$(2n\pi, 0)$ $\qquad\qquad$ $\theta = 2n\pi$

$((2n + 1)\pi, 2)$ $\qquad\qquad$ $(2n\pi, 0)$

Although both the numerator and denominator are 0 when $\theta = 2n\pi$, the curve has vertical tangents at these points. **See Figure 6.**

29. $\dfrac{dy}{dx} = \dfrac{dy/dt}{dx/dt} = \dfrac{-\text{sech } t \tanh t}{1 - \text{sech}^2 t} = -\dfrac{\text{sech } t \tanh t}{\tanh^2 t} = -\dfrac{1}{\sinh t}$

$\sinh t = 0, \quad t = 0, \quad (0, 1).$

33. $\dfrac{dy}{dx} = \dfrac{dy/dt}{dx/dt}$

$$\frac{d^2y}{dx^2} = \frac{d}{dx}\frac{dy}{dx} = \frac{\frac{d}{dt}\frac{dy}{dx}}{\frac{dx}{dt}} = \frac{\frac{dx}{dt}\frac{d^2y}{dt^2} - \frac{dy}{dt}\frac{d^2x}{dt^2}}{\left(\frac{dx}{dt}\right)^3}.$$

CHAPTER 13, SECTION 4 (pp. 648-650)

1. $x = \cos^3\theta$ $\qquad\qquad$ $y = \sin^3\theta$

$x' = -3\cos^2\theta \sin\theta, \quad y' = 3\sin^2\theta \sin\theta$

$$s = \int_0^{\pi/2} \sqrt{9\cos^4\theta \sin^2\theta + 9\sin^4\theta\cos^2\theta}\, d\theta$$

$$= \int_0^{\pi/2} 3\sin\theta\cos\theta\, d\theta = \frac{3\sin^2\theta}{2}\Big|_0^{\pi/2} = \frac{3}{2}.$$

5. $y = x^{3/2}, \quad y' = \frac{3}{2}x^{1/2}$

$$s = \int_0^{4/3} \sqrt{1 + \frac{9x}{4}}\, dx$$

$$= \frac{1}{18}\int_0^{4/3} 9\sqrt{4 + 9x}\, dx = \frac{1}{27}(4 + 9x)^{3/2}\Big|_0^{4/3}$$

$$= \frac{1}{27}(64 - 8) = \frac{56}{27}.$$

9. $x = \cosh y, \quad x' = \sinh y$

$$s = \int_0^1 \sqrt{1 + \sinh^2 y}\, dy = \int_0^1 \cosh y\, dy = \sinh y \Big|_0^1 = \sinh 1.$$

13. $x = \operatorname{Arccos}(1 - y) + \sqrt{2y - y^2}$

$$x' = \frac{1}{\sqrt{1 - (1 - y)^2}} + \frac{1 - y}{\sqrt{2y - y^2}} = \frac{2 - y}{\sqrt{2y - y^2}}$$

$$s = \int_{1/2}^1 \sqrt{1 + \frac{4 - 4y + y^2}{2y - y^2}}\, dy = \int_{1/2}^1 \sqrt{\frac{4 - 2y}{2y - y^2}}\, dy$$

$$= \int_{1/2}^1 \sqrt{\frac{2}{y}}\, dy = 2\sqrt{2y}\Big|_{1/2}^1 = 2\sqrt{2} - 2.$$

17. $d = \sqrt{(-1 - 2)^2 + (5 - 1)^2} = \sqrt{9 + 16} = \sqrt{25} = 5$

$m = -\frac{4}{3}$

$$d = \int_{-1}^2 \sqrt{1 + 16/9}\, dx = \frac{5}{3}\int_{-1}^2 dx = \frac{5}{3}x\Big|_{-1}^2 = \frac{5}{3}(2 + 1) = 5.$$

21. $y = R \sin \theta, \quad dy = R \cos \theta\, d\theta; \quad x = R \cos \theta,\ dx = -R \sin d\theta;$

$$y' = \frac{dy}{dx} = -\cot \theta, \quad \frac{d^2y}{dx^2} = \frac{dy'}{dx} = \frac{\csc^2\theta}{-R \sin \theta} = -\frac{\csc^3\theta}{R}$$

$$\kappa = \frac{-\csc^3\theta}{R(1 + \cot^2\theta)^{3/2}} = (-\frac{1}{R})\frac{\csc^3\theta}{(\csc^2\theta)^{3/2}} = -\frac{1}{R}.$$

25. $dy = 3a \sin^2 t \cos t\, dt, \quad dx = -3a \cos^2 t \sin t\, dt,$

$$\frac{dy}{dx} = y' = \frac{\sin t}{-\cos t} = -\tan t,$$

$$\frac{d^2y}{dx^2} = \frac{dy'}{dx} = \frac{-\sec^2 t}{-3a \cos^2 t \sin t} = \frac{1}{3a \cos^4 t \sin t}$$

$$\kappa = \frac{1}{3a \cos^4 t \sin t(1 + \tan^2 t)^{3/2}}$$

$$= \frac{1}{3a\cos^4 t \sin t(\sec^2 t)^{3/2}} = \frac{1}{3a\cos t \sin t}$$

$$= \frac{2}{(3a)(2\sin t\cos t)} = \frac{2}{3a\sin 2t}, \quad R = \frac{1}{|\kappa|} = \frac{3}{2}a|\sin 2t|.$$

29. $y' = x,\ y'' = 1,\ h = x - \frac{y'[1 + (y')^2]}{y''} = x - \frac{x(1 + x^2)}{1} = -x^3,$

$$k = y + \frac{1 + (y')^2}{y''} = y + \frac{1 + x^2}{1} = \frac{x^2}{2} + 1 + x^2 = \frac{3x^2}{2} + 1.$$

$$x^2 = h^{2/3};\ k = \frac{3}{2}h^{2/3} + 1,\ 2(k - 1) = 3h^{2/3},\ h^2 = \frac{8(k - 1)^3}{27}.$$

CHAPTER 13, SECTION 5 (pp. 655-656)

1. $x = a\cos^3\theta, \quad y = a\sin^3\theta$

$$\frac{dx}{d\theta} = -3a\cos^2\theta\sin\theta, \quad \frac{dy}{d\theta} = 3a\sin^2\theta\cos\theta$$

$$S_x = \int_0^{\pi/2} 2\pi y\sqrt{9a^2\cos^4\theta\sin^2\theta + 9a^2\sin^4\theta\cos^2\theta}\,d\theta$$

$$= \int_0^{\pi/2} 2\pi a\sin^3\theta\cdot 3a\sin\theta\cos\theta\sqrt{\cos^2\theta + \sin^2\theta}\,d\theta$$

$$= 6\pi a^2\int_0^{\pi/2}\sin^4\theta\cos\theta\,d\theta = \frac{6\pi a^2}{5}\sin^5\theta\Big|_0^{\pi/2} = \frac{6\pi a^2}{5}.$$

5. $y = \cosh x, \quad y' = \sinh x$

$$S_x = \int_{-1}^{1} 2\pi\cosh x\sqrt{1 + \sinh^2 x}\,dx = 2\pi\int_{-1}^{1}\cosh^2 x\,dx$$

$$= \pi\int_{-1}^{1}(1 + \cosh 2x)dx = \pi(x + \frac{1}{2}\sinh 2x)\Big|_{-1}^{1}$$

$$= \pi(x + \sinh x\cosh x)\Big|_{-1}^{1} = 2\pi(1 + \sinh 1\cosh 1).$$

9. $x = \sqrt{y}$, $x' = \dfrac{1}{2\sqrt{y}}$

$$S_y = \int_0^4 2\pi\sqrt{y}\,\sqrt{1 + \frac{1}{4y}}\,dy$$

$$= \pi \int_0^4 \sqrt{4y + 1}\,dy = \frac{\pi}{4}\,\frac{2}{3}\,(4y + 1)^{3/2}\Big|_0^4 = \frac{\pi}{6}\,(17\sqrt{17} - 1).$$

13. $y = \dfrac{x^3}{3}$, $y' = x^2$

$$S_y = \int_0^3 2\pi x\,\sqrt{1 + x^4}\,dx$$

$u = x^2$, $du = 2xdx$

$$= \pi \int_0^9 \sqrt{1 + u^2}\,du$$

$u = \tan\theta$, $du = \sec^2\theta\,d\theta$

$$= \pi \int_0^{\text{Arctan } 9} \sqrt{1 + \tan^2\theta}\,\sec^2\theta\,d\theta = \pi \int_0^{\text{Arctan } 9} \sec^3\theta d\theta$$

$$= \frac{\pi}{2}(\sec\theta\tan\theta + \ln|\sec\theta + \tan\theta|)\Big|_0^{\text{Arctan } 9}$$

$$= \frac{\pi}{2}\,[9\sqrt{82} + \ln(\sqrt{82} + 9)]\ .$$

17. $y = \sqrt{4 - x^2}$, $y' = \dfrac{-x}{\sqrt{4 - x^2}}$

$$S_x = \int_{-1/2}^{1/2} 2\pi\,\sqrt{4 - x^2}\,\sqrt{1 + \frac{x^2}{4 - x^2}}\,dx = \int_{-1/2}^{1/2} 4\pi\,dx$$

$$= 4\pi\,x\Big|_{-1/2}^{1/2} = 4\pi\ .$$

21. $y = \sin x\ ,\quad y' = \cos x$

$$S_x = \int_0^{\pi} 2\pi \sin x\sqrt{1+\cos^2 x}\ dx$$

$$\cos x = y\ ,\quad dy = -\sin x\ dx$$

$$= \int_1^{-1} -2\pi\sqrt{1+y^2}\ dy$$

$$y = \tan\theta\ ,\quad dy = \sec^2\theta\ d\theta$$

$$= \int_{\pi/4}^{-\pi/4} -2\pi \sec^3\theta\ d\theta$$

$$= -\pi[\sec\theta\tan\theta + \ln|\sec\theta + \tan\theta|]\Big|_{\pi/4}^{-\pi/4}$$

$$= -\pi(-\sqrt{2} + \ln|\sqrt{2}-1| - \sqrt{2} - \ln|\sqrt{2}+1|)$$

$$= 2\sqrt{2}\,\pi + \pi\ \ln\frac{\sqrt{2}+1}{\sqrt{2}-1} = 2\sqrt{2}\,\pi + \pi\ \ln(1+\sqrt{2})^2$$

$$= 2\sqrt{2}\,\pi + 2\pi\ \ln(1+\sqrt{2})\ .$$

25. $y = \frac{r}{h}x\ ,\quad y' = \frac{r}{h}$

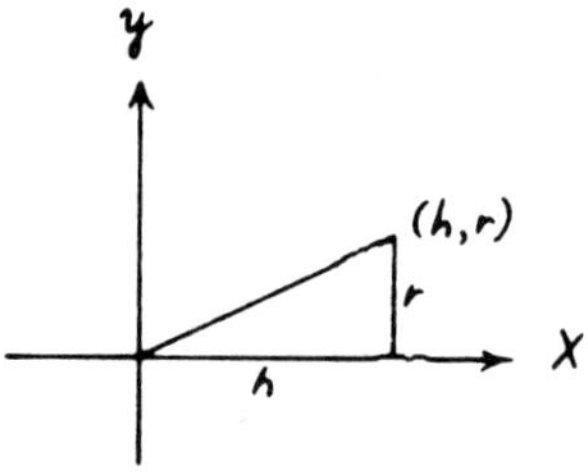

$$S_x = \int_0^h 2\pi y\sqrt{1+\frac{r^2}{h^2}}\ dx$$

$$= \int_0^h 2\pi\frac{r}{h}x\sqrt{\frac{r^2+h^2}{h^2}}\ dx$$

$$= \frac{2\pi r\sqrt{r^2+h^2}}{h^2}\int_0^h x\ dx$$

$$= \frac{\pi r\sqrt{r^2+h^2}}{h^2}x^2\Big|_0^h = \pi r\sqrt{r^2+h^2}\ .$$

29. $y = \frac{1}{x}$, $y' = -\frac{1}{x^2}$

$$S_x = \int_1^{+\infty} 2\pi y \sqrt{1 + (y')^2} = 2\pi \int_1^{+\infty} \frac{1}{x} \sqrt{1 + \frac{1}{x^4}}\, dx$$

$$= 2\pi \int_1^{+\infty} \frac{\sqrt{x^4 + 1}}{x^3}\, dx = 2\pi \lim_{k \to +\infty} \int_1^k \frac{\sqrt{x^4 + 1}}{x^3}\, dx$$

$$u = \sqrt{x^4 + 1} \qquad v' = \frac{1}{x^3}$$

$$u' = \frac{2x^3}{\sqrt{x^4 + 1}} \qquad v = -\frac{1}{2x^2}$$

$$= 2\pi \lim_{k \to +\infty} \left(-\frac{\sqrt{x^4 + 1}}{2x^2}\Big|_1^k + \int_1^k \frac{x\, dx}{\sqrt{x^4 + 1}} \right)$$

$$= 2\pi \lim_{k \to +\infty} \left(\frac{\sqrt{2}}{2} - \frac{1}{2} \sqrt{1 + \frac{1}{k^4}} + \int_1^k \frac{x\, dx}{\sqrt{x^4 + 1}} \right)$$

$$= \pi(\sqrt{2} - 1) + 2\pi \lim_{k \to +\infty} \int_1^k \frac{x\, dx}{\sqrt{x^4 + 1}}$$

$$\int_1^k \frac{x\, dx}{\sqrt{x^4 + 1}}$$

$$u = x^2, \quad du = 2x\, dx$$

$$= \frac{1}{2} \int_1^{k^2} \frac{du}{\sqrt{1 + u^2}}$$

$$u = \tan\theta, \quad du = \sec^2\theta\, d\theta$$

$$= \frac{1}{2} \int_{\pi/4}^{\text{Arctan } k^2} \frac{\sec^2\theta\, d\theta}{\sqrt{1 + \tan^2\theta}} = \frac{1}{2} \int_{\pi/4}^{\text{Arctan } k^2} \sec\theta\, d\theta$$

$$= \frac{1}{2} \ln|\sec\theta + \tan\theta|\Big|_{\pi/4}^{\text{Arctan } k^2}$$

$$= \frac{1}{2} \ln| \sqrt{1+k^4} + k^2| - \frac{1}{2} \ln(\sqrt{2} + 1)$$

$$\longrightarrow \ +\infty \quad \text{as} \quad k \longrightarrow +\infty$$

$$S_x = +\infty.$$

CHAPTER 13, SECTION 6 (pp. 661-662)

1. $y^{2/3} = a^{2/3} - x^{2/3}$

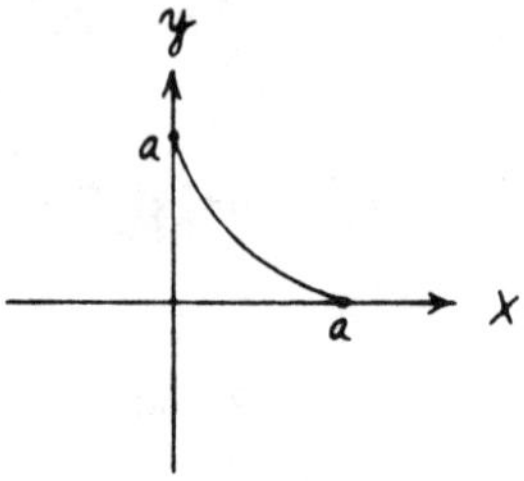

$$y = (a^{2/3} - x^{2/3})^{3/2},$$

$$y' = \frac{3}{2}(a^{2/3} - x^{2/3})^{1/2}(-\ \frac{2}{3}x^{-1/3})$$

$$= -x^{-1/3}(a^{2/3} - x^{2/3})^{1/2},$$

$$s = \int_0^a \sqrt{1 + \frac{a^{2/3} - x^{2/3}}{x^{2/3}}}\, dx$$

$$= \int_0^a \frac{a^{1/3}}{x^{1/3}}\, dx = \frac{3}{2}\, a^{1/3}x^{2/3}\Big|_0^a = \frac{3}{2}\, a,$$

$$M_y = \int_0^a x\, \frac{a^{1/3}}{x^{1/3}}\, dx = a^{1/3} \int_0^a x^{2/3} dx$$

$$= \frac{3}{5}\, a^{1/3} a^{5/3}\Big|_0^a = \frac{3}{5}\, a^2,$$

$$\bar{x} = \bar{y} = \frac{3a^2/5}{3a/2} = \frac{2a}{5}.$$

5. $y = 1 - \frac{x^2}{4}$, $y' = -\frac{x}{2}$,

$$s = \int_0^2 \sqrt{1 + \frac{x^2}{4}}\, dx$$

$x = 2 \tan\theta$, $dx = 2 \sec^2\theta\, d\theta$

$$= \int_0^{\pi/4} \sqrt{1 + \tan^2\theta}\; 2 \sec^2\theta\, d\theta$$

$$= 2 \int_0^{\pi/4} \sec^3\theta\, d\theta = (\sec\theta \tan\theta + \ln|\sec\theta + \tan\theta|)\Big|_0^{\pi/4}$$

$$= \sqrt{2} + \ln(\sqrt{2} + 1) ,$$

$$M_x = \int_0^2 \left(1 - \frac{x^2}{4}\right)\sqrt{1 + \frac{x^2}{4}}\, dx$$

$$= \int_0^2 \sqrt{1 + \frac{x^2}{4}}\, dx - \frac{1}{4}\int_0^2 x^2\sqrt{1 + \frac{x^2}{4}}\, dx$$

$$= s - \frac{1}{4}\int_0^{\pi/4} 4 \tan^2\theta\sqrt{1 + \tan^2\theta}\; 2 \sec^2\theta\, d\theta$$

$$= s - 2\int_0^{\pi/4} \tan^2\theta \sec^3\theta\, d\theta ,$$

$$\int \tan^2\theta \sec^3\theta\, d\theta = \int (\sec^2\theta - 1)\sec^3\theta\, d\theta$$

$$= \int \sec^5\theta\, d\theta - \int \sec^3\theta\, d\theta$$

$u = \sec^3\theta$ $\qquad v' = \sec^2\theta$

$u' = 3 \sec^3\theta \tan\theta$ $\qquad v = \tan\theta$

$$= \sec^3\theta \tan\theta - 3\int \sec^3\theta \tan^2\theta\, d\theta - \int \sec^3\theta\, d\theta$$

$$4\int \tan^2\theta \sec^3\theta\, d\theta = \sec^3\theta \tan\theta - \int \sec^3\theta\, d\theta$$

$$\int_0^{\pi/4} \tan^2\theta \sec^3\theta \, d\theta = \frac{1}{4}\sec^3\theta \tan\theta - \frac{1}{8}\sec\theta\tan\theta$$

$$- \frac{1}{8}\ln|\sec\theta + \tan\theta|\Big|_0^{\pi/4}$$

$$= \frac{\sqrt{2}}{2} - \frac{\sqrt{2}}{8} - \frac{1}{8}\ln(\sqrt{2}+1) = \frac{3\sqrt{2}}{8} - \frac{1}{8}\ln(\sqrt{2}+1) ,$$

$$M_x = \sqrt{2} + \ln(\sqrt{2}+1) - 2[\frac{3\sqrt{2}}{8} - \frac{1}{8}\ln(\sqrt{2}+1)]$$

$$= \frac{\sqrt{2}}{4} + \frac{5}{4}\ln(\sqrt{2}+1) ,$$

$$M_y = \int_0^2 x\sqrt{1+\frac{x^2}{4}}\,dx = 2\int_0^2 \frac{1}{2}x\sqrt{1+\frac{x^2}{4}}\,dx$$

$$= 2\cdot\frac{2}{3}(1+\frac{x^2}{4})^{3/2}\Big|_0^2 = \frac{4}{3}(2\sqrt{2}-1) = \frac{8\sqrt{2}-4}{3} ,$$

$$\bar{x} = \frac{M_y}{s} = \frac{(8\sqrt{2}-4)/3}{\sqrt{2}+\ln(\sqrt{2}+1)} ,$$

$$\bar{y} = \frac{M_x}{s} = \frac{(\sqrt{2}+5\ln(\sqrt{2}+1))/4}{\sqrt{2}+\ln(\sqrt{2}+1)} = \frac{\sqrt{2}+5\ln(\sqrt{2}+1)}{4(\sqrt{2}+\ln(\sqrt{2}+1))} .$$

9. $y = \cosh x$, $y' = \sinh x$,

$$S_x = \int_{-1}^1 2\pi y\sqrt{1+\sinh^2 x}\,dx = 2\pi\int_{-1}^1 \cosh^2 x\,dx$$

$$= \pi\int_{-1}^1 (1+\cosh 2x)\,dx = \pi(x+\frac{1}{2}\sinh 2x)\Big|_{-1}^1$$

$$= \pi(x+\sinh x\cosh x)\Big|_{-1}^1 = 2\pi(1+\sinh 1\cosh 1),$$

$$M_{yz} = \int_{-1}^1 2\pi x\cosh^2 x\,dx$$

$$u = 2\pi x \quad v' = \cosh^2 x$$

$$u' = 2\pi \quad v = \frac{1}{2}(x + \sinh x \cosh x)$$

$$= \pi x(x + \sinh x \cosh x)\Big|_{-1}^{1} - \pi \int_{-1}^{1} (x + \sinh x \cosh x)dx$$

$$= -\pi\left(\frac{x^2}{2} + \frac{\cosh^2 x}{2}\right)\Big|_{-1}^{1} = 0 ,$$

$\bar{x} = 0$.

13. $x^2 + y^2 - 2x = 0$, $y = \sqrt{2x - x^2}$, $y' = \dfrac{1 - x}{\sqrt{2x - x^2}}$,

$$S_y = \int_0^2 2\pi x \sqrt{1 + \frac{1 - 2x + x^2}{2x - x^2}}\, dx = 2\pi \int_0^2 \frac{x}{\sqrt{2x - x^2}}\, dx$$

$$= -\pi \int_0^2 \frac{-2x}{\sqrt{2x - x^2}}\, dx$$

$$= -\pi \int_0^2 \left(\frac{2 - 2x}{\sqrt{2x - x^2}} - \frac{2}{\sqrt{1 - (1 - 2x + x^2)}} \right) dx$$

$$= -\pi \int_0^2 \left(\frac{2 - 2x}{\sqrt{2x - x^2}} - 2\frac{1}{\sqrt{1 - (x - 1)^2}} \right) dx$$

$$= [\, -2\pi \sqrt{2x - x^2} + 2\pi \text{Arcsin}(x - 1)]\Big|_0^2$$

$$= 2\pi \cdot \frac{\pi}{2} - 2\pi(-\frac{\pi}{2}) = 2\pi^2 ,$$

$$M_{xz} = \int_0^2 \sqrt{2x - x^2}\, 2\pi x \sqrt{1 + \frac{1 - 2x + x^2}{2x - x^2}}\, dx = 2\pi \int_0^2 x dx$$

$$= \pi x^2\Big|_0^2 = 4\pi ,$$

$$\bar{y} = \frac{4\pi}{2\pi^2} = \frac{2}{\pi} .$$

17. $x = \frac{y^2}{4} , \quad \frac{dx}{dy} = \frac{y}{2}$

$$s = \int_{-2}^{2} \sqrt{1 + \frac{y^2}{4}}\, dy = 2 \int_0^2 \sqrt{1 + \frac{y^2}{4}}\, dy$$

$$y = 2 \tan \theta , \quad dy = 2 \sec^2\theta \, d\theta$$

$$= 2 \int_0^{\pi/4} \sqrt{1 + \tan^2\theta}\ 2 \sec^2\theta \, d\theta$$

$$= 4 \int_0^{\pi/4} \sec^3\theta \, d\theta = 2(\sec\theta \tan\theta + \ln|\sec\theta + \tan\theta|)\Big|_0^{\pi/4}$$

$$= 2[\sqrt{2} + \ln(\sqrt{2} + 1)] = 2\sqrt{2} + 2 \ln(\sqrt{2} + 1) ,$$

$$M_y = \int_{-2}^{2} x\sqrt{1 + \frac{y^2}{4}}\, dy = \frac{2}{4} \int_0^2 y^2\sqrt{1 + \frac{y^2}{4}}\, dy$$

$$y = 2 \tan \theta , \quad dy = 2 \sec^2\theta \, d\theta$$

$$= \frac{1}{2} \int_0^{\pi/4} 4 \tan^2\theta \sqrt{1 + \tan^2\theta}\ 2 \sec^2\theta \, d\theta$$

$$= 4 \int_0^{\pi/4} \tan^2\theta \sec^3\theta \, d\theta$$

(See Problem 5)

$$= \sec^3\theta \tan\theta - \frac{1}{2} \sec\theta \tan\theta - \frac{1}{2} \ln|\sec\theta + \tan\theta|\Big|_0^{\pi/4}$$

$$= [3\sqrt{2} - \ln(\sqrt{2} + 1)]/2$$

$$\bar{x} = \frac{M_y}{s} = \frac{3\sqrt{2} - \ln(\sqrt{2} + 2)}{4\sqrt{2} + 4 \ln(\sqrt{2} + 1)} , \quad \bar{y} = 0 \text{ by symmetry.}$$

21. $y = \sqrt{1 - x^2}$, $y' = \dfrac{-x}{\sqrt{1 - x^2}}$

$s = \pi$ and $\bar{y} = \dfrac{2}{\pi}$ (by Example 1, p. 625)

$S_x = \pi(2\pi \cdot \frac{2}{\pi}) = 4\pi$

$$S_x = \int_{-1}^{1} 2\pi\sqrt{1 - x^2}\sqrt{1 + \frac{x^2}{1 - x^2}}\,dx = 2\pi\int_{-1}^{1} dx$$

$$= 2\pi x\Big|_{-1}^{1} = 4\pi\ .$$

CHAPTER 13, REVIEW (pp. 662-664)

1. $t = x - 1$, $y = (x - 1)^2 + 4(x - 1) - 2$

$y = x^2 + 2x - 5.$

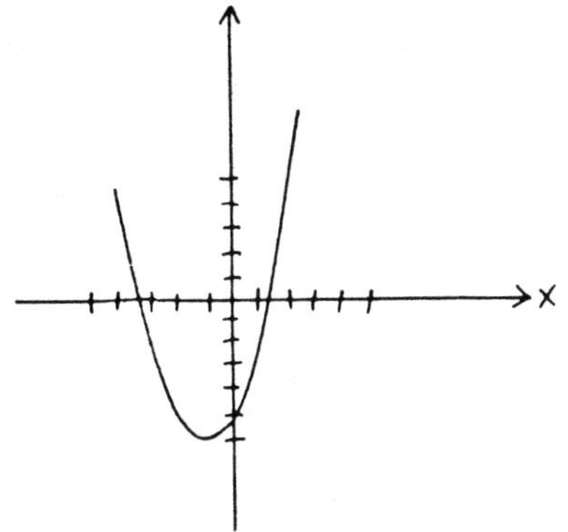

5. $y = \sin 2\theta = 2 \sin \theta \cos \theta$

$y^2 = 4 \sin^2\theta \cos^2\theta$

$= 4 \sin^2\theta(1 - \sin^2\theta)$

$y^2 = 4x^2(1 - x^2).$

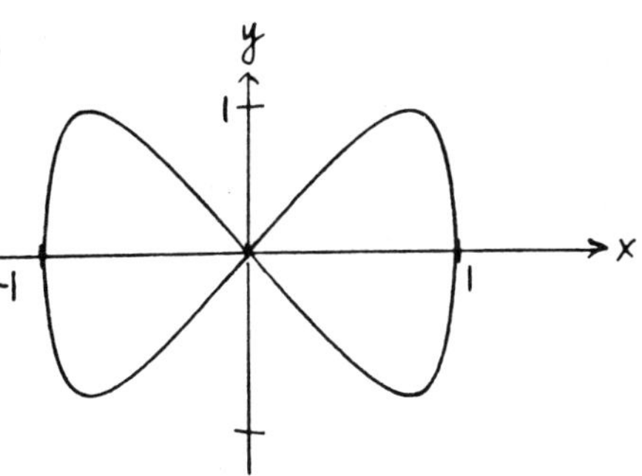

9. $\frac{dy}{dx} = \frac{dy/dt}{dx/dt} = \frac{3t^2}{4t} = \frac{3t}{4}$

$\frac{d^2y}{dx^2} = \frac{dy'/dt}{dx/dt} = \frac{3/4}{4t} = \frac{3}{16t}.$

13. $y = a \cosh \frac{x}{a}$, $y' = \sinh \frac{x}{a}$,

$$S_x = \int_{-a}^{a} 2\pi y \sqrt{1 + (y')^2}\, dx$$

$$= 2\pi \int_{-a}^{a} a \cosh \frac{x}{a} \sqrt{1 + \sinh^2 \frac{x}{a}}\, dx$$

$$= 2\pi a \int_{-a}^{a} \cosh^2 \frac{x}{a}\, dx = 4\pi a \int_{0}^{a} \cosh^2 \frac{x}{a}\, dx$$

$$= 2\pi a \int_{0}^{a} (1 + \cosh \frac{2x}{a}) dx$$

$$= 2\pi a(x + \frac{a}{2} \sinh \frac{2x}{a}) \Big|_{0}^{a} = 2\pi a(a + \frac{a}{2} \sinh 2)$$

$$= \pi a^2(2 + \sinh 2).$$

17. $\frac{dy}{dx} = \frac{dy/d\theta}{dx/d\theta} = \frac{-\cos \theta}{1 - \sin \theta}$

$\cos \theta = 0$	$1 - \sin \theta = 0$
$\theta = \frac{\pi}{2} + 2n\pi, \quad \frac{3\pi}{2} + 2n\pi$	$\sin \theta = 1$
	$\theta = \frac{\pi}{2} + 2n\pi$

$(\frac{\pi}{2} + 2n\pi, 0)$, $(\frac{3\pi}{2} + 2n\pi, 2)$, $(\frac{\pi}{2} + 2n\pi, 0)$.

21. $y = x^2 - \frac{\ln x}{8}, \quad y' = 2x - \frac{1}{8x},$

$$s = \int_1^2 \sqrt{1 + (2x - 1/8x)^2}\, dx = \int_1^2 \sqrt{4x^2 + \frac{1}{2} + \frac{1}{64x^2}}\, dx$$

$$= \int_1^2 (2x + \frac{1}{8x})dx = (x^2 + \frac{\ln x}{8})\Big|_1^2$$

$$= 4 + \frac{\ln 2}{8} - 1 = \frac{24 + \ln 2}{8},$$

$$M_x = \int_1^2 y\sqrt{1 + (y')^2}\, dx = \int_1^2 (x^2 - \frac{\ln x}{8})(2x + \frac{1}{8x})dx$$

$$= \int_1^2 (2x^3 + \frac{x}{8} - \frac{x \ln x}{4} - \frac{\ln x}{64x})\, dx$$

$u = \ln x \quad v' = \frac{x}{4}$

$u' = \frac{1}{x} \qquad v = \frac{x^2}{8}$

$$= (\frac{x^4}{2} + \frac{x^2}{16} - \frac{\ln^2 x}{128} - \frac{x^2 \ln x}{8})\Big|_1^2 + \int_1^2 \frac{x dx}{8}$$

$$= (\frac{x^4}{2} + \frac{x^2}{16} - \frac{\ln^2 x}{128} - \frac{x^2 \ln x}{8} + \frac{x^2}{16})\Big|_1^2$$

$$= (8 + \frac{1}{4} - \frac{\ln^2 2}{128} - \frac{\ln 2}{2} + \frac{1}{4}) - \frac{5}{8}$$

$$= \frac{1008 - 64 \ln 2 - \ln^2 2}{128},$$

$$M_y = \int_1^2 x\sqrt{1 + (y')^2}\, dx = \int_1^2 x(2x + \frac{1}{8x})dx = \int_1^2 (2x^2 + \frac{1}{8})dx$$

$$= (\frac{2x^3}{3} + \frac{x}{8})\Big|_1^2 = (\frac{16}{3} + \frac{1}{4}) - (\frac{2}{3} + \frac{1}{8}) = \frac{115}{24},$$

$$\bar{x} = \frac{M_y}{s} = \frac{115/24}{(24 + \ln 2)/8} = \frac{115}{3(24 + \ln 2)},$$

$$\bar{y} = \frac{M_x}{s} = \frac{\frac{1008 - 64 \ln 2 - \ln^2 2}{128}}{\frac{24 + \ln 2}{8}} = \frac{1008 - 64 \ln 2 - \ln^2 2}{16(24 + \ln 2)}.$$

25. $\sin \theta = \frac{\overline{OB}}{\overline{OT}} = \frac{\overline{OB}}{2a}$

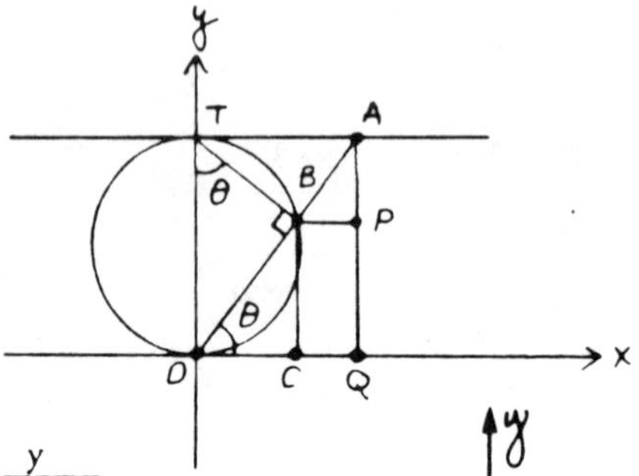

$\overline{OB} = 2a \sin \theta$

$\tan \theta = \frac{\overline{AQ}}{\overline{OQ}} = \frac{2a}{x}$

$x = 2a \cot \theta$

$\sin \theta = \frac{\overline{BC}}{\overline{OB}} = \frac{\overline{PQ}}{\overline{OB}} = \frac{y}{2a \sin \theta}$

$y = 2a \sin^2\theta.$

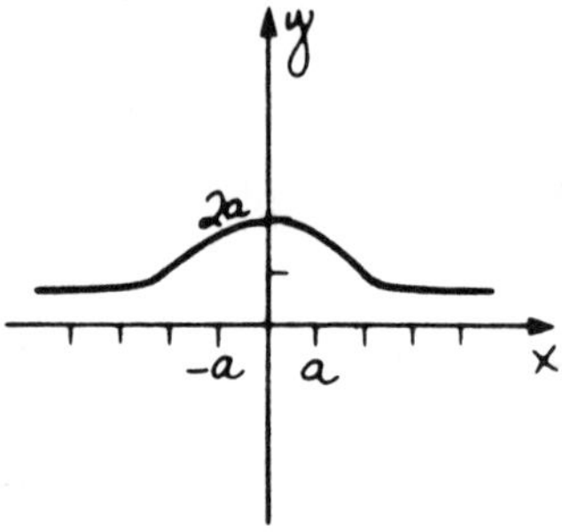

29. $y = x^4 + \frac{1}{32x^2}$, $y' = 4x^3 - \frac{1}{16x^3}$

$$s = \int_0^x \sqrt{1 + (4t^3 - \frac{1}{16t^3})}\, dt$$

$$= \int_0^x \sqrt{1 + 16t^6 - \frac{1}{2} + \frac{1}{256t^6}}\, dt$$

$$= \int_0^x (4t^3 + \frac{1}{16t^3})\, dt$$

$$= t^4 - \frac{1}{32t^2}\Big|_1^x = x^4 - \frac{1}{32x^2} - 1 + \frac{1}{32}$$

$$\frac{2881}{36} = x^4 - \frac{1}{32x^2} - \frac{31}{32},$$

$$32x^6 - \frac{23,329}{9}x^2 - 1 = 0, \ x = 3$$

$$y = 81 + \frac{1}{288} = \frac{23,329}{288}, \ (3, \frac{23,329}{288}).$$

Chapter 14
Polar Coordinates

CHAPTER 14, SECTION 1 (pp. 674-675)

1.

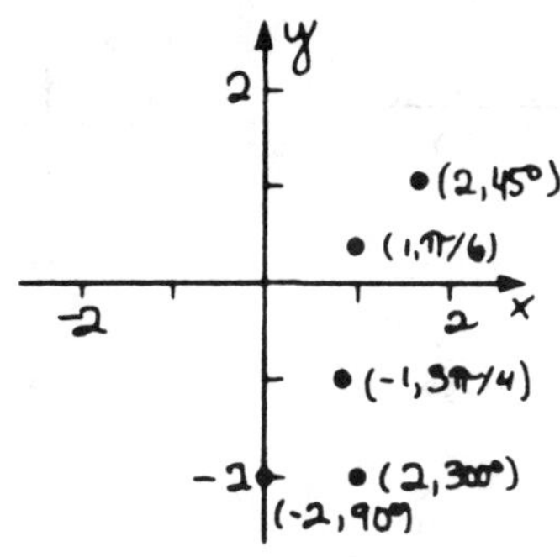

5. $r = \cos\theta$.

Symmetry about the x axis: $r = \cos(-\theta) = \cos\theta$.

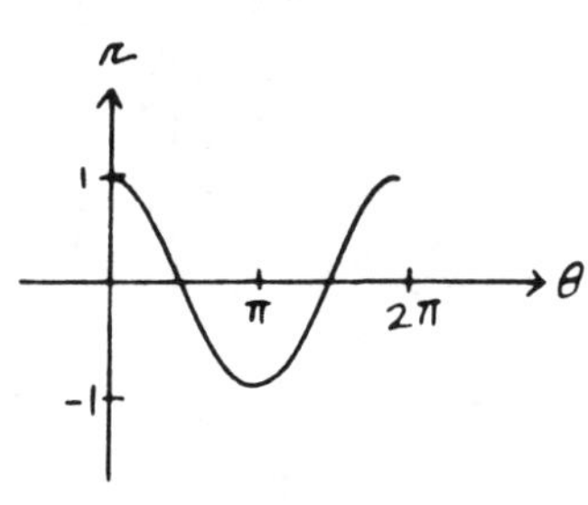

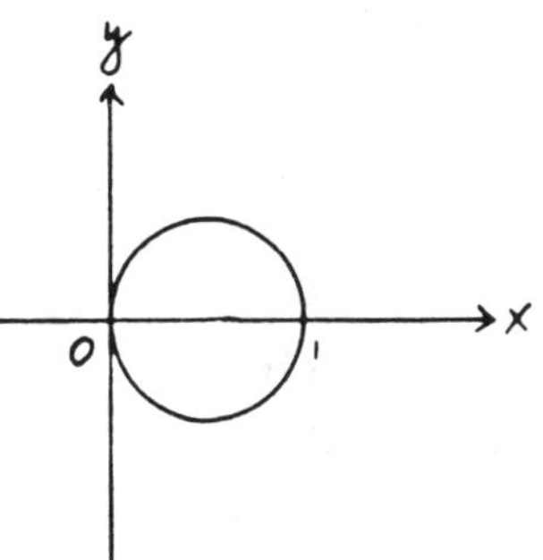

9. $r = 1 - \sin\theta$.

Symmetry about the y axis: $r = 1 - \sin(\pi - \theta) = 1 - \sin\theta$.

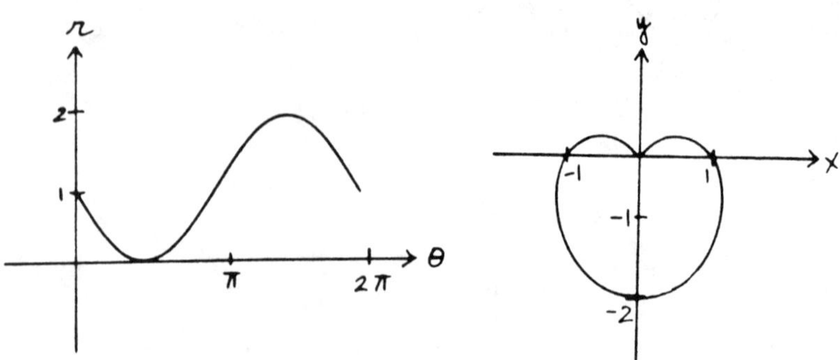

13. $r = \sin 3\theta$.

Symmetry about the y axis: $-r = \sin 3(-\theta) = \sin 3\theta$.

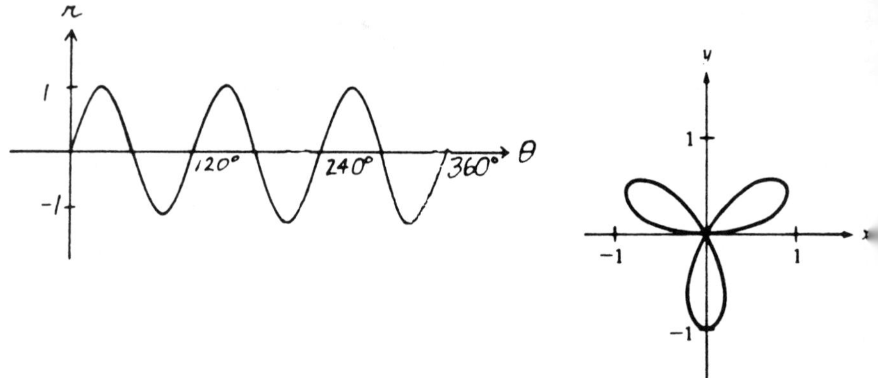

17. $r = 1 + 2\sin\theta$.

Symmetry about the y axis: $r = 1 + 2\sin(\pi - \theta) = 1 + 2\sin$

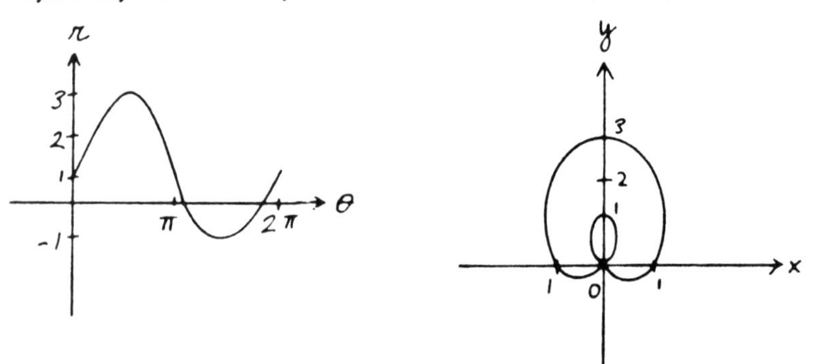

21. $r = \tan \theta$.

Symmetry about the pole: $r = \tan(\theta + \pi) = \dfrac{\tan \theta + \tan \pi}{1 - \tan \theta \tan \pi}$

$= \tan \theta$.

Symmetry about the x axis: $-r = \tan(\pi - \theta) = \dfrac{\tan \pi - \tan \theta}{1 + \tan \pi \tan \theta}$

$= -\tan \theta$.

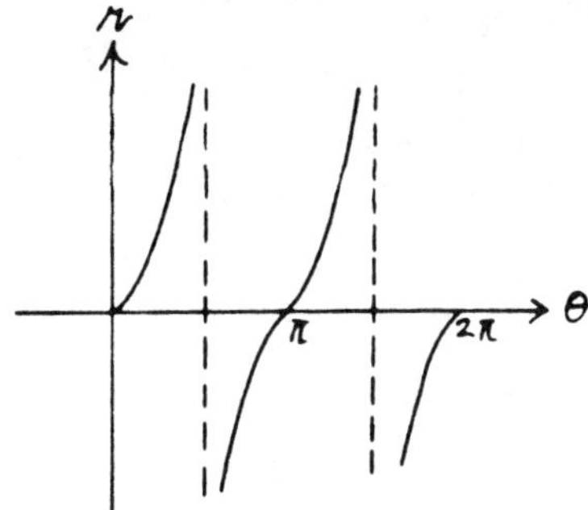

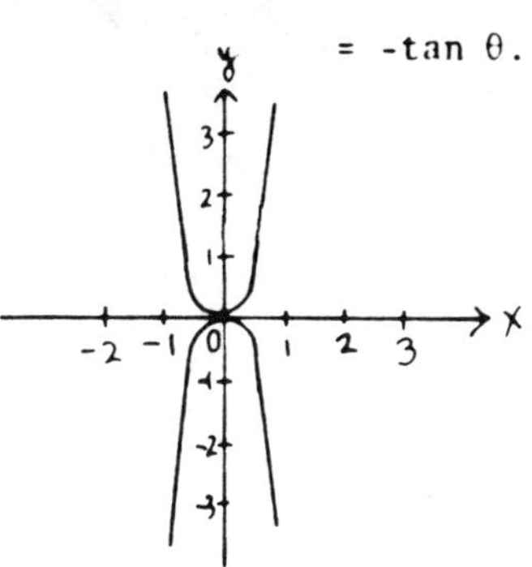

25. $r^2 = \cos 4\theta$.

Symmetry about the x axis: $r^2 = \cos 4(-\theta) = \cos 4\theta$.

Symmetry about the y axis: $(-r)^2 = \cos 4(-\theta)$, $r^2 = \cos 4\theta$.

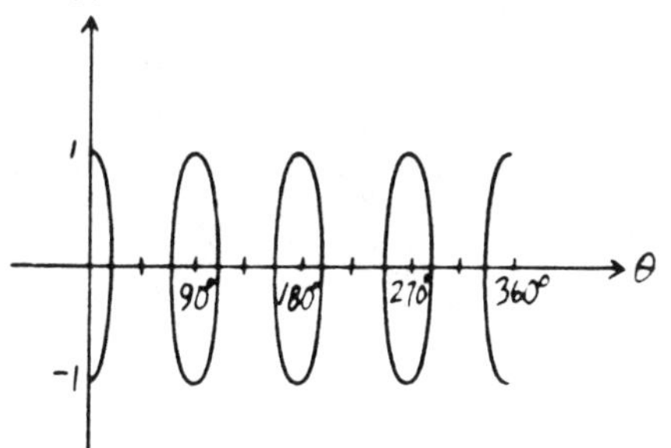

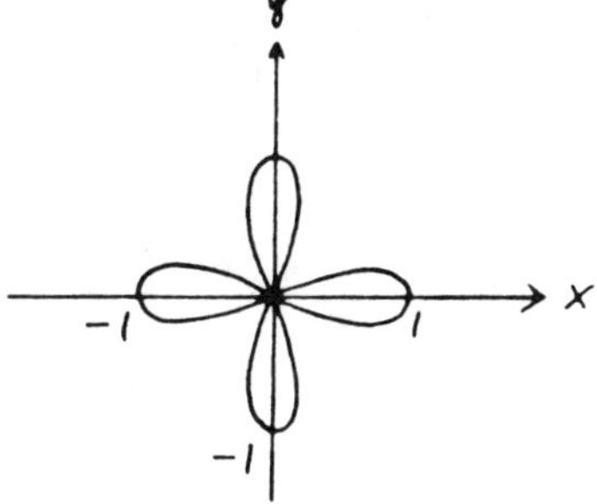

29. $r = \theta$.

Symmetry about the y axis: $-r = -\theta$, $r = \theta$.

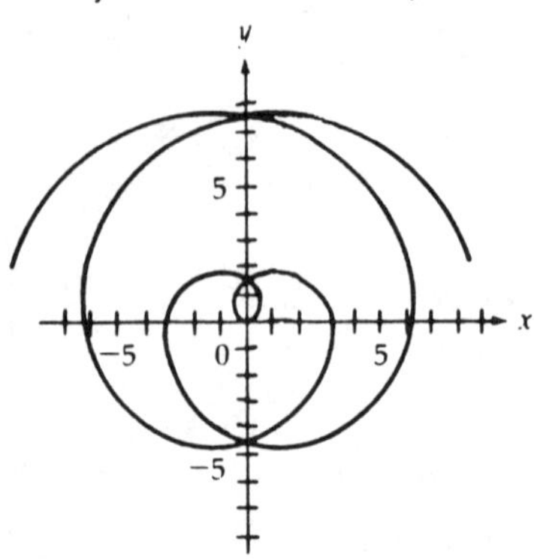

33. $r = \dfrac{2}{1 - 2\cos\theta}$.

Symmetry about the x axis: $r = \dfrac{2}{1 - 2\cos(-\theta)} = \dfrac{1}{1 - 2\cos\theta}$.

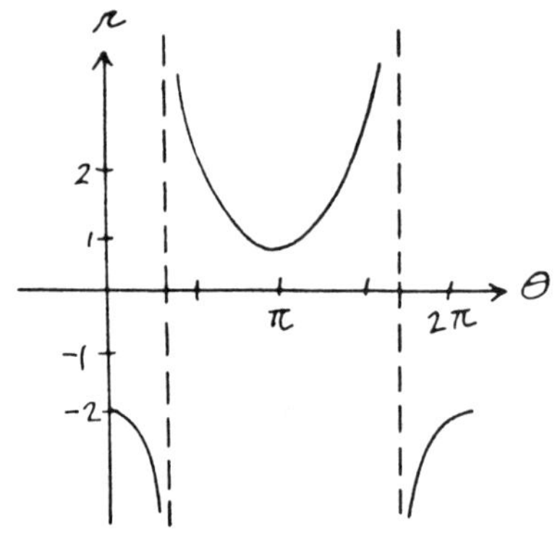

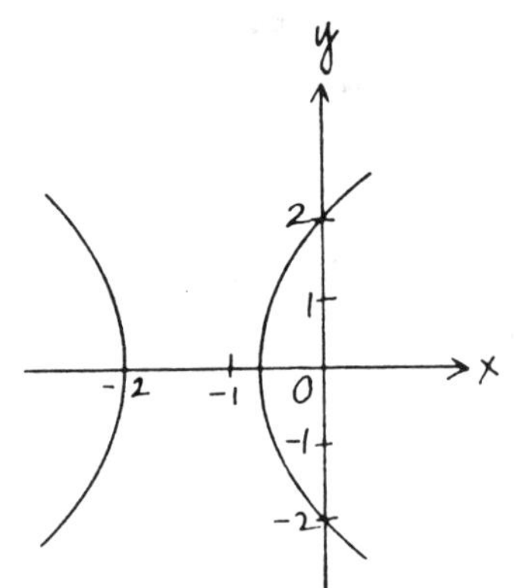

CHAPTER 14, SECTION 2 (pp. 681-682)

1. $x = r\cos\theta = (1)\cos\frac{\pi}{2} = 0$,

$y = r\sin\theta = (1)\sin\frac{\pi}{2} = 1$, $(1, \pi/2) = (0, 1)$.

$x = r\cos\theta = \sqrt{3}\cos\frac{\pi}{6} = \frac{3}{2}$,

$y = r\sin\theta = \sqrt{3}\sin\frac{\pi}{6} = \frac{\sqrt{3}}{2}$, $(\sqrt{3}, \pi/6) = (3/2, \sqrt{3}/2)$.

$x = r \cos \theta = -2 \cos 3\pi = 2,$

$y = r \sin \theta = -2 \sin 3\pi = 0, \quad (-2, 3\pi) = (2, 0).$

$x = r \cos \theta = \sqrt{2} \cos \frac{3\pi}{4} = -1,$

$y = r \sin \theta = \sqrt{2} \sin \frac{3\pi}{4} = 1, \quad (\sqrt{2}, 3\pi/4) = (-1, 1).$

$x = r \cos \theta = 3\sqrt{3} \cos \frac{5\pi}{3} = \frac{3\sqrt{3}}{2},$

$y = r \sin \theta = 3\sqrt{3} \sin \frac{5\pi}{3} = -\frac{9}{2}, \quad (3\sqrt{3}, 5\pi/3) = (3\sqrt{3}/2, -9/2).$

$x = r \cos \theta = -5 \cos \frac{7\pi}{6} = \frac{5\sqrt{3}}{2},$

$y = r \sin \theta = -5 \sin \frac{7\pi}{6} = \frac{5}{2}, \quad (-5, 7\pi/6) = (5\sqrt{3}/2, 5/2).$

$x = r \cos \theta = 0,$

$y = r \sin \theta = 0, \quad (0, 5\pi/4) = (0, 0).$

$x = r \cos \theta = 6 \cos 0 = 6,$

$y = r \sin \theta = 6 \sin 0 = 0, \quad (6, 0) = (6, 0).$

$x = r \cos \theta = -4 \cos \frac{7\pi}{4} = -2\sqrt{2},$

$y = r \sin \theta = -4 \sin \frac{7\pi}{4} = 2\sqrt{2}, \quad (-4, 7\pi/4) = (-2\sqrt{2}, 2\sqrt{2}).$

5. $x^2 + y^2 = 4, \quad r^2 = 4, \quad r = 2.$

9. $(x + y)^2 = x - y, \quad (r \cos \theta + r \sin \theta)^2 = r \cos \theta - r \sin \theta,$

$r(1 + 2 \sin \theta \cos \theta) = \cos \theta - \sin \theta,$

$r = \frac{\cos \theta - \sin \theta}{1 + 2 \sin \theta \cos \theta}.$

13. $r = a, \quad r^2 = a^2, \quad x^2 + y^2 = a^2.$

17. $r = 4 \sin \theta, \quad r^2 = 4r \sin \theta, \quad x^2 + y^2 - 4y = 0.$

21. $r = 2, \quad r = \cos \theta + 2, \quad 2 = \cos \theta + 2, \quad \cos \theta = 0,$

$\theta = 90°, 270°, \quad (2, 90°), (2, 270°).$

25. $r = \sin 2\theta$, $r = \sin \theta$, $\sin 2\theta = \sin \theta$,

$2 \sin \theta \cos \theta = \sin \theta$, $\sin \theta = 0$, $\cos \theta = \frac{1}{2}$,

$\theta = 0°, 180°, 60°, 300°$,

$(0, 0°) = (0, 180°)$, $(\frac{\sqrt{3}}{2}, 60°)$, $(-\frac{\sqrt{3}}{2}, 300°)$.

29. $r = \cos 2\theta$, $r = \cos^2\theta - \sin^2\theta$, $r^3 = r^2\cos^2\theta - r^2\sin^2\theta$,

$(r^2)^3 = (r^2\cos^2\theta - r^2\sin^2\theta)^2$, $(x^2 + y^2)^3 = (x^2 - y^2)^2$.

33. $r = 1 - \sin \theta$, $r = 1 - \cos \theta$,

$1 - \sin \theta = 1 - \cos \theta$, $\sin \theta = \cos \theta$,

$\tan \theta = 1$, $\theta = 45°, 225°$,

$(1 - 1/\sqrt{2}, 45°)$,

$(1 + 1/\sqrt{2}, 225°)$.

From the figure: $(0, 90°) = (0, 0°)$.

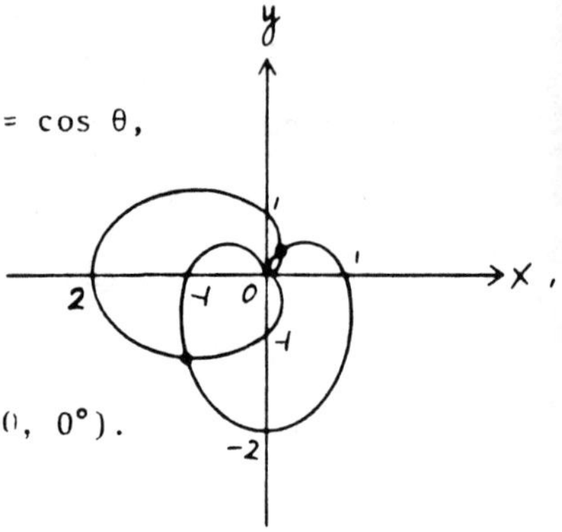

37. $[(x + 1)^2 + y^2][(x - 1)^2 + y^2] = 1$

$(r^2 + 2r \cos \theta + 1)(r^2 - 2r \cos \theta + 1) = 1$

$(r^2 + 1)^2 - 4r^2 \cos^2\theta = 1$

$r^4 + 2r^2 = 4r^2 \cos^2\theta$

$r^2 = -2 + 4 \cos^2\theta$

$= -2 + 2 + 2 \cos 2\theta$

$r^2 = 2 \cos 2\theta$.

CHAPTER 14, SECTION 3 (pp. 685-687)

1. $r = \frac{6}{1 + 3 \cos \theta}$, $e = 3$, $p = 2$,

hyperbola, F: $(0, 0)$, D: $x = 2$.

5. $r = \dfrac{4}{1 + \sin\theta}$, $e = 1$, $p = 4$,

parabola, F: (0, 0), D: $y = 4$.

9. $r = \dfrac{ep}{1 + e\cos\theta} = \dfrac{8/3}{1 + 2/3\cos\theta} = \dfrac{8}{3 + 2\cos\theta}$.

13. $r = \dfrac{ep}{1 + e\cos\theta} = \dfrac{49/4}{1 + 7/4\cos\theta} = \dfrac{49}{4 + 7\cos\theta}$.

17.

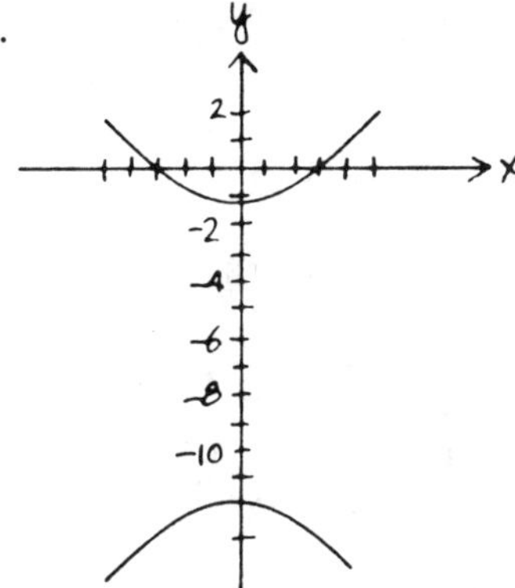

21.

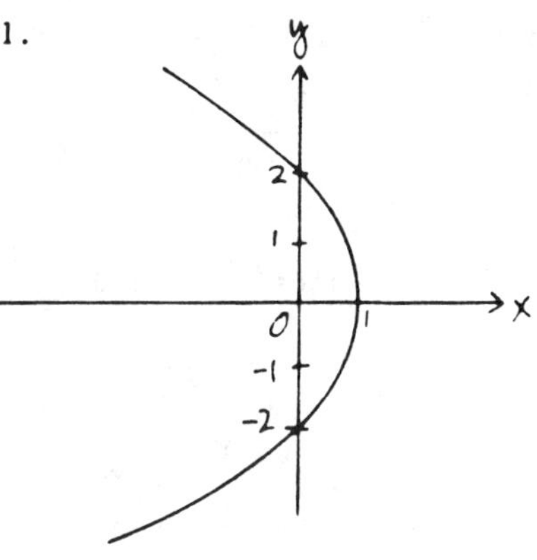

25. $a + c = 4500$, $a - c = 4100$

$2a = 8600$ $2c = 400$

$a = 4300$ $c = 200$

$e = \dfrac{c}{a} = \dfrac{200}{4300} = \dfrac{2}{43}$, D: $x = -\dfrac{a}{e} = \dfrac{-4300}{2/43} = -92{,}450$,

$r = \dfrac{ep}{1 - e\cos\theta} = \dfrac{4300}{1 - \frac{2}{43}\cos\theta} = \dfrac{184{,}900}{43 - 2\cos\theta}$.

29. The line through (x_1, y_1) and parallel to the given line is $Ax + By - (Ax_1 + By_1) = 0$. Both lines can be put into the normal form by dividing by $\pm\sqrt{A^2 + B^2}$. Thus, we have

$$\frac{A}{\pm\sqrt{A^2 + B^2}} x + \frac{B}{\pm\sqrt{A^2 + B^2}} y + \frac{C}{\pm\sqrt{A^2 + B^2}} = 0$$

and

$$\frac{A}{\pm\sqrt{A^2 + B^2}} x + \frac{B}{\pm\sqrt{A^2 + B^2}} y - \frac{Ax_1 + By_1}{\pm\sqrt{A^2 + B^2}} = 0.$$

By choosing the same sign in both cases, the polar coordinates of Q_1 and Q_2 are

$$Q_1: \left(\frac{C}{\pm\sqrt{A^2 + B^2}}, \alpha\right), \qquad Q_2: \left(-\frac{Ax_1 + By_1}{\pm\sqrt{A^2 + B^2}}, \alpha\right).$$

Thus, the distance between them is

$$d = \left|\frac{C}{\pm\sqrt{A^2 + B^2}} - \left(-\frac{Ax_1 + By_1}{\pm\sqrt{A^2 + B^2}}\right)\right| = \frac{|Ax_1 + By_1 + C|}{\sqrt{A^2 + B^2}}.$$

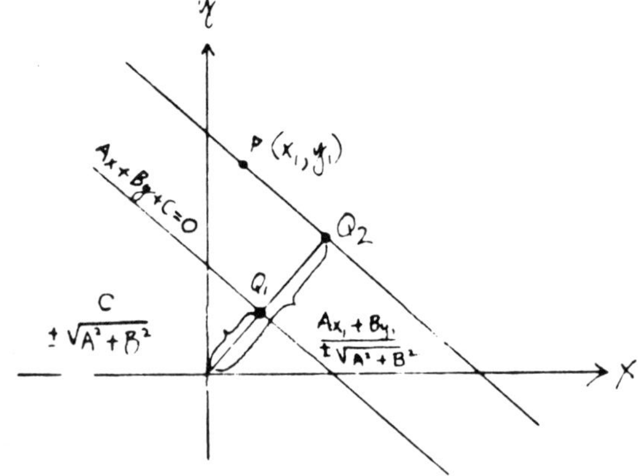

CHAPTER 14, SECTION 4 (pp. 690-692)

1. $r = 1 + \cos\theta$, $r' = -\sin\theta$,

$$\frac{dy}{dx} = \frac{r'\sin\theta + r\cos\theta}{r'\cos\theta - r\sin\theta} = \frac{-\sin^2\theta + (1+\cos\theta)\cos\theta}{-\sin\theta\cos\theta - (1+\cos\theta)\sin\theta}$$

$$= \frac{\cos\theta + \cos^2\theta - \sin^2\theta}{-\sin\theta - 2\sin\theta\cos\theta} = -\frac{\cos\theta + \cos 2\theta}{\sin\theta + \sin 2\theta}.$$

5. $r = \cos 3\theta$, $r' = -3\sin 3\theta$,

$$\frac{dy}{dx} = \frac{r'\sin\theta + r\cos\theta}{r'\cos\theta - r\sin\theta} = \frac{-3\sin 3\theta\sin\theta + \cos 3\theta\cos\theta}{-3\sin 3\theta\cos\theta - \cos 3\theta\sin\theta}$$

$$= \frac{-2\sin 3\theta\sin\theta + \cos 4\theta}{-2\sin 3\theta\sin\theta - \sin 4\theta}.$$

9. $$\frac{dy}{dx} = \frac{r'\sin\theta + r\cos\theta}{r'\cos\theta - r\sin\theta} = \frac{3\sin\theta\cos\theta + 3\sin\theta\cos\theta}{3\cos^2\theta - 3\sin^2\theta}$$

$$= \frac{2\sin\theta\cos\theta}{\cos^2\theta - \sin^2\theta} = \frac{\sin 2\theta}{\cos 2\theta} = \tan 2\theta = -\sqrt{3}.$$

13. $$\frac{dy}{dx} = \frac{r'\sin\theta + r\cos\theta}{r'\cos\theta - r\sin\theta} = \frac{3\sin\theta\cos 3\theta + \cos\theta\sin 3\theta}{3\cos\theta\cos 3\theta - \sin\theta\sin 3\theta}$$

$$= \frac{3\cdot 1/2\cdot 0 + \sqrt{3}/2\cdot 1}{3\cdot\sqrt{3}/2\cdot 0 - 1/2\cdot 1} = -\sqrt{3}.$$

17. $r = -1$, $r' = -2\sin 2\theta = 0$, $(-1, \pi/2) = (0, -1)$,

$$\frac{dy}{dx} = \frac{r'\sin\theta + r\cos\theta}{r'\cos\theta - r\sin\theta} = \frac{0\cdot 1 - 1\cdot 0}{0\cdot 0 - (-1)\cdot 1} = 0, \quad y + 1 = 0.$$

21. $r = \dfrac{3}{1+\sin\theta}$, $r' = \dfrac{-3\cos\theta}{(1+\sin\theta)^2}$,

$$\frac{dy}{dx} = \frac{r'\sin\theta + r\cos\theta}{r'\cos\theta - r\sin\theta} = \frac{\dfrac{-3\sin\theta\cos\theta}{(1+\sin\theta)^2} + \dfrac{3\cos\theta}{1+\sin\theta}}{\dfrac{-3\cos^2\theta}{(1+\sin\theta)^2} - \dfrac{3\sin\theta}{1+\sin\theta}}$$

$$= \frac{-\sin\theta\cos\theta + \cos\theta(1+\sin\theta)}{-\cos^2\theta - \sin\theta(1+\sin\theta)}$$

$$= \frac{-\cos\theta}{\sin^2\theta + \cos^2\theta + \sin\theta} = \frac{-\cos\theta}{1+\sin\theta}.$$

25. $r = \sin 2\theta$,

$$\frac{dy}{dx} = \frac{r'\sin\theta + r\cos\theta}{r'\cos\theta - r\sin\theta} = \frac{2\sin\theta\cos 2\theta + \cos\theta\sin 2\theta}{2\cos\theta\cos 2\theta - \sin\theta\sin 2\theta}.$$

$2\sin\theta\cos 2\theta + \cos\theta\sin 2\theta = 0$,

$2\sin\theta(\cos^2\theta - \sin^2\theta) + 2\sin\theta\cos^2\theta = 0$,

$2\sin\theta(3\cos^2\theta - 1) = 0$,

$\sin\theta = 0$, $3\cos^2\theta - 1 = 0$,

$\theta = 0, \pi$, $\cos^2\theta = \frac{1}{3}$,

(0, 0) neither, $\cos\theta = \pm\frac{1}{\sqrt{3}} = \pm 0.577$,

$(0, \pi)$ neither, $\theta = 54.8°, 125.2°, 234.8°, 305,2°$,

$(2\sqrt{2}/3, 54.8°)$ max, $(-2\sqrt{2}/3, 125.2°)$ min,

$(2\sqrt{2}/3, 234.8°)$ min, $(-2\sqrt{2}/3, 305.2°)$ max.

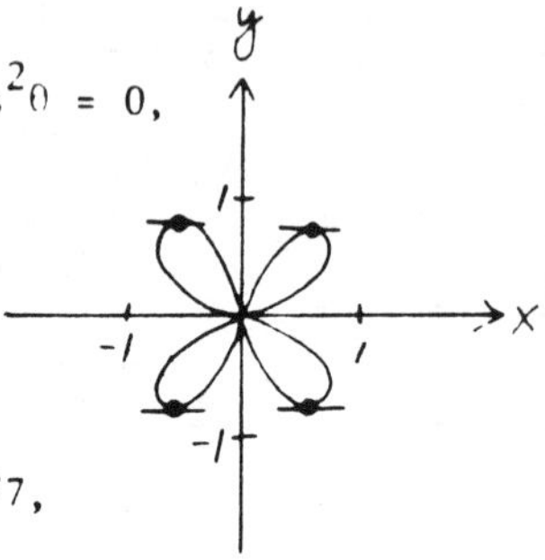

y

1

-1

1

-1

x

29. $f(\theta_0) = 0$, $f'(\theta_0) \neq 0$, $\dfrac{dy}{dx} = \dfrac{f'(\theta)\sin\theta + f(\theta)\cos\theta}{f'(\theta)\cos\theta - f(\theta)\sin\theta}$

$$= \frac{f'(\theta_0)\sin\theta_0}{f'(\theta_0)\cos\theta_0} = \tan\theta_0.$$

33. $r = \cos\theta, \quad r = 1 - \cos\theta,$

$$\tan\psi_1 = \frac{f_1(\theta)}{f_1'(\theta)} = \frac{\cos\theta}{-\sin\theta}$$

$$\tan\psi_2 = \frac{f_2(\theta)}{f_2'(\theta)} = \frac{1 - \cos\theta}{\sin\theta}.$$

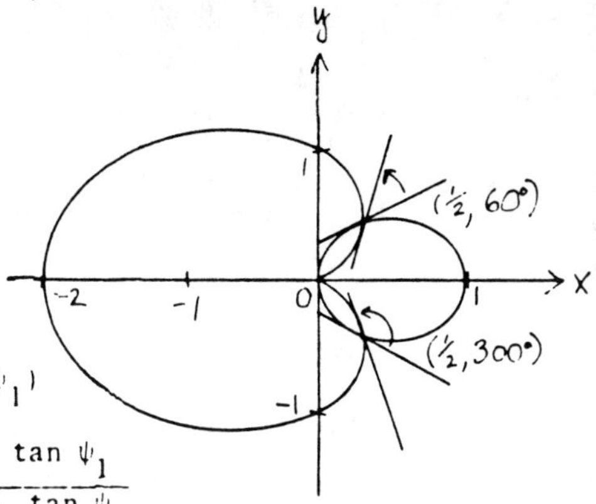

At $(1/2, 60°)$,

$$\tan\psi_1 = \frac{1/2}{-\sqrt{3}/2} = -\frac{1}{\sqrt{3}},$$

$$\tan\psi_2 = \frac{1 - 1/2}{\sqrt{3}/2} = \frac{1}{\sqrt{3}}$$

$$\tan(\theta_2 - \theta_1) = \tan(\psi_2 - \psi_1)$$

$$= \frac{\tan\psi_2 - \tan\psi_1}{1 + \tan\psi_1 \tan\psi_2}$$

$$= \frac{\frac{1}{\sqrt{3}} + \frac{1}{\sqrt{3}}}{1 - \frac{1}{3}} = \frac{2/\sqrt{3}}{2/3} = \sqrt{3}$$

$\theta_2 - \theta_1 = 60°.$

At $(1/2, 300°)$, $\tan\psi_1 = \dfrac{1/2}{\sqrt{3}/2} = \dfrac{1}{\sqrt{3}}.$

$$\tan\psi_2 = \frac{1 - 1/2}{-\sqrt{3}/2} = -\frac{1}{\sqrt{3}}$$

$\tan(\theta_2 - \theta_1) = \tan(\psi_2 - \psi_1) = -\sqrt{3}$

$\theta_2 - \theta_1 = 120°.$

At $(0, 90°) = (0, 0°)$, $\theta_2 - \theta_1 = 90°$ from the graph.

1. $A = 2\int_0^{\pi/2} \frac{1}{2} r^2\, d\theta$

$= \int_0^{\pi/2} 4^2\sin^2\theta\, d\theta$

$= 8\int_0^{\pi/2} (1 - \cos 2\theta)\, d\theta$

$= 4(2\theta - \sin 2\theta)\Big|_0^{\pi/2} = 4\pi.$

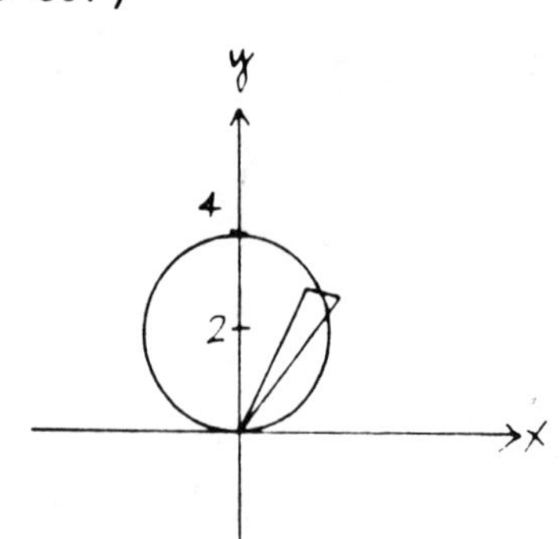

5. $A = 2\int_0^{\pi/2} \frac{1}{2} r^2\, d\theta$

$= \int_0^{\pi/2} \sin\theta\, d\theta$

$= -\cos\theta\Big|_0^{\pi/2}$

$= 1.$

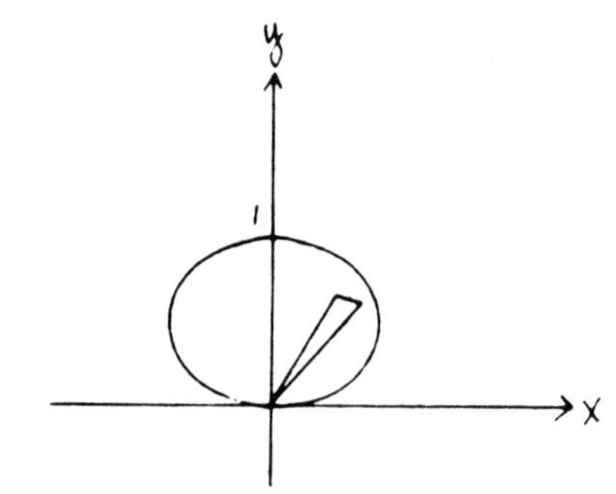

9. $A = 2\int_0^{\pi/2} \frac{1}{2} r^2 d\theta$

$= \int_0^{\pi/2} 9\sin 2\theta\, d\theta$

$= -\frac{9}{2}\cos 2\theta\Big|_0^{\pi/2}$

$= \frac{9}{2} + \frac{9}{2}$

$= 9.$

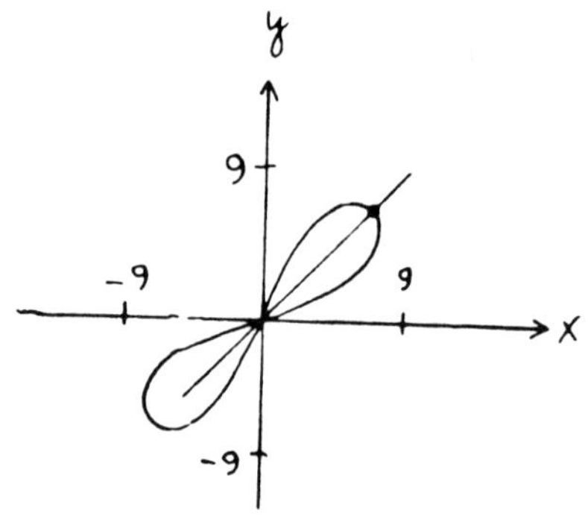

13. $A = 2\int_{\pi/4}^{5\pi/4} \frac{1}{2} r_1^2 d\theta$

$= \int_{\pi/4}^{5\pi/4} (1 + \cos\theta)^2 d\theta$

$= \int_{\pi/4}^{5\pi/4} (1 + 2\cos\theta + \cos^2\theta)\, d\theta$

$= \int_{\pi/4}^{5\pi/4} (1 + 2\cos\theta + \frac{1 + \cos 2\theta}{2})\, d\theta$

$= (\frac{3}{2}\theta + 2\sin\theta + \frac{1}{4}\sin 2\theta)\Big|_{\pi/4}^{5\pi/4}$

$= \frac{15\pi}{8} - \sqrt{2} + \frac{1}{4} - \frac{3\pi}{8} - \sqrt{2} - \frac{1}{4}$

$= \frac{3\pi}{2} - 2\sqrt{2}.$

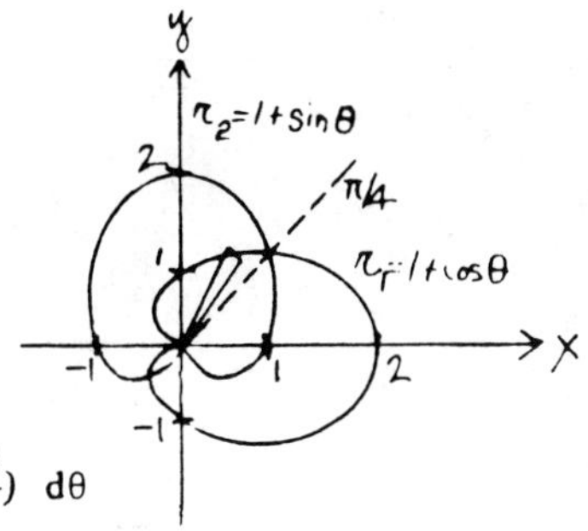

17. $r = \sin\theta, \quad r = \sin 2\theta = 2\sin\theta\cos\theta$

$2\sin\theta\cos\theta - \sin\theta = 0$

$\sin\theta(2\cos\theta - 1) = 0$

$\sin\theta = 0, \quad \cos\theta = \frac{1}{2}$

$\theta = 0, \pi, \frac{\pi}{3}, \frac{5\pi}{3}$

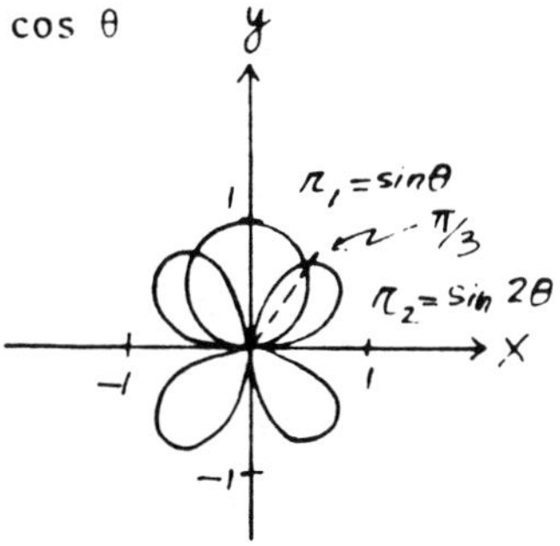

$A = 2\left[\int_0^{\pi/3} \frac{1}{2} r_1^2 d\theta + \int_{\pi/3}^{\pi/2} \frac{1}{2} r_2^2 d\theta\right]$

$= \int_0^{\pi/3} \sin^2\theta\, d\theta + \int_{\pi/3}^{\pi/2} \sin^2 2\theta\, d\theta$

$= \int_0^{\pi/3} \frac{1 - \cos 2\theta}{2} d\theta + \int_{\pi/3}^{\pi/2} \frac{1 - \cos 4\theta}{2} d\theta$

$= (\frac{\theta}{2} - \frac{1}{4}\sin 2\theta)\Big|_0^{\pi/3} + (\frac{\theta}{2} - \frac{1}{8}\sin 4\theta)\Big|_{\pi/3}^{\pi/2}$

$= \frac{\pi}{6} - \frac{\sqrt{3}}{8} + \frac{\pi}{4} - \frac{\pi}{6} - \frac{\sqrt{3}}{16}$

$= \frac{\pi}{4} - \frac{3\sqrt{3}}{16}.$

21. $\frac{dy}{dx} = \frac{r \cos \theta + r' \sin \theta}{-r \sin \theta + r' \cos \theta}$

$$s = \int_{x_1}^{x_2} \sqrt{1 + (y')^2}\, dx$$

$$= \int_{\theta_1}^{\theta_2} \sqrt{1 + \left(\frac{r \cos \theta + r' \sin \theta}{-r \sin \theta + r' \cos \theta}\right)^2} (-r \sin \theta + r' \cos \theta) d\theta$$

$$= \int_{\theta_1}^{\theta_2} \sqrt{(-r \sin \theta + r' \cos \theta)^2 + (r \cos \theta + r' \sin \theta)^2} d\theta$$

$$= \int_{\theta_1}^{\theta_2} \sqrt{\begin{array}{c} r^2 \sin^2 \theta - 2rr' \sin \theta \cos \theta + (r')^2 \cos^2 \theta + r^2 \cos^2 \theta \\ + 2rr' \sin \theta \cos \theta + (r')^2 \sin^2 \theta \end{array}}$$

$$= \int_{\theta_1}^{\theta_2} \sqrt{r^2 + (r')^2}\, d\theta.$$

24. $r = 1 + \cos \theta, \quad r' = -\sin \theta$

$$s = \int_0^{2\pi} \sqrt{1 + 2 \cos \theta + \cos^2 \theta + \sin^2 \theta}\, d\theta$$

$$= \int_0^{2\pi} \sqrt{2 + 2 \cos \theta}\, d\theta = \sqrt{2} \int_0^{2\pi} \frac{\sqrt{(1 + \cos \theta)(1 - \cos \theta)}}{\sqrt{1 - \cos \theta}}\, d\theta$$

$$= \sqrt{2} \int_0^{\pi} \frac{\sin \theta}{\sqrt{1 - \cos \theta}}\, d\theta - \sqrt{2} \int_{\pi}^{2\pi} \frac{\sin \theta}{\sqrt{1 - \cos \theta}}\, d\theta$$

$$= 2\sqrt{2} \sqrt{1 - \cos \theta}\Big|_0^{\pi} - 2\sqrt{2} \sqrt{1 - \cos \theta}\Big|_{\pi}^{2\pi}$$

$$= 4 + 4 = 8.$$

CHAPTER 14, REVIEW (pp. 697-698)

1. $r = 1 + 3\cos\theta$.

Symmetry about the x axis: $r = 1 + 3\cos(-\theta) = 1 + 3\cos\theta$.

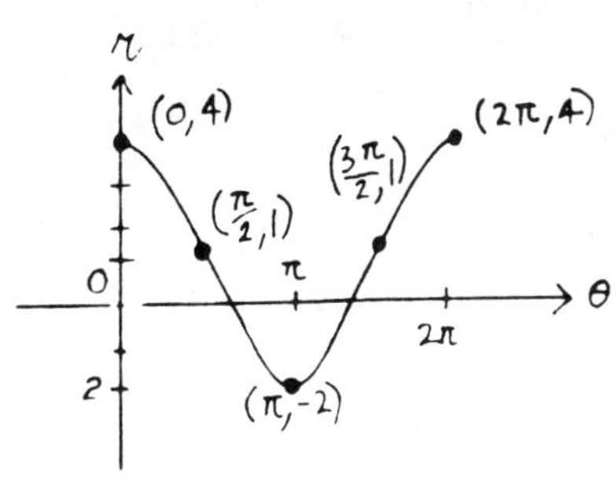

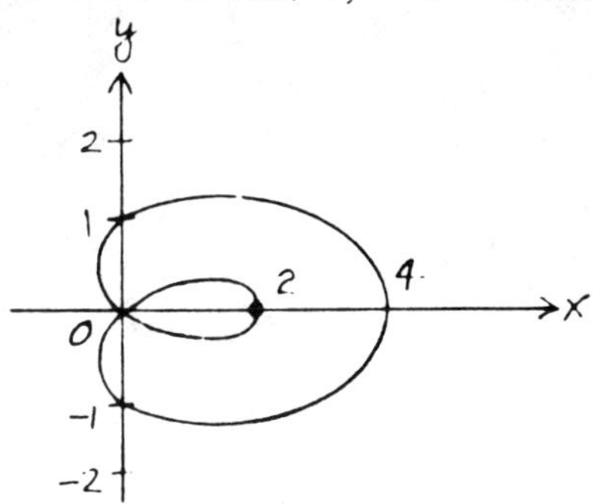

5. $r^2 = \sin 2\theta$.

Symmetry about the pole: $(-r)^2 = \sin 2\theta$, $r^2 = \sin 2\theta$.

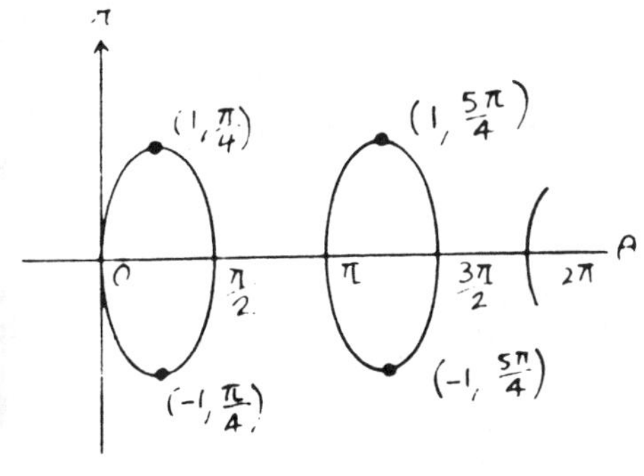

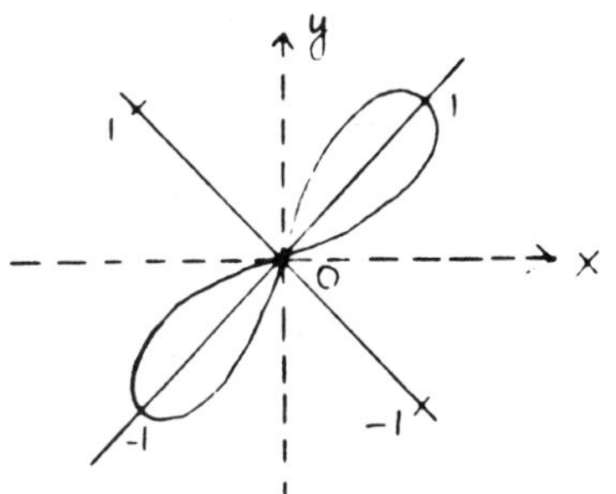

9. $r = \sin\theta$, $r = 1 + 2\sin\theta$,

$\sin\theta = 1 + 2\sin\theta$,

$\sin\theta = -1$,

$\theta = \frac{3\pi}{2}$,

$(-1, 3\pi/2)$, $(0, 0) = (0, 7\pi/6)$.

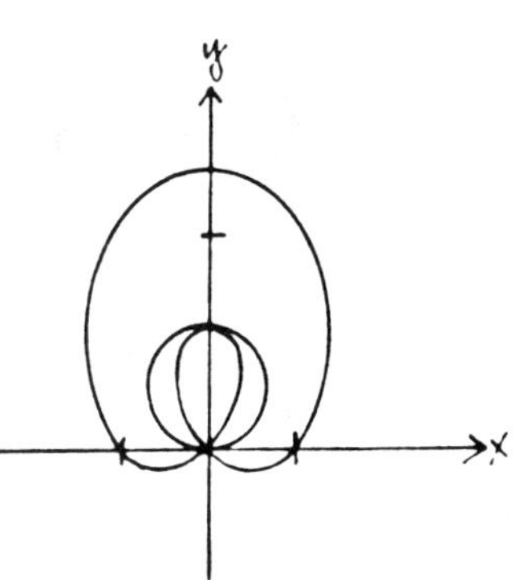

13. (a) $x = r\cos\theta = (-1)\cos\frac{\pi}{2} = 0$

$y = r\sin\theta = (-1)\sin\frac{\pi}{2} = -1$, $(0, -1)$

$x = r\cos\theta = 2\sqrt{2}(-\frac{1}{\sqrt{2}}) = -2$

$y = r\sin\theta = 2\sqrt{2}(\frac{1}{\sqrt{2}}) = 2$, $(-2, 2)$

$x = r\cos\theta = 6(-\frac{\sqrt{3}}{2}) = -3\sqrt{3}$

$y = r\sin\theta = 6(-\frac{1}{2}) = -3$, $(-3\sqrt{3}, -3)$

$x = r\cos\theta = -2\sqrt{3}(-\frac{1}{2}) = \sqrt{3}$

$y = r\sin\theta = -2\sqrt{3}(\frac{\sqrt{3}}{2}) = -3$, $(\sqrt{3}, -3)$

(b) $r^2 = x^2 + y^2 = 4 + 4 = 8$

$\tan\theta = \frac{y}{x} = \frac{2}{-2} = -1$, $(2\sqrt{2}, 3\pi/4)$

$r^2 = x^2 + y^2 = 25$

$\tan\theta = \frac{y}{x} = \frac{0}{-5} = 0$, $(-5, 0) = (5, \pi)$

$r^2 = x^2 + y^2 = 1 + 3 = 4$

$\tan\theta = \frac{y}{x} = -\sqrt{3}$, $(2, -\pi/3)$

$r^2 = x^2 + y^2 = 100 + 16 = 116$

$\tan\theta = \frac{y}{x} = \frac{2}{5}$, $(2\sqrt{29}, \text{Arctan } 2/5)$.

17. $A = 2\int_{-\pi/2}^{\pi/2} \frac{1}{2} r^2 d\theta$

$= \int_{-\pi/2}^{\pi/2} (2 + \sin\theta)^2 d\theta$

$= \int_{-\pi/2}^{\pi/2} (4 + 4\sin\theta + \frac{1 - \cos 2\theta}{2}) d\theta$

$$= \int_{-\pi/2}^{\pi/2} \left(\frac{9}{2} + 4 \sin \theta - \frac{1}{4} \cdot 2 \cos 2\theta\right) d\theta$$

$$= \left(\frac{9}{2}\theta - 4 \cos \theta - \frac{1}{4} \sin 2\theta\right)\Big|_{-\pi/2}^{\pi/2}$$

$$= \frac{9\pi}{4} + \frac{9\pi}{4}$$

$$= \frac{9\pi}{2}.$$

21. $F = (0, 0)$

D: $y = -2$

$e = 1$.

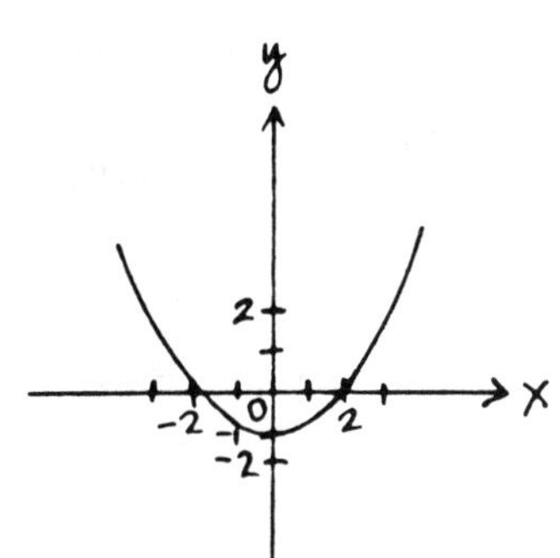

25. $r = \sin \theta + \cos \theta$, $r' = \cos \theta - \sin \theta$,

$$\frac{dy}{dx} = \frac{r \cos \theta + r' \sin \theta}{-r \sin \theta + r' \cos \theta}$$

$$= \frac{(\sin \theta + \cos \theta)\cos \theta + (\cos \theta - \sin \theta)\sin \theta}{-(\sin \theta + \cos \theta)\sin \theta + (\cos \theta - \sin \theta)\cos \theta}$$

$$= \frac{\cos^2\theta - \sin^2\theta + 2 \sin \theta \cos \theta}{\cos^2\theta - \sin^2\theta - 2 \sin \theta \cos \theta} = \frac{\cos 2\theta + \sin 2\theta}{\cos 2\theta - \sin 2\theta}$$

$\cos 2\theta = -\sin 2\theta$,

$\tan 2\theta = -1$

$2\theta = \frac{3\pi}{4}, \frac{7\pi}{4}, \frac{11\pi}{4}, \frac{15\pi}{4}$

$\theta = \frac{3\pi}{8}, \frac{7\pi}{8}, \frac{11\pi}{8}, \frac{15\pi}{8}$

$(1.307, 3\pi/8)$ max

$(-0.541, 7\pi/8)$ min.

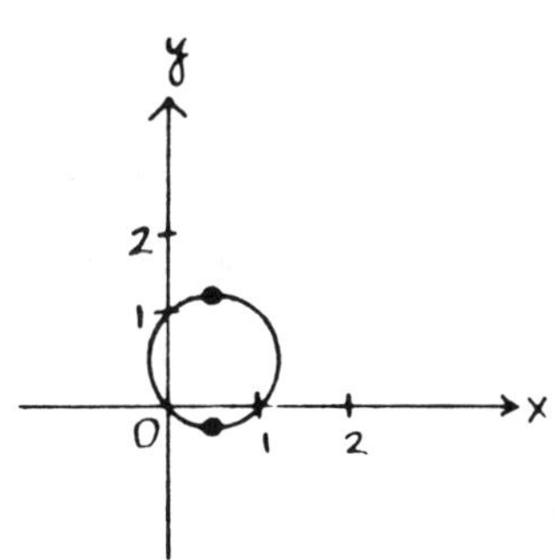

29. $a(1 + \cos\theta) = a(1 - \cos\theta)$

$$2\cos\theta = 0$$

$$\theta = \frac{\pi}{2}, \frac{3\pi}{2}$$

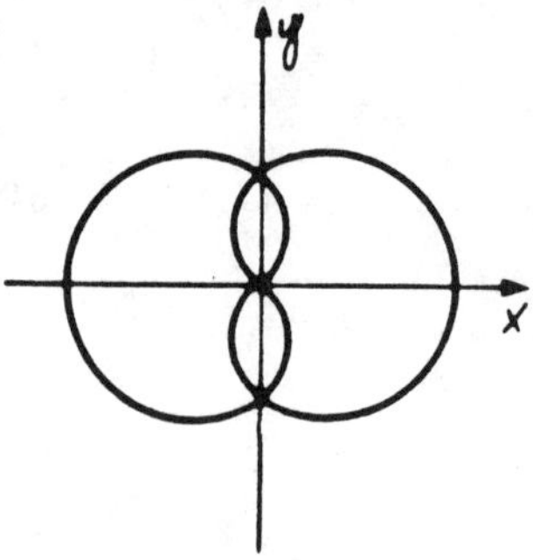

$(1, \frac{\pi}{2})$, $(1, \frac{3\pi}{2})$, $(0, 0)$

See Section 4, problems 30-31.

$$\tan\psi_1 = \frac{a(1 + \cos\theta)}{-a\sin\theta} = -\csc\theta - \cot\theta$$

$$\tan\psi_2 = \frac{a(1 - \cos\theta)}{a\sin\theta} = \csc\theta - \cot\theta$$

$(1, \frac{\pi}{2})$: $\tan\psi_1 = -1 - 0 = -1$, $\tan\psi_2 = 1 - 0 = 1$

$-1 \cdot 1 = -1$, perpendicular

$(1, \frac{3\pi}{2})$: $\tan\psi_1 = -(-1) = 1$, $\tan\psi_2 = -1$

$1(-1) = -1 = -1$, perpendicular

$(0, \pi) = (0, 0)$: $\psi_1 = \frac{\pi}{2}$, $\psi_2 = \frac{\pi}{2}$, parallel.

Chapter 15
Limits and Continuity: The Epsilon-Delta Approach

CHAPTER 15, SECTION 1 (pp. 704-705)

1. Limit point.

5. Not a limit point.

9. Limit point.

13. Suppose $0 < |x - 1| < \delta$,

$|(x - 3) + 2| = |x - 1| < \delta$, $\delta = \varepsilon$.

17. Suppose $0 < |x| < \delta$.

$|x^2| < \delta^2$, $\delta = \sqrt{\varepsilon}$.

21. Suppose $0 < |x - 1| < \delta$ and $\delta \leqq 1$.

$|x - 1| < 1$

$-1 < x - 1 < 1$

$0 < x < 2$

$|x| < 2$

$|x^2 - x| = |x| \cdot |x - 1| < 2\delta$

$\delta = \min\{\varepsilon/2, 1\}$.

25. Suppose $0 < |x - 2| < \delta$ and $\delta \leqq 1$.

$|x - 2| < 1$

$-1 < x - 2 < 1$

$1 < x < 3$

$|x| < 3$

$$|x^3 - 8| = |x^2 + 2x + 4| \cdot |x - 2|$$
$$\leqq (|x|^2 + 2|x| + 4)|x - 2|$$
$$< (3^2 + 2\cdot 3 + 4)\delta = 19\delta$$

$\delta = \min\{\varepsilon/19, 1\}$.

29. Suppose $0 < |x - 1| < \delta$ and $\delta \leqq \frac{1}{2}$.

$$|x - 1| < \frac{1}{2}$$

$$-\frac{1}{2} < x - 1 < \frac{1}{2}$$

$$\frac{1}{2} < x < \frac{3}{2} \quad \text{and} \quad \frac{3}{2} < x + 1 < \frac{5}{2}$$

$$|x| > \frac{1}{2} \qquad |x + 1| < \frac{5}{2}$$

$$\frac{1}{|x|} < 2$$

$$\left|\frac{1}{x^2} - 1\right| = \left|\frac{1 - x^2}{x^2}\right| = \frac{|1 + x| \cdot |1 - x|}{|x|^2}$$

$$= \frac{1}{|x|^2} |x + 1| \cdot |x - 1| < 2^2 \cdot \frac{5}{2}\delta = 10\delta$$

$$\delta = \min\{\varepsilon/10, 1/2\} .$$

CHAPTER 15, SECTION 2 (pp. 709-710)

1. Choose $\varepsilon = 1$. If $\delta > 0$, let $1 - \delta < x < 1$.

$$-\delta < x - 1 < 0 \qquad x - 3 < -2$$

$$0 < |x - 1| < \delta \qquad |x - 3| > 2$$

$$|(x + 1) - 4| > 1 = \varepsilon$$

$$|f(x) - 4| \nless \varepsilon.$$

5. Choose $\varepsilon = 1$. If $\delta > 0$, let $0 < x < \delta$.

$$0 < |x| < \delta \qquad f(x) = 1 = \varepsilon$$

$$|f(x)| \nless \varepsilon.$$

9. Choose $\varepsilon = 1$. If $\delta > 0$, let $0 < x < \min\{\delta,1\}$.

$0 < |x| < \delta$ $\qquad |x^2| < 1$

$$\left|\frac{1}{x^2}\right| > 1 = \varepsilon$$

$$|f(x)| \nless \varepsilon.$$

13. False. Choose $\varepsilon = 1$. If $\delta > 0$, let $0 < x < \min\{1,\delta\}$.

$0 < |x| < \delta$ $\qquad 0 < x < 1$

$$\frac{1}{x} > 1 = \varepsilon$$

$$|f(x)| \nless \varepsilon.$$

17. Choose $\varepsilon = 2$. If $\delta > 0$, let $2 < x < 2 + \delta$ (x irrational).

$0 < x - 2 < \delta$ $\qquad f(x) = -1$

$0 < |x - 2| < \delta$ $\qquad f(x) - 1 = -2$

$$|f(x) - 1| = 2 = \varepsilon$$

$$|f(x) - 1| \nless \varepsilon.$$

21. Suppose $\lim_{x\to 0} f(x) = b$ and $b \geq 1$. Choose $\varepsilon = \frac{1}{2}$.

If $\delta > 0$, let $0 < x < \min\{\delta, 1/\sqrt{2}\}$.

$0 < |x| < \delta$ $\qquad 0 < x < \frac{1}{\sqrt{2}}$

$$0 < f(x) < \frac{1}{2}$$

$$f(x) - b < -\frac{1}{2} = -\varepsilon$$

$$|f(x) - b| > \varepsilon$$

$$|f(x) - b| \nless \varepsilon.$$

Suppose $b < 1$. Choose $\varepsilon = \frac{1-b}{2}$. If $\delta > 0$, let $-\delta < x < 0$.

$0 < |x| < \delta$ $\qquad f(x) = 1$

$$f(x) - b = 1 - b > \frac{1-b}{2} = \varepsilon$$

$$|f(x) - b| \nless \varepsilon.$$

24. Suppose $\lim\limits_{x \to 0} \frac{1}{x^2} = b$ and $b \geqq 0$.

Choose $\varepsilon = 1$. If $\delta > 0$, let $0 < x < \min\{\delta, \frac{1}{\sqrt{b+1}}\}$,

$0 < x < \delta.$ $\qquad 0 < x < \frac{1}{\sqrt{b+1}}$

$0 < |x| < \delta$

$$\frac{1}{x^2} > b + 1$$

$$\frac{1}{x^2} - b > 1 = \varepsilon$$

$$\left|\frac{1}{x^2} - b\right| \nless \varepsilon.$$

Suppose $b < 0$. Choose $\varepsilon = -b$. If $\delta > 0$, let $0 < x < \delta$.

$0 < |x| < \delta$ $\qquad \frac{1}{x^2} > 0$

$$\frac{1}{x^2} - b > -b = \varepsilon$$

$$\left|\frac{1}{x^2} - b\right| \nless \varepsilon.$$

CHAPTER 15, SECTION 3 (pp. 717-718)

1. Suppose $0 < x < \delta$. Then $f(x) = x$ and $|f(x)| = x$. Hence $|f(x) - 0| < \delta$. Let $\delta = \varepsilon$. Then $|f(x) - 0| < \varepsilon$ whenever $0 < x < \delta$.

5. Suppose $0 < |x| < \delta$. $|\frac{1}{x}| > \frac{1}{\delta}$, $\frac{1}{x^2} > \frac{1}{\delta^2}$, $\frac{-1}{x^2} < \frac{-1}{\delta^2}$.

Choose δ such that $-1/\delta^2 = N$ or $\delta = 1/\sqrt{-N}$ when N is negative; choose any δ when $N \geqq 0$.

9. Suppose $x > N$ (N positive). $\frac{1}{x^2} < \frac{1}{N^2}$. Choose N such that $1/N^2 = \varepsilon$ or $N = 1/\sqrt{\varepsilon}$.

13. Suppose $x < N$. $e^x < e^N$. Choose N such that $e^N = \varepsilon$ or $N = \ln \varepsilon$.

17. Choose $N = 1$. If $\delta > 0$, let $-\delta < x < 0$.
$-\delta^3 < x^3$, $\frac{1}{x^3} < \frac{-1}{\delta^3} < N$, $\frac{1}{x^3} \not> N$.

21. Choose $\varepsilon = 1$. If N is a number, let $x > \max\{N,1\}$.
$x > N$, $x > 1$. $(x + 1) - 1 > 1 = \varepsilon$

$$|\frac{x^2 - 1}{x - 1} - 1| \not< \varepsilon.$$

25. Suppose $x > M$.
$x^2 > M^2$.
Choose M such that $N = M^2$ or $M = \sqrt{N}$ if N is positive; choose any M if N is negative.

CHAPTER 15, SECTION 4 (pp. 725-726)

1. Suppose $|x| < \delta$. Then $|f(x)| = |x| < \delta$, $\delta = \varepsilon$.

5. Suppose $|x| < \delta$. Then $|f(x) - 1| = |\frac{x^2 - 1}{x - 1} - 1|$
$= |(x + 1) - 1| = |x| < \delta$, $\delta = \varepsilon$.

9. Suppose $|x - 1| < \delta$ and $\delta < 1$.
Then $|f(x) - 1| = |1 - 1| = 0 < \varepsilon$; δ can be any number less than 1.

13. Suppose $|x - 1| < \delta$ and $\delta \leqq 1/2$. Then

$$|x - 1| < \frac{1}{2} \qquad |x| > \frac{1}{2}$$

$$-\frac{1}{2} < x - 1 < \frac{1}{2} \qquad \left|\frac{1}{x} - 1\right| = \left|\frac{1 - x}{x}\right| = \frac{1}{|x|}|x - 1| < 2\delta$$

$$\frac{1}{2} < x < \frac{3}{2} \qquad \delta = \min\{\varepsilon/2, 1/2\} .$$

17. Choose $\varepsilon = \frac{1}{2}$. If $\delta > 0$, let $0 < x < \min\{\delta, 1/2\}$.

$$0 < x < \delta \qquad 0 < x < \frac{1}{2}$$

$$|x| < \delta \qquad -1 < x - 1 < -\frac{1}{2}$$

$$|x - 1| > \frac{1}{2} = \varepsilon$$

$$|f(x) - 1| \nless \varepsilon.$$

21. $f(x) = \dfrac{x^2 + x - 2}{x - 1} = \dfrac{(x - 1)(x + 2)}{x - 1} = x + 2$ for $x \neq 1$

Define $f(1) = 3$. Then $f(x) = x + 2$, which is continuous.

25. Suppose $f(x)$ is a polynomial, $f(a) < 0$, and $f(b) > 0$. By Problem 22, f is continuous on $[a,b]$. By the intermediate-value theorem, $f(x) = 0$ for some x between a and b.

29. Suppose not. The velocity v cannot be less than v_{av} for all $a \leqq t \leqq b$, nor can it be greater than v_{av} for all $a \leqq t \leqq b$. Hence there must be times t_0 and t_1 such that $v(t_0) < v_{av} < v(t_1)$. Since the velocity of a moving object is continuous, there is a time t_2 between t_0 and t_1 such that $v(t_2) = v_{av}$.

CHAPTER 15, REVIEW (pp. 726-727)

1. True. Suppose $0 < |x - 2| < \delta$. $-\delta < x - 2 < \delta$,

   $-2\delta < 2x - 4 < 2\delta$, $|f(x) - 5| = |(2x + 1) - 5| = |2x - 4| < 2\delta$.

   Choose $\delta = \varepsilon/2$. Then $|f(x) - 5| < \varepsilon$ whenever

   $0 < |x - 2| < \delta$.

5. True. Suppose $N \geq 1$. Let $0 < x - 3 < \delta$ for some positive δ. Then $\frac{1}{x-3} > \frac{1}{\delta}$. Choose $\delta = 1/N$. Then

   $\frac{1}{x-3} > N$ whenever $0 < x - 3 < \delta$.

9. True. Let $\varepsilon > 0$. Suppose, for some $\delta > 0$,

   $0 < |x + 2| < \delta$, $-\delta < x + 2 < \delta$, $-4 - \delta < x - 2 < -4 + \delta$.

   Let us choose $\delta \leqq 1$. Then

   $-5 \leqq -4 - \delta < x - 2 < -4 + \delta \leqq -3$, $\qquad |x - 2| < 5$

   $|f(x) - 4| = |x^2-4| = |(x+2)(x-2)| = |x+2||x - 2| < 5\delta$.

   Choose $\delta = \min\{1,\varepsilon/5\}$. Then $|f(x) - 4| < \varepsilon$ whenever

   $0 < |x + 2| < \delta$.

13. $\sin u = 1$ when $u = \pi/2 + 2\pi n = (4n + 1)\pi/2$

    $\sin \frac{1}{x} = 1$ when $x = 2/\pi(4n + 1)$

    $\sin u = -1$ when $u = -\pi/2 + 2\pi n = (4n - 1)\pi/2$

    $\sin \frac{1}{x} = -1$ when $x = 2/\pi(4n - 1)$

    Assume $\lim_{x \to 0} \sin(1/x) = L \neq 1$. Let $\varepsilon = |1 - L|/2$.

    For any $\delta > 0$, let $|x| < \delta$ and $x = 2/\pi(4n + 1) < \delta$ for some integer n. Then $\sin(1/x) = 1$ and

    $$|\sin(1/x) - L| = |1 - L| \not< \varepsilon.$$

    Thus if the limit exists, it must be 1. Assume $L = 1$.

    Let $\varepsilon = 1$. For any $\delta > 0$, let $|x| < \delta$ and

    $x = 2/(4n-1)\pi < \delta$ for some integer n. Then

    $\sin(1/x) = -1$ and

$$|\sin(1/x) - L| = |-1 - 1| = 2 \nless \varepsilon .$$

Hence $\lim_{x \to 0} \sin(1/x)$ does not exist.

17. $f(x) = \dfrac{x^2 + 2x - 8}{x + 4} = \dfrac{(x + 4)(x - 2)}{x + 4} = x - 2$ for $x \neq -4$.

Define $f(-4) = -6$. Then $f(x) = x - 2$ for all x and is continuous.

21. Let $\varepsilon = f(a)/2$. Then there is a $\delta > 0$ such that $|f(x) - f(a)| < \varepsilon$ whenever $|x - a| < \delta$.

$$|f(x) - f(a)| < f(a)/2$$

$$-f(a)/2 < f(x) - f(a) < f(a)/2$$

$$f(a)/2 < f(x) < 3f(a)/2$$

Letting $b = a - \delta$ and $c = a + \delta$, $f(x) > 0$ for all x in (b,c).

Chapter 16
Indeterminate Forms

CHAPTER 16, SECTION 1 (pp. 736-737)

1. $\lim_{x\to 1} \frac{x^2 - 1}{x - 1} = \lim_{x\to 1} \frac{2x}{1} = 2.$

5. $\lim_{x\to a} \frac{\sqrt[3]{x} - \sqrt[3]{a}}{x - a} = \lim_{x\to a} \frac{1/3x^{2/3}}{1} = \frac{1}{3a^{2/3}}.$

9. $\lim_{x\to 1^-} \frac{\ln x}{\sqrt{1 - x}} = \lim_{x\to 1^-} \frac{1/x}{-1/2\sqrt{1 - x}} = \lim_{x\to 1^-} - \frac{2\sqrt{1 - x}}{x} = 0.$

13. $\lim_{x\to +\infty} \frac{\ln x}{x} = \lim_{x\to +\infty} \frac{1/x}{1} = 0.$

17. $$\begin{aligned}\lim_{x\to 0} \frac{\tan x - \sin x}{x^3} &= \lim_{x\to 0} \frac{\sec^2 x - \cos x}{3x^2} \\ &= \lim_{x\to 0} \frac{2\sec^2 x \tan x + \sin x}{6x} \\ &= \lim_{x\to 0} \frac{2\sec^4 x + 4\sec^2 x \tan^2 x + \cos x}{6} \\ &= \frac{2 + 0 + 1}{6} = \frac{1}{2}.\end{aligned}$$

21. $$\begin{aligned}\lim_{x\to 0} \frac{\ln(\tan x)}{\ln(\tan 2x)} &= \lim_{x\to 0} \frac{(\sec^2 x)/\tan x}{(2\sec^2 2x)/\tan 2x} \\ &= \lim_{x\to 0} \frac{\sec^2 x \tan 2x}{2\sec^2 2x \tan x} \\ &= \lim_{x\to 0} \frac{2\sec^2 x \sec^2 2x + 2\sec^2 x \tan x \tan 2x}{8\sec^2 2x \tan 2x \tan x + 2\sec^2 2x \sec^2 x} \\ &= \frac{2}{2} = 1.\end{aligned}$$

25. $\lim_{x\to 0} \frac{\sin x - x \cos x}{x - \sin x} = \lim_{x\to 0} \frac{x \sin x}{1 - \cos x}$

$= \lim_{x\to 0} \frac{x \cos x + \sin x}{\sin x}$

$= \lim_{x\to 0} \frac{-x \sin x + 2 \cos x}{\cos x} = 2.$

29. $\lim_{x\to 0^+} \frac{e^{-1/x}}{x} = \lim_{u\to +\infty} \frac{e^{-u}}{1/u} = \lim_{u\to +\infty} \frac{u}{e^u} = \lim_{u\to +\infty} \frac{1}{e^u} = 0.$

33. $\int_1^{+\infty} \frac{\ln x}{x^2}\, dx = \lim_{k\to +\infty} \int_1^k \frac{\ln x}{x^2}\, dx$

$u = \ln x \qquad v' = 1/x^2$

$u' = 1/x \qquad v = -1/x$

$= \lim_{k\to +\infty} \left[-\frac{\ln x}{x}\Big|_1^k + \int_1^k \frac{1}{x^2}\, dx \right]$

$= \lim_{k\to +\infty} -\frac{1 + \ln x}{x}\Big|_1^k = -\lim_{k\to +\infty} (1 - \frac{1 + \ln k}{k})$

$= 1 - \lim_{k\to +\infty} \frac{1/k}{1} = 1.$

CHAPTER 16, SECTION 2 (pp. 741-742)

1. $\lim_{x\to 0^+} x^2 \ln x = \lim_{x\to 0^+} \frac{\ln x}{1/x^2} = \lim_{x\to 0} \frac{1/x}{-2/x^3} = \lim_{x\to 0} -\frac{x^2}{2} = 0.$

5. $\lim_{x\to\pi/2^-} \sec x \cos 5x = \lim_{x\to\pi/2^-} \frac{\cos 5x}{\cos x}$

$$= \lim_{x\to\pi/2^-} \frac{-5 \sin 5x}{-\sin x} = \frac{-5}{-1} = 5.$$

9. $\lim_{x\to+\infty} e^{-x}\ln x = \lim_{x\to+\infty} \frac{\ln x}{e^x} = \lim_{x\to+\infty} \frac{1/x}{e^x} = 0.$

13. $\lim_{x\to+\infty} x \ln(1 + \frac{a}{x}) = \lim_{x\to+\infty} \frac{\ln(1 + a/x)}{1/x} = \lim_{x\to+\infty} \frac{\frac{-a/x^2}{1 + a/x}}{-1/x^2}$

$$= \lim_{x\to+\infty} \frac{a}{1 + a/x} = a.$$

17. $\lim_{x\to 1^+} (\frac{1}{\ln x} - \frac{1}{x - 1}) = \lim_{x\to 1^+} \frac{x - 1 - \ln x}{(x - 1)\ln x}$

$$= \lim_{x\to 1^+} \frac{1 - 1/x}{(x - 1)/x + \ln x}$$

$$= \lim_{x\to 1^+} \frac{1/x^2}{1/x^2 + 1/x} = \lim_{x\to 1^+} \frac{1}{1 + x} = \frac{1}{2}.$$

21. $\lim_{x\to 0} \left(\frac{1}{x(1 + x)} - \frac{\ln(1 + x)}{x^2}\right)$

$$\lim_{x\to 0} \frac{\ln(1 + x)}{x^2} = \lim_{x\to 0} \frac{1/(1 + x)}{2x} = \begin{cases} +\infty & \text{if } x > 0 \\ -\infty & \text{if } x < 0 \end{cases},$$

$$\lim_{x\to 0} \left(\frac{1}{x(1 + x)} - \frac{\ln(1 + x)}{x^2}\right) = \lim_{x\to 0} \frac{x - (1 + x)\ln(1 + x)}{x^2(1 + x)},$$

$$= \lim_{x\to 0} \frac{-\ln(1 + x)}{2x + 3x^2}$$

$$= \lim_{x\to 0} \frac{-1/(1 + x)}{2 + 6x} = -\frac{1}{2}.$$

25. $\lim_{x\to 0} x \ln x = \lim_{u\to+\infty} \frac{1}{u} \ln \frac{1}{u} = \lim_{u\to+\infty} \frac{-\ln u}{u} = \lim_{u\to+\infty} \frac{-1/u}{1} = 0.$

29. $$\int_0^{+\infty} \frac{dx}{x(x+1)} = \lim_{\varepsilon\to 0^+} \int_\varepsilon^1 \frac{dx}{x(x+1)} + \lim_{k\to+\infty} \int_1^k \frac{dx}{x(x+1)}$$

$$= \lim_{\varepsilon\to 0^+} \int_\varepsilon^1 \left(\frac{1}{x} - \frac{1}{x+1}\right) dx + \lim_{k\to+\infty} \int_1^k \left(\frac{1}{x} - \frac{1}{x+1}\right) dx$$

$$= \lim_{\varepsilon\to 0^+} \left| \ln \frac{x}{x+1} \right| \Big|_\varepsilon^1 + \lim_{k\to+\infty} \ln \left| \frac{x}{x+1} \right| \Big|_1^k$$

$$= \lim_{\varepsilon\to 0^+} \left(\ln \frac{1}{2} - \ln \frac{\varepsilon}{\varepsilon+1}\right) + \lim_{k\to+\infty} \left(\ln \frac{k}{k+1} - \ln\right.$$

$$= \ln \frac{1}{2} + \infty - \ln \frac{1}{2} = +\infty.$$

CHAPTER 16, SECTION 3 (pp. 745-746)

1. $\lim_{x\to 0^+} x^2 \ln x = \lim_{x\to 0^+} \frac{\ln x}{1/x^2} = \lim_{x\to 0^+} \frac{1/x}{-2/x^3} = \lim_{x\to 0^+} -\frac{x^2}{2} = 0,$

$\lim_{x\to 0^+} x^{x^2} = e^0 = 1.$

5. $$\lim_{x\to+\infty} x \ln\left(1 + \frac{1}{x}\right) = \lim_{x\to+\infty} \frac{\ln(1 + 1/x)}{1/x}$$

$$= \lim_{x\to+\infty} \frac{(-1/x^2)/(1 + 1/x)}{-1/x^2} = \lim_{x\to+\infty} \frac{1}{1 + 1/x} =$$

$\lim_{x\to+\infty} \left(1 + \frac{1}{x}\right)^x = e^1 = e.$

9. $\lim_{x\to\pi/2^-} \cos x \ln \tan x = \lim_{x\to\pi/2^-} \frac{\ln \tan x}{\sec x} = \lim_{x\to\pi/2^-} \frac{\sec^2 x/\tan x}{\sec x \tan x}$

$= \lim_{x\to\pi/2^-} \frac{\sec x}{\tan^2 x} = \lim_{x\to\pi/2^-} \frac{1/\cos x}{\sin^2 x/\cos^2 x}$

$= \lim_{x\to\pi/2^-} \frac{\cos x}{\sin^2 x} = 0,$

$\lim_{x\to\pi/2^-} (\tan x)^{\cos x} = e^0 = 1.$

13. $\lim_{x\to 1} \frac{1}{1 - x} \ln x = \lim_{x\to 1} \frac{\ln x}{1 - x} = \lim_{x\to 1} \frac{1/x}{-1} = -1,$

$\lim_{x\to 1} x^{1/(1-x)} = e^{-1} = \frac{1}{e}.$

17. $\lim_{x\to 0} \frac{1}{x} \ln(1 - x) = \lim_{x\to 0} \frac{\ln(1 - x)}{x} = \lim_{x\to 0} \frac{-1/(1 - x)}{1} = -1,$

$\lim_{x\to 0} (1 - x)^{1/x} = e^{-1} = \frac{1}{e}.$

21. $\lim_{x\to 0^+} x \ln(-\ln x) = \lim_{x\to 0^+} \frac{\ln(-\ln x)}{1/x} = \lim_{x\to 0^+} \frac{1/(x \ln x)}{-1/x^2}$

$= \lim_{x\to 0^+} \frac{-x}{\ln x} = 0,$

$\lim_{x\to 0^+} (-\ln x)^x = e^0 = 1.$

25. $\lim_{x\to 0} \frac{b}{x} \ln(1 + ax) = b \lim_{x\to 0} \frac{\ln(1 + ax)}{x} = b \lim_{x\to 0} \frac{a}{1 + ax} = ab,$

$\lim_{x\to 0} (1 + ax)^{b/x} = e^{ab}.$

29. $\lim_{x\to 0^+} x^x = L$

$$\ln L = \lim_{x\to 0^+} \ln x^x = \lim_{x\to 0^+} x \ln x = \lim_{x\to 0^+} \frac{\ln x}{1/x}$$

$$= \lim_{x\to 0^+} \frac{1/x}{-1/x^2} = \lim_{x\to 0^+} (-x) = 0,$$

$L = e^0 = 1$

$\lim_{x\to +\infty} x^x = +\infty$

$\ln y = x \ln x$

$\frac{y'}{y} = x \cdot \frac{1}{x} + \ln x = 1 + \ln x$

$y' = y(1 + \ln x) = 0$

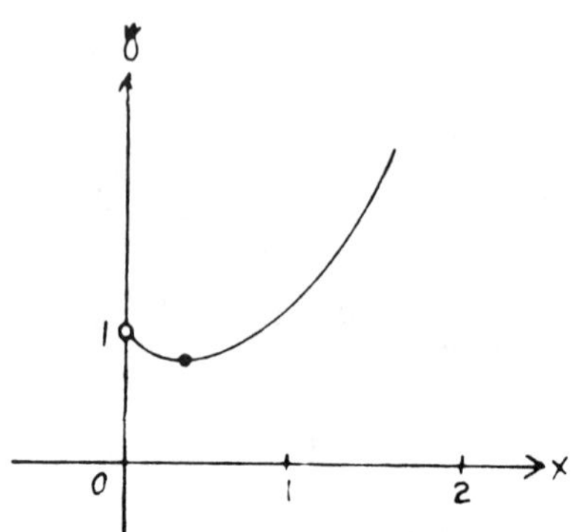

$\ln x = -1$

$x = e^{-1} = \frac{1}{e}$

$(1/e,\ e^{-1/e})$.

CHAPTER 16, REVIEW (p. 746)

1. $\lim_{x\to 0} \frac{\sin x}{1 + \cos x} = \frac{0}{2} = 0.$

5. $\lim_{x\to 0} (1 - e^x)\cot 3x = \lim_{x\to 0} \frac{1 - e^x}{\tan 3x} = \lim_{x\to 0} \frac{-e^x}{3 \sec^2 3x} = -\frac{1}{3}.$

9. $\lim_{x\to 0} \frac{e^x - e^{-x}}{\cot 2x} = 0.$

13. $\lim_{x\to 0} (e^x - e^{-x})^x = y$,

$$\ln y = \lim_{x\to 0} \ln(e^x - e^{-x})^x = \lim_{x\to 0} x \ln(e^x - e^{-x})$$

$$= \lim_{x\to 0} \frac{\ln(e^x - e^{-x})}{1/x} = \lim_{x\to 0} \frac{\frac{e^x + e^{-x}}{e^x - e^{-x}}}{-1/x^2} = \lim_{x\to 0} \frac{x^2(e^x + e^{-x})}{-(e^x - e^{-x})}$$

$$= \lim_{x\to 0} \frac{x^2(e^x - e^{-x}) + 2x(e^x + e^{-x})}{-e^{-x} - e^x} = \frac{0}{-2} = 0$$

$y = e^0 = 1$.

17. $$\lim_{x\to 0} \sin ax \ln ax = \lim_{x\to 0} \frac{\ln ax}{\csc ax} = \lim_{x\to 0} \frac{1/x}{-a \csc ax \cot ax}$$

$$= \lim_{x\to 0} \frac{1/x}{-a \cos ax/\sin^2 ax} = \lim_{x\to 0} \frac{\sin^2 ax}{-ax \cos ax}$$

$$= \lim_{x\to 0} \frac{2a \sin ax \cos ax}{a^2 x \sin ax - a \cos ax} = 0.$$

21. $\lim_{x\to 1^+} (\frac{1}{x - 1})^{\sin \pi x} = y$,

$$\ln y = \lim_{x\to 1^+} \ln(\frac{1}{x - 1})^{\sin \pi x} = \lim_{x\to 1^+} [-\sin \pi x \ln(x - 1)]$$

$$= \lim_{x\to 1^+} \frac{-\ln(x - 1)}{\csc \pi x} = \lim_{x\to 1^+} \frac{-\frac{1}{x - 1}}{-\pi \csc \pi x \cot \pi x}$$

$$= \lim_{x\to 1^+} \frac{\frac{1}{x - 1}}{\frac{\pi \cos \pi x}{\sin^2 \pi x}} = \lim_{x\to 1^+} \frac{\sin^2 \pi x}{\pi(x - 1)\cos \pi x}$$

$$= \lim_{x\to 1^+} \frac{2\pi \sin \pi x \cos \pi x}{-\pi^2(x - 1)\sin \pi x + \pi \cos \pi x} = \frac{0}{-\pi} = 0,$$

$y = e^0 = 1$.

Chapter 17
Infinite Series

CHAPTER 17, SECTION 1 (pp. 757-759)

1. 1, 4, 9, 16.

5. $\frac{1}{\ln 2}, -\frac{1}{\ln 3}, \frac{1}{\ln 4}, -\frac{1}{\ln 5}$.

9. $s_n = \frac{n}{n+1}$, $n = 1,2,3,\cdots$.

13. $s_n = (2n - 1)(2n + 1)$, $n = 1,2,3,\cdots$.

17. $\lim_{n\to+\infty} \frac{1}{n} = 0$; converges to 0.

21. $\lim_{n\to+\infty} \frac{n^2 - 1}{n^2} = \lim_{n\to+\infty} (1 - \frac{1}{n^2}) = 1$; converges to 1.

25. $\lim_{n\to+\infty} \frac{n}{3^n} = \lim_{n\to+\infty} \frac{1}{3^n \ln 3} = 0$; converges to 0.

29. Since $\{s_n\}$ is bounded, there is a number k such that $s_n \geq k$ for every integer n. Let $S = \{x | x \leq s_n$ for every integer $n\}$. This set is not empty since k is in it. Let u be the largest number in S.

Let $\varepsilon > 0$, and assume that

$$s_n \geq u + \varepsilon$$

for every positive integer n. Then $u + \varepsilon$ is in S. But $u + \varepsilon > u$, contradicting the statement that u is the largest number in S. Thus, the assumption is wrong, and there is a number N such that $s_N < u + \varepsilon$. If $n > N$, then $s_n < s_N < u + \varepsilon$ and $s_n > u$. Thus, there is a number N such that, if $n > N$, $|s_n - u| < \varepsilon$; and $\{s_n\}$ converges to u.

CHAPTER 17, SECTION 2 (pp. 768-770)

1. $\frac{1}{1} + \frac{1}{2^2} + \frac{1}{3^2} + \frac{1}{4^2} + \cdots = 1 + \frac{1}{4} + \frac{1}{9} + \frac{1}{16} + \cdots.$

5. $\frac{-1}{1!} + \frac{1}{2!} - \frac{1}{3!} + \frac{1}{4!} - \cdots = -1 + \frac{1}{2} - \frac{1}{6} + \frac{1}{24} - \cdots.$

9. $\sum_{n=1}^{\infty} 2^{n-1}$

13. $0.2222\cdots = \frac{2}{10} + \frac{2}{100} + \frac{2}{1000} + \cdots = \frac{2/10}{1 - 1/10} = \frac{2}{9}.$

Other method: let $N = 0.2222\cdots$, then $10N = 2.2222\cdots$ and $10N - N = 2.0000\cdots$, or $9N = 2$, so $N = \frac{2}{9}$.

17. $0.9999\cdots = \frac{9}{10} + \frac{9}{100} + \frac{9}{1000} + \cdots = \frac{9/10}{1 - 1/10} = 1.$

Or: $10N - N = 9.999\cdots - 0.999\cdots$, $9N = 9$, $N = 1$.

21. $a = \frac{1}{4}$, $r = \frac{1}{4}$; converges. $s = \frac{a}{1 - r} = \frac{1/4}{1 - 1/4} = \frac{1}{3}.$

25. $\lim_{n\to+\infty} \frac{n}{n + 1} = 1 \neq 0$; diverges.

29. $\frac{2n - 1}{[(n - 1)^2 + 1](n^2 + 1)} = \frac{1}{(n - 1)^2 + 1} - \frac{1}{n^2 + 1},$

$s_n = 1 - \frac{1}{n^2 + 1} \to 1$. Converges to 1.

33. The series $\frac{1}{2} + \frac{1}{2^2} + \frac{1}{2^3} + \cdots$ converges to 1 (see Problem 32).

The series $\frac{1}{3} + \frac{1}{3^2} + \frac{1}{3^3} + \cdots$ converges to $\frac{1}{2}$ (see Problem 32).

By Theorem 6, the original series converges to $1 - \frac{1}{2} = \frac{1}{2}.$

37. Since $\sum_{n=N}^{\infty} a_n$ converges, $\lim_{n\to+\infty} t_n$ exists, where $t_n = a_N + a_{N+1} + \cdots + a_{N+n-1}$. The k-th partial sum of $\sum_{n=1}^{\infty} a_n$ is $s_k = a_1 + a_2 + \cdots + a_k$. Letting $k = N + n - 1$ $(k \geq N - 1)$, $s_k = s_{N+n-1} = a_1 + a_2 + \cdots + a_{N-1} + t_n$. Since $\lim_{n\to+\infty} t_n$ exists, $\lim_{n\to+\infty} s_{N+n-1}$ exists and $\lim_{k\to+\infty} s_k$ exists. Thus, $\sum_{n=1}^{\infty} a_n$ converges.

CHAPTER 17, SECTION 3 (pp. 778-779)

1. Converges, since it is a p-series with $p = 2 > 1$.

5. $\sum_{n=1}^{\infty} \frac{\ln(n + 2)}{n + 2} = \sum_{n=3}^{\infty} \frac{\ln n}{n}$, $\lim_{n\to+\infty} \frac{(\ln n)/n}{1/n} = \lim_{n\to+\infty} \ln n = +\infty$. $\sum \frac{\ln n}{n}$ is of a greater order of magnitude than $\sum \frac{1}{n}$. Since $\sum \frac{1}{n}$ diverges, the given series diverges.

9. $\lim_{n\to+\infty} \frac{n/(n + 1)2^n}{1/2^n} = \lim_{n\to+\infty} \frac{n}{n + 1} = 1$. $\sum \frac{n}{(n + 1)2^n}$ and $\sum \frac{1}{2^n}$ are of the same order of magnitude. Since $\sum \frac{1}{2^n}$ converges (geometric series, $r = \frac{1}{2} < 1$), the given series converges.

13. $\lim_{n\to+\infty} \frac{2n/(n + 1)(n + 2)}{1/n} = \lim_{n\to+\infty} \frac{2}{(1 + 1/n)(1 + 2/n)} = 2$. Hence, $\sum \frac{2n}{(n + 1)(n + 2)}$ and $\sum \frac{1}{n}$ are of the same order of magnitude. Since $\sum \frac{1}{n}$ diverges (harmonic series), the given series diverges.

17. $\lim_{n\to+\infty} \frac{n!}{n^2} = \lim_{n\to+\infty} \frac{n-1}{n}(n-2)! = +\infty$. Diverges.

21. $\lim_{n\to+\infty} \frac{1/n^n}{1/n^2} = \lim_{n\to+\infty} \frac{1}{n^{n-2}} = 0$.

Hence, $\sum \frac{1}{n^n}$ is of a lesser order of magnitude than $\sum \frac{1}{n^2}$.

Since $\sum \frac{1}{n^2}$ converges (p-series, $p = 2 > 1$), the given series converges.

25. $\lim_{n\to+\infty} \frac{(n+1)2^{n+1}/3^n}{(\sqrt{2})^n 2^n/3^n} = \lim_{n\to+\infty} \frac{2(n+1)}{(\sqrt{2})^n} = \lim_{n\to+\infty} \frac{2}{(\sqrt{2})^n \ln\sqrt{2}} = 0$.

Thus, $\sum \frac{(n+1)2^{n+1}}{3^n}$ is of a lesser order of magnitude than $\sum \frac{(\sqrt{2})^n 2^n}{3^n} = \sum (\frac{2\sqrt{2}}{3})^n$. Since $\sum (\frac{2\sqrt{2}}{3})^n$ converges (geometric series, $r = \frac{2\sqrt{2}}{3} < 1$), the given series converges.

29. Let $\sum a_n = \sum (-n)$ and $\sum b_n = \sum \frac{1}{n^2}$. $\sum b_n$ converges and $a_n < b_n$ for all n, but $\sum a_n$ diverges.

33. Suppose $\sum a_n$ is of a greater order of magnitude than $\sum b_n$. Then

$$\lim_{n\to+\infty} \frac{a_n}{b_n} = +\infty,$$

or

$$\lim_{n\to+\infty} \frac{b_n}{a_n} = 0.$$

If $\sum a_n$ converges, then the second part of this theorem assures us that $\sum b_n$ also converges. Since this is a contradiction, $\sum a_n$ must diverge.

CHAPTER 17, SECTION 4 (pp. 786-787)

1. $\lim_{n\to+\infty} \frac{n+2}{(n+1)2^{n+1}} \frac{n \cdot 2^n}{n+1} = \lim_{n\to+\infty} \frac{n(n+2)}{2(n+1)^2} = \frac{1}{2}$; converges.

5. $\lim_{n\to+\infty} \frac{3^{n+1}}{(n+1)!} \frac{n!}{3^n} = \lim_{n\to+\infty} \frac{3}{n+1} = 0$; converges.

9. $\lim_{n\to+\infty} \frac{n(n+2)}{(n+1)(n+3)} = 1$; diverges.

13. $\int_1^\infty \frac{dx}{2x+3} = \lim_{n\to+\infty} \int_1^n \frac{dx}{2x+3} = \lim_{n\to+\infty} \frac{1}{2} \ln(2x+3)\Big|_1^n$

$= \lim_{n\to+\infty} [\frac{1}{2} \ln(2n+3) - \frac{1}{2} \ln 5] = +\infty$; diverges.

17. $\int_2^\infty \frac{dx}{x \ln x} = \lim_{n\to+\infty} \int_2^n \frac{dx}{x \ln x} = \lim_{n\to+\infty} \ln \ln x\Big|_2^n$

$= \lim_{n\to+\infty} (\ln \ln n - \ln \ln 2) = +\infty$; diverges.

21. $\int_1^\infty \ln \frac{x+2}{x}\, dx$

$= \lim_{k\to+\infty} \int_1^k [\ln(x+2) - \ln x]dx$

$= \lim_{k\to+\infty} [(x+2)\ln(x+2) - (x+2) - x \ln x + x]\Big|_1^k$

$= \lim_{k\to+\infty} [(k+2)\ln(k+2) - k \ln k - 2] - [3 \ln 3 - 2]$

$= \lim_{k\to+\infty} [(\ln \frac{(k+2)^{k+2}}{k^k} - 2) - (3 \ln 3 - 2)]$

$= \lim_{k\to+\infty} \ln(k+2)^2(1 + \frac{2}{k})^k - 3 \ln 3$

$= \lim_{k\to+\infty} \ln(k+2)^2[(1 + \frac{2}{k})^{k/2}]^2 - 3 \ln 3 = +\infty$; diverges.

25. $\lim_{n\to+\infty} \frac{1\cdot 3\cdot 5 \cdots (2n+1)}{2\cdot 5\cdot 8 \cdots (3n+2)} \frac{2\cdot 5\cdot 8 \cdots (3n-1)}{1\cdot 3\cdot 5 \cdots (2n-1)}$

$= \lim_{n\to+\infty} \frac{2n+1}{3n+2} = \frac{2}{3}$; converges.

29. $\sum_{n=1}^{\infty} (\sqrt{n^2+1} - n) = \sum_{n=1}^{\infty} \frac{1}{\sqrt{n^2+1} + n}$. This is of the same order of magnitude as $\sum \frac{1}{n}$ since

$$\lim_{n\to+\infty} \frac{1/n}{1/(\sqrt{n^2+1}+n)} = \lim_{n\to+\infty} \frac{\sqrt{n^2+1}+n}{n}$$

$$= \lim_{n\to+\infty} (\sqrt{1+1/n^2} + 1) = 2.$$

Since $\sum \frac{1}{n}$ diverges, this series diverges.

33. $S - s_N = \sum_1^{\infty} a_n - \sum_N a_n = a_{N+1} + a_{N+2} + a_{N+3} + \cdots$

$$= a_N[\frac{a_{N+1}}{a_N} + (\frac{a_{N+2}}{a_{N+1}})(\frac{a_{N+1}}{a_N}) + (\frac{a_{N+3}}{a_{N+2}})(\frac{a_{N+2}}{a_{N+1}})(\frac{a_{N+1}}{a_N}) + \cdots]$$

$$\le a_N(u + u^2 + u^3 + \cdots) = (a_N)(\frac{u}{1-u})$$

because $u + u^2 + u^3 + \cdots$ is a convergent geometric series. Since $n \ge N$, $s_n \ge s_N$, $S - s_n \le S - s_N \le \frac{a_N u}{1-u}$.

CHAPTER 17, SECTION 5 (pp. 795-797)

1. $\lim_{n\to+\infty} \frac{1}{2n} = 0$. $\frac{1}{2(n+1)} < \frac{1}{2n}$; converges. $\sum \frac{1}{2n}$ diverges.

   The original series converges conditionally.

5. $\lim_{n\to+\infty} \frac{1}{\sqrt{n}} = 0$, $\frac{1}{\sqrt{n+1}} < \frac{1}{\sqrt{n}}$; converges. $\sum \frac{1}{\sqrt{n}}$ diverges.

   The original series converges conditionally.

9. $\sum \frac{1}{n!}$ converges (by the ratio test).

The original series converges absolutely.

13. $\lim_{n\to+\infty} \frac{n!}{n^2 2^n} = \frac{n(n-1)}{n \cdot n} \cdot \frac{(n-2)\cdots 2 \cdot 1}{2 \cdots 2 \cdot 2} \cdot \frac{1}{4}$

$> \frac{n(n-1)}{8n^2} \cdot \frac{n-2}{2} \to +\infty$ as $n \to \infty$; diverges.

17. $\lim_{n\to+\infty} \frac{1}{n \ln n} = 0$, $\frac{1}{(n+1)\ln(n+1)} < \frac{1}{n \ln n}$; converges.

$\sum \frac{1}{n \ln n}$ diverges (by the integral test).

The original series converges conditionally.

21. $\sum_{n=1}^{\infty} \frac{(-1)^{n-1}}{n^2} \approx \frac{1}{1} - \frac{1}{4} + \frac{1}{9} \approx 0.9$.

25. Consider the series

$(-1)^{n+2} b_{n+1} + (-1)^{n+3} b_{n+2} + (-1)^{n+4} b_{n+3} + \cdots + (-1)^{n+m+1} b_{n+m}$

Let $t_m = (-1)^{n+2} b_{n+1} + (-1)^{n+3} b_{n+2} + \cdots + (-1)^{n+m+1} b_{n+m}$.

If s_n represents the n-th partial sum of the original series $\sum a_i$, then

$$s_{n+m} = s_n + t_m \quad \text{or} \quad t_m = s_{n+m} - s_n$$

and

$$\lim_{m\to+\infty} t_m = s - s_n.$$

Suppose $(-1)^{n+2} b_{n+1} > 0$; then, by the argument of p. 726, $0 < t_m < b_{n+1}$. Furthermore, $t_m < b_{n+1} - b_{n+2} + b_{n+3} < b_{n+1}$ for all $m \geq 3$. Thus,

$$0 < \lim_{m\to+\infty} t_m < b_{n+1}$$

and $|s - s_n| < |b_{n+1}|$. Now suppose $(-1)^{n+2}b_{n+1} < 0$. By a similar argument,

$$-b_{n+1} < \lim_{m\to+\infty} t_m < 0$$

and $|s - s_n| < |b_{n+1}|$.

29. Take positive terms of the given series (beginning with 1 and going in descending order) until the partial sum is greater than 1. Then take negative terms until the partial sum is less than 1. Then take positive terms until the partial sum exceeds 2 and negative terms until the partial sum is less than 2. Repeat using 3, 4, etc. The resulting series diverges to $+\infty$.

CHAPTER 17, SECTION 6 (pp. 802-804)

1. $\lim_{n\to+\infty} \left|\frac{(n + 1)x^{n+1}}{nx^n}\right| = \lim_{n\to+\infty} \left|\frac{(n + 1)x}{n}\right| = |x| < 1;\ -1 < x < 1.$

   $x = 1$: $\sum n$ diverges. $x = -1$: $\sum (-1)^n n$ diverges.

   Converges for $-1 < x < 1$.

5. $\lim_{n\to+\infty} \left|\frac{(n + 1)^2 x^{n+1}}{2^{n+1}} \frac{2^n}{n^2 x^n}\right| = \lim_{n\to+\infty} \left|\frac{(n + 1)^2 x}{2n^2}\right| = \left|\frac{x}{2}\right| < 1;$

   $-2 < x < 2.$

   $x = 2$: $\sum n^2$ diverges. $x = -2$: $\sum (-1)^n n^2$ diverges.

   Converges for $-2 < x < 2$.

9. $\lim_{n\to+\infty} \left|\frac{(n + 1)^2 2^{n+1} x^{n+1}}{n^2 2^n x^n}\right| = \lim_{n\to+\infty} \left|\frac{2(n + 1)^2 x}{n^2}\right| = |2x| < 1;$

   $-\frac{1}{2} < x < \frac{1}{2}.$

   $x = \frac{1}{2}$: $\sum n^2$ diverges. $x = -\frac{1}{2}$: $\sum (-1)n^2$ diverges.

   Converges for $-\frac{1}{2} < x < \frac{1}{2}$.

13. $\lim_{n\to+\infty} \left|\frac{(x+3)^{n+1}}{2^{n+1}} \frac{2^n}{(x+3)^n}\right| = \lim_{n\to+\infty} \left|\frac{x+3}{2}\right| = \left|\frac{x+3}{2}\right| < 1;$

$-1 < \frac{x+3}{2} < 1, \quad -2 < x+3 < 2, \quad -5 < x < -1.$

$x = -1$: $\sum 1$ diverges. $x = -5$: $\sum (-1)^n$ diverges.

Converges for $-5 < x < -1$.

17. $\lim_{n\to+\infty} \left|\frac{(n+1)(2x-3)^{n+1}}{(n+2)^2} \frac{(n+1)^2}{n(2x-3)^n}\right| = \lim_{n\to+\infty} \left|\frac{(n+1)^3(2x -}{n(n+2)^2}\right.$

$= |2x-3| < 1;$

$-1 < 2x - 3 < 1, \quad 2 < 2x < 4, \quad 1 < x < 2.$

$x = 2$: $\sum \frac{n}{(n+1)^2}$ diverges. $x = 1$: $\sum \frac{(-1)^n n}{(n+1)^2}$ converges

Converges for $1 \leq x < 2$.

21. $\lim_{n\to+\infty} \left|\frac{(x+3)^{n+1}}{(n+1)(n+2)} \frac{n(n+1)}{(x+3)^n}\right| = \lim_{n\to+\infty} \left|\frac{n(x+3)}{n+2}\right| = |x+3|$

$-1 < x + 3 < 1, \quad -4 < x < -2.$

$x = -2$: $\sum \frac{1}{n(n+1)}$ converges. $x = -4$: $\sum \frac{(-1)^n}{n(n+1)}$ conver

Converges for $-4 \leq x \leq -2$.

25. $\lim_{n\to+\infty} \left|\frac{2^{n+1}\sin^{n+1}x}{2^n \sin^n x}\right| = \lim_{n\to+\infty} 2|\sin x| = 2|\sin x| < 1;$

$|\sin x| < \frac{1}{2}, \quad -\frac{\pi}{6} + n\pi < x < \frac{\pi}{6} + n\pi.$

$x = \frac{\pi}{6}$: $\sum_{n=0}^{\infty} 2^n \frac{1}{2^n} = \sum_{n=0}^{\infty} 1$ diverges.

$x = -\frac{\pi}{6}$: $\sum_{n=0}^{\infty} 2^n \frac{(-1)^n}{2^n} = \sum_{n=1}^{\infty} (-1)^n$ diverges.

Converges for $-\frac{\pi}{6} + n\pi < x < \frac{\pi}{6} + n\pi$.

29.

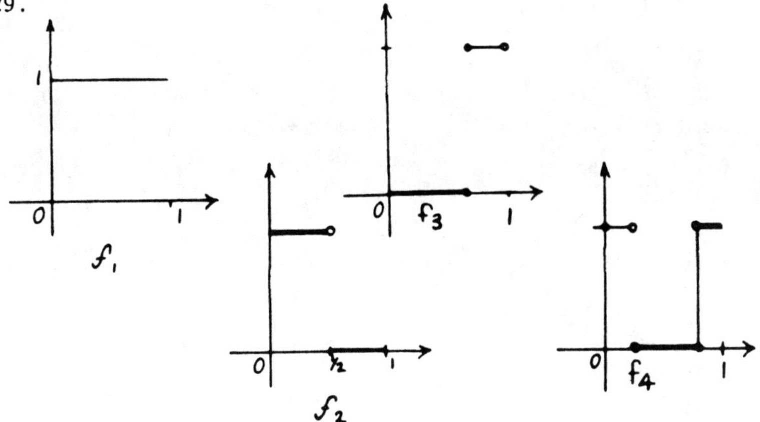

CHAPTER 17, SECTION 7 (pp. 810-811)

1. $f(x) = e^x \qquad f(1) = e$

$f'(x) = e^x \qquad f'(1) = e$

$\vdots \qquad\qquad \vdots$

$$e^x = e + e(x - 1) + \frac{e}{2!}(x - 1)^2 + \frac{e}{3!}(x - 1)^3 + \cdots$$

$$= \sum_{n=0}^{\infty} \frac{e}{n!}(x - 1)^n$$

$$\lim_{n\to+\infty} \left|\frac{e(x - 1)^{n+1}}{(n + 1)!} \frac{n!}{e(x - 1)^n}\right| = \lim_{n\to+\infty} \left|\frac{x - 1}{n + 1}\right| = 0.$$

Converges for all x.

5. $f(x) = \sin x \qquad f(\pi/4) = 1/\sqrt{2}$

$f'(x) = \cos x \qquad f'(\pi/4) = 1/\sqrt{2}$

$f''(x) = -\sin x \qquad f''(\pi/4) = -1/\sqrt{2}$

$f'''(x) = -\cos x \qquad f'''(\pi/4) = -1/\sqrt{2}$

$f^{(4)}(x) = \sin x \qquad f^{(4)}(\pi/4) = 1/\sqrt{2}$

$\vdots \qquad \vdots$

$$\sin x = \frac{1}{\sqrt{2}} + \frac{1}{\sqrt{2}}(x - \frac{\pi}{4}) - \frac{1}{\sqrt{2}}\frac{(x - \pi/4)^2}{2!} - \frac{1}{\sqrt{2}}\frac{(x - \pi/4)^3}{3!} + \cdots$$

$$\lim_{n\to+\infty} \left|\frac{(x - \pi/4)^{n+1}}{\sqrt{2}(n + 1)!}\,\frac{\sqrt{2}n!}{(x - \pi/4)^n}\right| = \lim_{n\to+\infty} \left|\frac{x - \pi/4}{n + 1}\right| = 0.$$

Converges for all x.

9. $f(x) = \cosh x \qquad f(0) = 1$

$f'(x) = \sinh x \qquad f'(0) = 0$

$f''(x) = \cosh x \qquad f''(0) = 1$

$\vdots \qquad \vdots$

$$\cosh x = 1 + \frac{x^2}{2!} + \frac{x^4}{4!} + \cdots + \frac{x^{2n}}{(2n)!} + \cdots$$

$$\lim_{n\to+\infty} \left|\frac{x^{2n+2}}{(2n + 2)!}\,\frac{(2n)!}{x^{2n}}\right| = \lim_{n\to+\infty} \left|\frac{x^2}{(2n + 1)(2n + 2)}\right| = 0.$$

Converges for all x.

13. $f(x) = \text{Arctan } x \qquad f(0) = 0$

$f'(x) = \dfrac{1}{x^2 + 1} \qquad f'(0) = 1$

$f''(x) = \dfrac{-2x}{(x^2 + 1)^2} \qquad f''(0) = 0$

$f'''(x) = \dfrac{2(3x^2 - 1)}{(x^2 + 1)^3} \qquad f'''(0) = -2$

$f^{(4)}(x) = \dfrac{-24x(x^2 - 1)}{(x^2 + 1)^4} \qquad f^{(4)}(0) = 0$

$f^{(5)}(x) = \dfrac{24(5x^4 - 10x^2 + 1)}{(x^2 + 1)^5} \qquad f^{(5)}(0) = 24$

$\vdots \qquad \vdots$

$$\text{Arctan } x = x - \frac{2x^3}{3!} + \frac{24x^5}{5!} - \cdots$$

$$= x - \frac{x^3}{3} + \frac{x^5}{5} - \cdots + (-1)^{n+1}\frac{x^{2n-1}}{2n-1} + \cdots$$

$$\lim_{n\to+\infty}\left|\frac{x^{2n+1}}{2n+1}\,\frac{2n-1}{x^{2n-1}}\right| = \lim_{n\to+\infty}\left|\frac{2n-1}{2n+1}x^2\right| = |x^2| < 1;$$

$$-1 < x < 1.$$

$x = 1$: $\sum \frac{(-1)^{n+1}}{2n-1}$ converges. $x = -1$: $\sum \frac{(-1)^n}{2n-1}$ converges.

Converges for $-1 \leq x \leq 1$.

17. $f(x) = x^{-1/2}$ $\quad f(1) = 1$

$f'(x) = -\frac{1}{2}x^{-3/2}$ $\quad f'(1) = -\frac{1}{2}$

$f''(x) = \frac{1\cdot 3}{2^2}x^{-5/2}$ $\quad f''(1) = \frac{1\cdot 3}{2^2}$

$f'''(x) = -\frac{1\cdot 3\cdot 5}{2^3}x^{-7/2}$ $\quad f'''(1) = -\frac{1\cdot 3\cdot 5}{2^3}$

$\vdots$ $\quad\vdots$

$$\frac{1}{\sqrt{x}} = 1 - \frac{1}{2}(x-1) + \frac{1\cdot 3}{2^2}\frac{(x-1)^2}{2!} - \frac{1\cdot 3\cdot 5}{2^3}\frac{(x-1)^3}{3!} + \cdots$$

$$= 1 - \frac{1}{2}(x-1) + \frac{1\cdot 3}{2\cdot 4}(x-1)^2 - \frac{1\cdot 3\cdot 5}{2\cdot 4\cdot 6}(x-1)^3 + \cdots.$$

21. $f(x) = \ln x$ does not have a Maclaurin's series expansion because there is no interval containing 0 in which f and all its derivatives exist.

25. $\frac{1}{1+x^2} = 1 - x^2 + x^4 - x^6 + \cdots$

$$\int (1 - x^2 + x^4 - x^6 + \cdots)dx = C + x - \frac{x^3}{3} + \frac{x^5}{5} - \frac{x^7}{7} + \cdots$$

Arctan 0 = 0; C = 0

$$\text{Arctan } x = x - \frac{x^3}{3} + \frac{x^5}{5} - \frac{x^7}{7} + \cdots.$$

29. $(x + 8)^{1/3} = 2(1 + \frac{x}{8})^{1/3}$

$$= 2[1 + \frac{1}{3}(\frac{x}{8}) + \frac{1/3(-2/3)}{2!}(\frac{x}{8})^2 + \frac{1/3(-2/3)(-5/3)}{3!}(\frac{x}{8})^3 + \cdots]$$

$$= 2(1 + \frac{x}{24} - \frac{x^2}{576} + \frac{5x^3}{41{,}472} + \cdots)$$

$$= 2 + \frac{x}{12} - \frac{x^2}{288} + \frac{5x^3}{20{,}736} + \cdots$$

Converges for $|x| < 8$.

33. Let $y = f(x) = \ln(1 + x)$ and $x = 1 + e^{-2}$. Since

$$\ln(1 + x) = \sum_{n=1}^{\infty} (-1)^{n+1} \frac{x^n}{n},$$

$$\ln(1 + 1 + e^{-2}) = \sum_{n=1}^{\infty} (-1)^{n+1} \frac{(1 + e^{-2})^n}{n} = -2 + \ln(2e^2 + 1)$$

CHAPTER 17, SECTION 8 (pp. 818-819)

1. $e^x = 1 + x + \frac{x^2}{2!} + R_2, \quad e \doteq 1 + 1 + \frac{1}{2} = 2.5.$

$R_2 = \frac{x^3}{3!} e^c$, where $0 < c < 1$, $< \frac{1}{3!} 3 = 0.5$.

5. $6° \doteq .10472, \quad \cos x = 1 - \frac{x^2}{2!} + \frac{x^4}{4!} + R_4,$

$\cos 6° \doteq 1 - \frac{(.10472)^2}{2!} + \frac{(.10472)^4}{4!} = 0.99452.$

$R_4 = \frac{x^5}{5!} \sin c$, where $0 < c < .10472$, $< \frac{(.2)^5}{5!} \doteq 0.000003$.

From Theorem 12, $R_4 < (.10472)^{6/6}! \doteq 0.000000002$.

9. $6° = 0.1047, \quad \tan x = x + \frac{x^3}{3} + R_4,$

$$\tan 6° \doteq .1047 + \frac{.1047^3}{3} = 0.1051.$$

$$R_4 = \frac{x^5}{5!}(16 \sec^6 c + 88 \sec^4 c \tan^2 c + 16 \sec^2 c \tan^4 c),$$

$$\text{where } 0 < c < .1047, < \frac{.1047^5}{5!} \cdot 17 = 0.000002.$$

13. $10° = 0.1745, \quad R_n = \frac{x^{n+1}}{(n+1)!} f^{(n+1)}(c)$, where $0 < c < 0.1745$,

$$< \frac{(0.2)^{n+1}}{(n+1)!} < 0.00005 \text{ when } n = 4.$$

$$\sin x = x - \frac{x^3}{3!} + R_4,$$

$$\sin 10° \doteq 0.1745 - \frac{(0.1745)^3}{3!} = 0.1736.$$

17. $R_n = \frac{(x - \pi/6)^{n+1}}{(n+1)!} f^{(n+1)}(c)$, where $\frac{\pi}{6} < c < 0.61087$,

$$< \frac{(0.1)^{n+1}}{(n+1)!} < 0.00005.$$

When $n = 4$, $\sin x = \frac{1}{2} + \frac{\sqrt{3}}{2}(x - \frac{\pi}{6}) - \frac{1}{2}\frac{(x - \pi/6)^2}{2!}$

$$- \frac{\sqrt{3}}{2}\frac{(x - \pi/6)^3}{3!} + \frac{1}{2}\frac{(x - \pi/6)^4}{4!} + R_4,$$

$$\sin 35° \doteq \frac{1}{2} + \frac{\sqrt{3}}{2}(0.0873) - \frac{1}{2}\frac{(0.0873)^2}{2!}$$

$$- \frac{\sqrt{3}}{2}\frac{(0.0873)^3}{3!} + \frac{1}{2}\frac{(0.0873)^4}{4!}$$

$$= 0.5736.$$

21. $R_n = \frac{(x - 16)^{n+1}}{(n+1)!} f^{(n+1)}(c)$, where $16 < c < 17$,

$$|R_n| < \frac{(1)^{n+1}}{(n+1)!} \cdot \frac{1 \cdot 3 \cdot 5 \cdots (2n-1)}{2^{n+1} 4^{2n+1}} < 0.01.$$

When $n = 1$, $\sqrt{x} = 4 + \frac{1}{8}(x - 16) + R_1$,

$$\sqrt{17} \doteq 4 + \frac{1}{8} = 4.125.$$

CHAPTER 17, SECTION 9 (pp. 823-825)

1. $f(x) = 1 + x + x^2 + x^3 + \cdots,$

$g(x) = f'(x) = 1 + 2x + 3x^2 + 4x^3 + \cdots.$

Both series converge for $-1 < x < 1$.

5. $f(x) = 1 - x^2 + x^4 - x^6 + \cdots$ converges for $-1 < x < 1$,

$$g(x) = \int f(x)\,dx = C + x - \frac{x^3}{3} + \frac{x^5}{5} - \frac{x^7}{7} + \cdots.$$

Since $g(0) = 0$, $C = 0$.

$g(x) = x - \frac{x^3}{3} + \frac{x^5}{5} - \frac{x^7}{7} + \cdots$ converges for $-1 \leq x \leq 1$.

9. $\cos 2x = 1 - \frac{2^2}{2!}x^2 + \frac{2^4}{4!}x^4 - \frac{2^6}{6!}x^6 + \cdots,$

$$\cos^2 x = \frac{1}{2} + \frac{1}{2}\cos 2x = 1 - \frac{2}{2!}x^2 + \frac{2^3}{4!}x^4 - \frac{2^5}{6!}x^6 + \cdots.$$

13. $$\int e^{-x^2}dx = \int \left(1 - x^2 + \frac{x^4}{2!} - \frac{x^6}{3!} + \cdots\right) dx$$

$$= C + x - \frac{x^3}{3} + \frac{x^5}{5\cdot 2!} - \frac{x^7}{7\cdot 3!} + \cdots.$$

17. $$\lim_{x\to 0} \frac{1 - \cos x}{\sin x} = \lim_{x\to 0} \frac{\frac{x^2}{2!} - \frac{x^4}{4!} + \frac{x^6}{6!} - \cdots}{x - \frac{x^3}{3!} + \frac{x^5}{5!} - \frac{x^7}{7!} + \cdots}$$

$$= \lim_{x\to 0} \frac{\frac{x}{2!} - \frac{x^3}{4!} + \frac{x^5}{6!} - \cdots}{1 - \frac{x^2}{3!} + \frac{x^4}{5!} - \frac{x^6}{7!} + \cdots} = 0.$$

21. $\int \sqrt{1 + x^3}\, dx = \int (1 + \frac{1}{2} x^3 - \frac{1}{2^2 \cdot 2!} x^6 + \frac{1 \cdot 3}{2^3 \cdot 3!} x^9$

$- \frac{1 \cdot 3 \cdot 5}{2^4 \cdot 4!} x^{12} + \cdots)\, dx$

$= C + x + \frac{1}{2} \frac{x^4}{4} - \frac{1}{2^2 \cdot 2!} \frac{x^7}{7} + \frac{1 \cdot 3}{2^3 \cdot 3!} \frac{x^{10}}{10}$

$- \frac{1 \cdot 3 \cdot 5}{2^4 \cdot 4!} \frac{x^{13}}{13} + \cdots .$

25. $\lim_{x \to 0} \frac{e^x - 1}{\sin x} = \lim_{x \to 0} \frac{x + \frac{x^2}{2!} + \frac{x^3}{3!} + \frac{x^4}{4!} + \cdots}{x - \frac{x^3}{3!} + \frac{x^5}{5!} - \frac{x^7}{7!} + \cdots}$

$= \lim_{x \to 0} \frac{1 + \frac{x}{2!} + \frac{x^2}{3!} + \frac{x^3}{4!} + \cdots}{1 - \frac{x^2}{3!} + \frac{x^4}{5!} - \frac{x^6}{7!} + \cdots} = 1.$

29. $\sin x = x - \frac{x^3}{3!} + \frac{x^5}{5!} - \cdots,$

$\sin(\cos x) = \cos x - \frac{1}{3!} \cos^3 x + \frac{1}{5!} \cos^5 x - \cdots$

$= \cos x - \frac{1}{3!}(1 - \sin^2 x)\cos x$

$+ \frac{1}{5!}(1 - \sin^2 x)^2 \cos x - \cdots,$

$\int_0^{.1} \sin(\cos x)dx = \int_0^{.1} \cos x\, dx - \frac{1}{3!} \int_0^{.1} (1 - \sin^2 x)\cos x\, dx$

$+ \frac{1}{5!} \int_0^{.1} (1 - 2 \sin^2 x + \sin^4 x)\cos x\, dx - \cdots$

$= (\sin x - \frac{1}{6} \sin x + \frac{1}{18} \sin^3 x + \frac{1}{120} \sin x$

$- \frac{1}{180} \sin^3 x + \frac{1}{600} \sin^5 x + \cdots)\Big|_0^{.1}$

$= 0.09983 - 0.01664 + 0.00005 + 0.00083$

$= 0.084.$

CHAPTER 17, REVIEW (pp. 825-827)

1. $s_n = n \cdot 2^{n-1}, \quad n = 1, 2, 3, \cdots.$

5. $\lim_{n\to+\infty} \frac{n^2}{4n + 5} = \lim_{n\to+\infty} \frac{2n}{4} = +\infty.$ The sequence diverges.

9. $\frac{2}{4n^2 - 1} = \frac{1}{2n - 1} - \frac{1}{2n + 1}, \quad s_n = 1 - \frac{1}{2n + 1} \to 1.$

The series converges to 1.

13. $\lim_{n\to+\infty} \frac{n^2/(4n^2 - 1)^2}{1/n^2} = \lim_{n\to+\infty} \frac{n^4}{(4n^2 - 1)^2} = \lim_{n\to+\infty} \frac{1}{(4 - 1/n^2)^2} = \frac{1}{16}$

Hence, $\sum \frac{n^2}{(4n^2 - 1)^2}$ and $\sum \frac{1}{n^2}$ are of the same order of magnitude. Since $\sum \frac{1}{n^2}$ converges, the given series converges.

17. $\lim_{n\to+\infty} \frac{(n + 1)^3/2^{n+1}}{n^3/2^n} = \lim_{n\to+\infty} \frac{(1 + 1/n)^3}{2} = \frac{1}{2}.$

Thus, the given series converges absolutely.

21. $s_n = -2 + \frac{(n + 1)^2 + 1}{n + 1} \to +\infty.$ Diverges.

25. $\int_2^{\infty} \frac{dx}{x \ln x^4} = \lim_{k\to+\infty} \int_2^k \frac{dx}{4x \ln x} = \lim_{k\to+\infty} \frac{1}{4} \ln \ln x \Big|_2^k$

$= \lim_{k\to+\infty} \frac{1}{4}(\ln \ln k - \ln \ln 2) = +\infty.$ Diverges.

29. Since $\lim_{n\to+\infty} \frac{n}{n + 1} = 1 \neq 0$, the given series diverges.

33. $\lim_{n\to+\infty} \left|\frac{(n+1)!x^{n+1}/(n+1)^4 4^{n+1}}{n!x^n/n^4 4^n}\right| = \lim_{n\to+\infty} \left|\frac{n^4 x}{(n+1)^3 \cdot 4}\right|$

$= +\infty, \quad (x \neq 0).$

Converges for $x = 0$.

37. $f(x) = \frac{1}{1-x}$ $\qquad f(2) = -1$

$f'(x) = \frac{1}{(1-x)^2}$ $\qquad f'(2) = 1$

$f''(x) = \frac{2!}{(1-x)^3}$ $\qquad f''(2) = -2!$

$f'''(x) = \frac{3!}{(1-x)^4}$ $\qquad f'''(2) = 3!$

$$\frac{1}{1-x} = -1 + (x-2) - (x-2)^2 + (x-2)^3 - \cdots + (-1)^{n+1}(x-2)^n + \cdots,$$

$$\lim_{n\to+\infty} \left|\frac{(x-2)^{n+1}}{(x-2)^n}\right| = \lim_{n\to+\infty} |x-2| = |x-2| < 1,$$

$-1 < x - 2 < 1, \quad 1 < x < 3,$

$x = 1$: $\sum (-1)^{n+1}(-1)^n = \sum (-1)$ diverges.

$x = 3$: $\sum (-1)^{n+1}$ diverges. Converges for $1 < x < 3$.

41. $\cos x = 1 - \frac{x^2}{2!} + \frac{x^4}{4!} + R_5$. $12° = 0.2094$.

$$\cos 12° \doteq 1 - \frac{(0.2094)^2}{2} + \frac{(0.2094)^4}{24}$$

$$= 1 - 0.0219 + 0.0000 = 0.9781.$$

$$R_5 = \frac{f^{(6)}(c)}{6!}(0.2094)^6 < \frac{(0.2094)^6}{720} = (1.2)(10^{-7}).$$

45. $\displaystyle\int \frac{dx}{1 + x^4} = \int (1 - x^4 + x^8 - x^{12} + x^{16} - \cdots)\, dx$

$$= C + x - \frac{x^5}{5} + \frac{x^9}{9} - \frac{x^{13}}{13} + \cdots.$$

49. $\displaystyle\int_1^{\infty} \text{Arccot } x\, dx = \lim_{k\to+\infty} \int_1^k \text{Arccot } x\, dx$

$$u = \text{Arccot } x \qquad v' = 1$$

$$u' = \frac{-1}{1 + x^2} \qquad v = x$$

$$= \lim_{k\to+\infty} \{x \text{ Arccot } x\Big|_1^k + \int_1^k \frac{x\, dx}{1 + x^2}\}$$

$$= \lim_{k\to+\infty} \{x \text{ Arccot } x - \frac{1}{2} \ln(1 + x^2)\}\Big|_1^k$$

$$= \lim_{k\to+\infty} \{k \text{ Arccot } k + \frac{1}{2} \ln(1 + k^2) - \frac{\pi}{4} - \frac{1}{2} \ln 2$$

$$= +\infty.$$

Chapter 18
Vectors in the Plane

CHAPTER 18, SECTION 1 (pp. 835-837)

1. $(-2 - 4,\ 1 - 3) = (-6, -2)$.

5. $|v| = \sqrt{3^2 + (-1)^2} = \sqrt{10}$, $\quad \dfrac{v}{|v|} = (\dfrac{3}{\sqrt{10}}, -\dfrac{1}{\sqrt{10}})$.

9. $x = 3 + \frac{1}{4}(7 - 3) = 4$, $\quad y = 4 + \frac{1}{4}(0 - 4) = 3$; $\quad P = (4, 3)$.

13. $3 = x - 1$, $\quad x = 4$; $\quad -1 = y - 4$, $\quad y = 3$; $\quad B = (4, 3)$.

17. $3 = x - 4$, $\quad x = 7$; $\quad -1 = y + 2$, $\quad y = -3$; $\quad B = (7, -3)$.

21. $u + v = (3 + 1,\ -1 + 2)$
$= (4, 1)$.

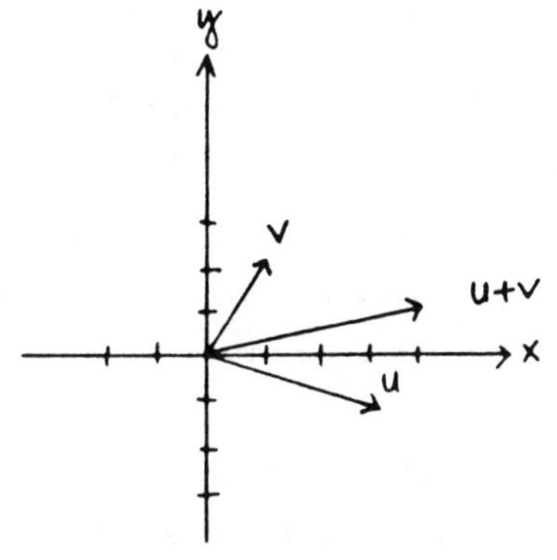

25. $A = (x_1, y_1)$, $\quad B = (x_2, y_2)$

$$\frac{x_1 + x_2}{2} = 4 \qquad \frac{y_1 + y_2}{2} = 1$$

$$x_2 - x_1 = 3 \qquad y_2 - y_1 = 5$$

$$x_1 = \frac{5}{2},\ x_2 = \frac{11}{2} \qquad y_1 = -\frac{3}{2},\ y_2 = \frac{7}{2}$$

$A = (\frac{5}{2}, -\frac{3}{2})$, $\quad B = (\frac{11}{2}, \frac{7}{2})$.

29. $2 = x + \frac{1}{2}(4 - x)$ $\qquad$ $-5 = y + \frac{1}{2}(-3 - y)$

$x = 0$ $\qquad$ $y = -7$

$A = (0, -7)$.

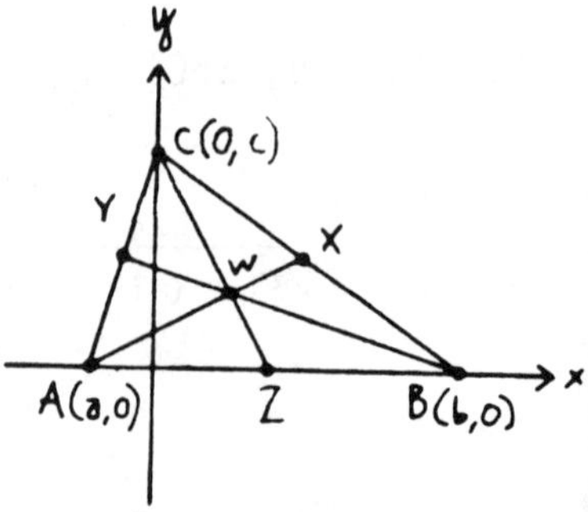

33. $X = (\frac{b}{2}, \frac{c}{2})$, $\quad Y = (\frac{a}{2}, \frac{c}{2})$,

$Z = (\frac{a + b}{2}, 0)$.

$W_1 = \frac{2}{3}$ of the way from A to X,

$W_2 = \frac{2}{3}$ of the way from B to Y,

$W_3 = \frac{2}{3}$ of the way from C to Z.

W_1: $x = a + \frac{2}{3}(\frac{b}{2} - \frac{a}{2}) = \frac{(a + b)}{3}$, $\quad y = 0 + \frac{2}{3}(\frac{c}{2} - 0) = \frac{c}{3}$.

W_2: $x = b + \frac{2}{3}(\frac{a}{2} - b) = \frac{a + b}{3}$, $\quad y = 0 + \frac{2}{3}(\frac{c}{2} - 0) = \frac{c}{3}$.

W_3: $x = 0 + \frac{2}{3}(\frac{a + b}{2} - 0) = \frac{a + b}{3}$, $\quad y = c + \frac{2}{3}(0 - c) = \frac{c}{3}$.

37. (a) Let $\overrightarrow{AB}$ be a representative of u and $\overrightarrow{BC}$ a representative of v. If A, B, and C are non-collinear, let D be a point such that ABCD is a parallelogram. Since AD and BC are equal and parallel, $\overrightarrow{AD}$ is a representative of v. Similarly, $\overrightarrow{DC}$ is a representative of u. Thus, $\overrightarrow{AC}$ is a representative of both u + v and v + u. u + v = v + u.

If A, B, and C are collinear, let D be the point on AC such that AD = BC. Again, $\overrightarrow{AD}$ is a representative of v and $\overrightarrow{DC}$ a representative of u; u + v = v + u.

(b) Let $\overrightarrow{AB}$ be a representative of u, $\overrightarrow{BC}$ a representative of v, and $\overrightarrow{CD}$ a representative of w. Then $\overrightarrow{BD}$ is a representative of v + w and $\overrightarrow{AC}$ a representative of u + v. Thus, $\overrightarrow{AD}$ is a representative of both u + (v + w) and (u + v) + w. u + (v + w) = (u + v) + w.

(c) $(ab)v$ is a vector of length $|ab||v|$; $a(bv)$ is a vector length $|a|(|b||v|) = |ab||v|$. $(ab)v$ has the same direction as v if ab is positive and opposite if ab is negative. bv has the same direction as v if b is positive and opposite if b is negative; furthermore, $a(bv)$ has the same direction as bv if a is positive and opposite if a is negative. Thus, $a(bv)$ has the same direction as v if ab is positive and opposite if a is negative. Thus, $(ab)v = a(bv)$.

(d) $(a + b)v$ is a vector of length $|a + b|\cdot|v|$. av is a vector of length $|a|\cdot|v|$ and bv a vector of length $|b|\cdot|v|$. Since av and bv have either the same direction (if a and b have the same sign) or the opposite direction (if a and b have opposite signs), the length of $av + bv$ is $|a + b|\cdot|v|$. Furthermore, $(a + b)v$ and $av + bv$ both have the same or opposite direction as v, depending upon whether $a + b$ is positive or negative. Thus, $(a + b)v = av + bv$.

(e) Since a representative of the zero vector is a point, any representative of v is a representative of $v + 0$. Thus, $v + 0 = v$.

(f) The length of $0v$ is $|0|\cdot|v| = 0$. Thus, $0v = 0$.

(g) The length of $a0$ is $|a|\cdot|0| = 0$. Thus, $a0 = 0$.

(h) $|av| = |a|\cdot|v|$ follows directly from the definition of av.

(i) If $\overrightarrow{AB}$ is a representative of u and $\overrightarrow{BC}$ a representative of v, then $\overrightarrow{AC}$ is a representative of $u + v$. Since the lengths of these segments are the absolute values of the vectors they represent, then $|u + v| < |u| + |v|$ if A, B, and C are non-collinear. If A, B, and C are collinear and B separates A and C, then $|u + v| = |u| + |v|$, while $|u + v| < |u| + |v|$ if B does not separate A and C. In all cases, $|u + v| \leq |u| + |v|$.

(j) Let $\overrightarrow{AB}$ be a representative of u and BC a representative of v; let AB' be a representative of au and B'C' a representative of av. The line AB and the line AB' are coincident; the line BC and the line B'C' are parallel or coincident. Thus, $\angle ABC = \angle AB'C'$. Since $AB' = a \cdot AB$ and $B'C' = a \cdot BC$, $AB'/B'C' = AB/BC$. Thus, ΔABC and $\Delta AB'C'$ are similar; $AC' = a \cdot AC$ and AC and AC' are either parallel or coincident. Since $\overrightarrow{AC}$ is a representative of u + v and $\overrightarrow{AC'}$ a representative of au + av, $a(u + v) = au + av$.

CHAPTER 18, SECTION 2 (pp. 843-844)

1. $\cos\theta = \dfrac{u \cdot v}{|u| \cdot |v|} = \dfrac{3 \cdot 1 - 1 \cdot 2}{\sqrt{9 + 1}\sqrt{1 + 4}} = \dfrac{1}{\sqrt{50}} = \dfrac{\sqrt{2}}{10} = 0.1414;\ \theta = 82°.$

5. $\cos\theta = \dfrac{u \cdot v}{|u||v|} = \dfrac{2 \cdot 1 - 1 \cdot 2}{|u||v|} = 0;\ \theta = 90°.$

9. $u \cdot v = 1 \cdot 2 - 1 \cdot 1 = 1 \neq 0$, not orthogonal.

13. $u \cdot v = 1 \cdot 3 - 1 \cdot 4 = -1$, not orthogonal.

17. $w = \dfrac{u \cdot v}{|v|^2} v = \dfrac{2 \cdot 1 - 1 \cdot 1}{1 + 1}(i + j) = \dfrac{1}{2} i + \dfrac{1}{2} j.$

21. $w = \dfrac{u \cdot v}{|v|^2} v = \dfrac{1 \cdot 2 - 1 \cdot 1}{4 + 1}(2i + j) = \dfrac{2}{5} i + \dfrac{1}{5} j.$

25. $u \cdot v = 3 \cdot 1 - 1 \cdot a = 0,\ \ a = 3.$

29. $\cos\theta = \dfrac{u \cdot v}{|u||v|} = \dfrac{1 \cdot a + 1(-1)}{\sqrt{1 + 1}\sqrt{a^2 + 1}} = \dfrac{a - 1}{\sqrt{2a^2 + 2}} = \pm 1$

$\dfrac{a^2 - 2a + 1}{2a^2 + 2} = 1,\ \ a^2 - 2a + 1 = 2a^2 + 2,$

$a^2 + 2a + 1 = 0,\ \ (a + 1)^2 = 0,\ \ a = -1.$

Alternate solution: u and v are parallel if their components are proportional, $\frac{a}{1} = \frac{-1}{1}$, $a = -1$.

33. $\cos \frac{\pi}{3} = \frac{u \cdot v}{|u| \cdot |v|} = \frac{a \cdot 1 + 2(-1)}{\sqrt{a^2 + 4}\sqrt{1 + 1}}$

$\frac{1}{2} = \frac{a - 2}{\sqrt{2(a^2 + 4)}}$, $\quad 2a^2 + 8 = 4a^2 - 16a + 16$,

$2a^2 - 16a + 8 = 0$, $\quad a^2 - 8a + 4 = 0$,

$a = \frac{8 \pm \sqrt{64 - 16}}{2} = 4 \pm 2\sqrt{3}$

$a = 4 + 2\sqrt{3}$ ($a = 4 - 2\sqrt{3}$ is extraneous).

37. $u = i + 4j$, $\quad v = 2i - j$, $\quad w = i - 5j$

$p_1(v \text{ on } u) = \frac{v \cdot u}{|u|^2} u = \frac{2 \cdot 1 - 1 \cdot 4}{1 + 16}(i + 4j) = -\frac{2}{17} i - \frac{8}{17} j$

$p_2(w \text{ on } u) = \frac{w \cdot u}{|u|^2} u = \frac{1 \cdot 1 - 5 \cdot 4}{1 + 16}(i + 4j) = -\frac{19}{17} i - \frac{76}{17} j$.

41. Since the zero vector is orthogonal to every other vector and $0 \cdot v = 0$, that case is trivial. Suppose now that neither u nor v is the zero vector. If u and v are orthogonal, then

$$\frac{u \cdot v}{|u| \cdot |v|} = \cos 90° = 0$$

Thus, $u \cdot v = 0$. On the other hand, if $u \cdot v = 0$, $\cos \theta = 0$ and $\theta = 90°$.

45. $u = \frac{C}{A} i - \frac{C}{B} j$ (from Problem 44)

$$\cos \theta = \frac{u \cdot v}{|u||v|} = \frac{(C/A)B - (C/B)(-A)}{\sqrt{C^2/A^2 + C^2/B^2}\sqrt{B^2 + A^2}}$$

$$= \frac{(B^2C + A^2C)/AB}{\sqrt{C^2(A^2 + B^2)/A^2B^2}\sqrt{A^2 + B^2}} = \pm 1$$

Therefore, v is parallel to $Ax + By + C = 0$.

CHAPTER 18, SECTION 3 (pp. 850-852)

1. $r'(t) = 3t^2 i + 2tj$,

   $r'(2) = 12i + 4j$.

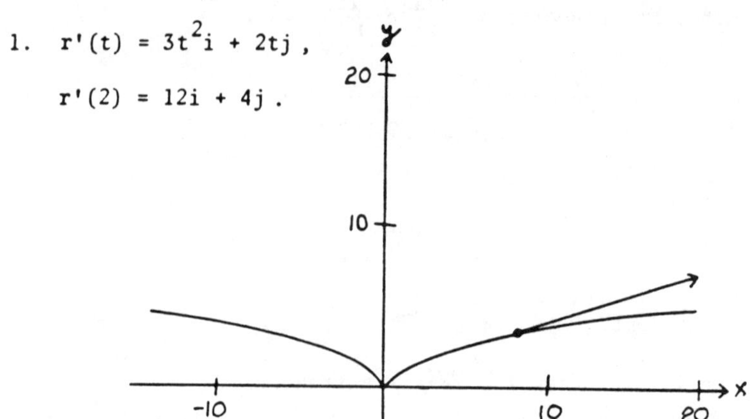

5. $r'(t) = 2ti + 2tj$,

   $r''(t) = 2i + 2j$.

9. $r'(t) = i + \cos tj$,

   $r''(t) = -\sin tj$.

13. $r' = \sinh ti + \cosh tj$,

    $r'' = \cosh ti + \sinh tj$,

    $$\frac{ds}{dt} = |r'| = \sqrt{\sinh^2 t + \cosh^2 t}$$

    $$= \sqrt{\cosh 2t},$$

    $$\frac{d^2s}{dt^2} = \frac{2 \sinh 2t}{2\sqrt{\cosh 2t}} = \frac{\sinh 2t}{\sqrt{\cosh 2t}}.$$

    At $t = 0$, $r' = j$, $r'' = i$,

    $$\frac{ds}{dt} = 1, \frac{d^2s}{dt^2} = 0.$$

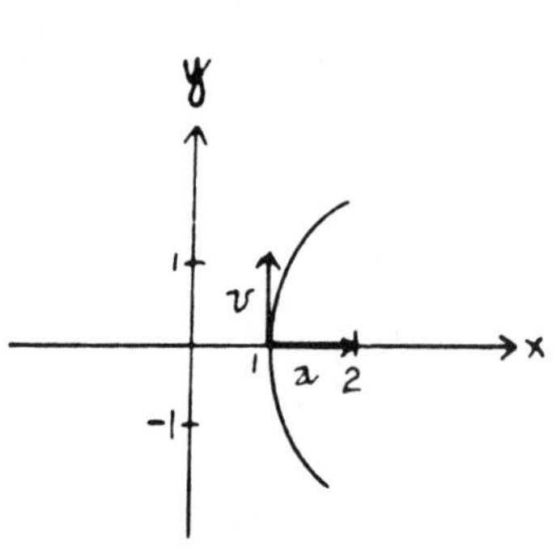

17. $\theta = 4t$

$x = OM + CQ$

$= 2t + \frac{1}{4}\cos(\frac{\pi}{2} - 4t)$

$= 2t + \frac{1}{4}\sin 4t$, $y = MC + QP$

$= \frac{1}{2} + \frac{1}{4}\sin(\frac{\pi}{2} - 4t)$

$= \frac{1}{2} + \frac{1}{4}\cos 4t$

$r = (2t + \frac{1}{4}\sin 4t)i + (\frac{1}{2} + \frac{1}{4}\cos 4t)j$

$r' = (2 + \cos 4t)i - (\sin 4t)j$

$r'' = -(4\sin 4t)i - (4\cos 4t)j.$

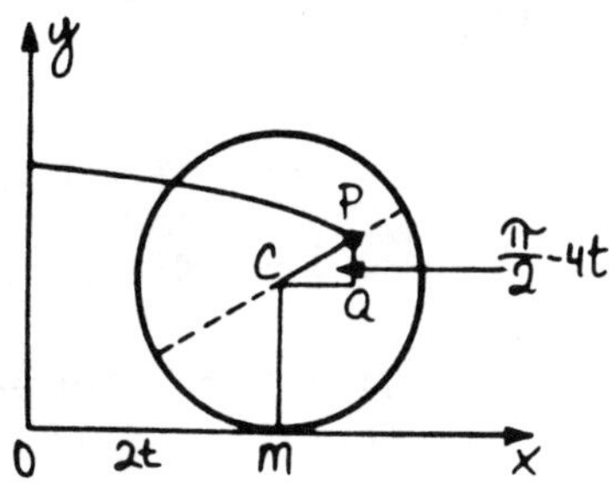

21. $x = 25\cos 60°t = \frac{25t}{2}$

$y = 25\sin 60°t - 4.9t^2$

$= \frac{25\sqrt{3}}{2}t - 4.9t^2$

At Q, $y = x\tan 30° = \frac{x}{\sqrt{3}}$,

$\frac{25\sqrt{3}}{2}t - 4.9t^2 = \frac{25}{2\sqrt{3}}t$,

$t = 2.9457\text{ s}$

$x = \frac{25}{2}t = 36.82$, $OQ = x\sec 30° = 42.52\text{ m}.$

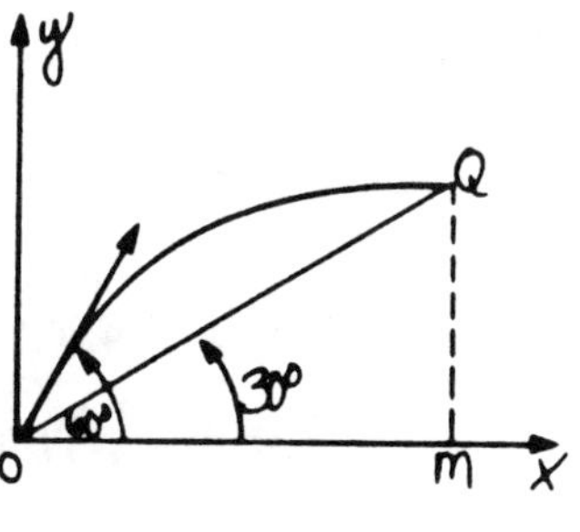

25. $s' = \frac{1}{2}\omega = 4$, $\omega = 8$.

From Example 2, $r = \frac{1}{2}(8\,t - \sin 8t)i + \frac{1}{2}(1 - \cos 8t)j$,

$v = (4 - 4\cos 8t)i + 4\sin 8t\,j$

$a = 32\sin 8\,t\,i + 32\cos 8t\,j$

At $t = T$, $r_0 = (4T - \frac{1}{2}\sin 8T)i + (\frac{1}{2} - \frac{1}{2}\cos 8T)j$,

$v_0 = (4 - 4\cos 8T)i + (4\sin 8T)j$

$r = (4t - \frac{1}{2}\sin 8t)i + \frac{1}{2}(1 - \cos 8t)j$, $t < T$,

$r = [4T - \frac{1}{2}\sin 8T + (4 - 4\cos 8T)(t - T)]i$

$+ [\frac{1}{2} - \frac{1}{2}\cos 8T + 4(t - T)\sin 8T - 4.9(t - T)^2]j$,

$t > T$.

For fixed $t > T$, $y = 4(t - T)\sin 8T - 4.9(t - T)^2$

$+ \frac{1}{2} - \frac{1}{2}\cos 8T$

$y' = 4\sin 8T - 9.8(t - T) = 0$, $t - T = \frac{4\sin 8T}{9.8}$,

For each T, maximum $y = \frac{1}{2} - \frac{1}{2}\cos 8T + 4(\frac{4\sin 8T}{9.8})\sin 8T$

$- 4.9(\frac{4\sin 8T}{9.8})^2$

$= \frac{1}{2} - \frac{1}{2}\cos 8T + \frac{8}{9.8}\sin^2 8T.$

$\frac{dy}{dT} = 4\sin 8T + \frac{8}{9.8}2\sin 8T\cos 8T \cdot 8 = 0$,

$\cos 8T = -\frac{9.8}{32}$, $T = \pm 0.2353 + \frac{n\pi}{4}$, n a positive integer

CHAPTER 18, SECTION 4 (pp. 859-861)

1. $v_0 = \sqrt{\frac{2GM}{R}} = \sqrt{\frac{2 \times 6.7 \times 10^{-11} \times 7.2 \times 10^{22}}{1.7 \times 10^6}}$

$= 100\sqrt{\frac{2 \times 6.7 \times 72}{1.7}} = 2382$ m/s.

5. $r = 2\theta$, $s' = \sqrt{(r')^2 + r^2(\theta')^2} = \sqrt{(\frac{dr}{d\theta})^2 + r^2}\ \frac{d\theta}{dt} = 2$,

$\theta' = \frac{2}{\sqrt{4 + 4\theta^2}} = \frac{1}{\sqrt{1 + \theta^2}}$, $r' = 2\theta' = \frac{2}{\sqrt{1 + \theta^2}}$

$v = r'i_r + r\theta' i_\theta = \frac{2}{\sqrt{1 + \theta^2}} i_r + \frac{2\theta}{\sqrt{1 + \theta^2}} i_\theta$

$r'' = \frac{d}{d\theta}(r')\theta' = \frac{2\theta}{(1 + \theta^2)^2}$, $\theta'' = \frac{d}{d\theta}(\theta')\theta' = -\frac{\theta}{(1 + \theta^2)^2}$

$a = (r'' - r\theta'^2)i_r + (r\theta'' + 2r'\theta')i_\theta$

$= (-\frac{2\theta}{(1 + \theta^2)^2} - \frac{2\theta}{1 + \theta^2})i_r + (-\frac{2\theta^2}{(1 + \theta^2)^2} + \frac{4}{(1 + \theta^2)})i_\theta$

$= \frac{4\theta + 2\theta^3}{(1 + \theta^2)^2} i_r + \frac{2\theta^2 + 4}{(1 + \theta^2)^2} i_\theta$, where

$2t = 2\int \sqrt{1 + \theta^2}\, d\theta = \theta\sqrt{1 + \theta^2} + \ln(\theta + \sqrt{1 + \theta^2})$.

9. $r' = 0$, $r'' > 0$ at P_0

$v = r' i_r + r\theta' i_\theta = r\theta' i_\theta = r\theta'(-\sin\theta i + \cos\theta j)$.

Position vector is $r \cos\theta i + r \sin\theta j$.

$(-r\theta' \sin\theta i + r\theta' \cos\theta j) \cdot (r\cos\theta i + r\sin\theta j)$

$= -r^2\theta' \sin\theta \cos\theta + r^2\theta' \cos\theta \sin\theta = 0$.

Since $r \neq 0$, the two vectors are perpendicular.

13. From page 795, $r = \dfrac{C^2/GM}{1 + (AC^2/GM)\cos\theta}$, where $C = \dfrac{\ell}{m}$, so $e = \dfrac{AC^2}{GM}$ is the eccentricity of the path. If $e = 1$, the path is a parabola; if $e > 1$, the path is a hyperbola.

CHAPTER 18, SECTION 5 (pp. 868-870)

1. $4 + 1 + 1 + x = 0$, $x = -6$, $3 - 2 + 1 + y = 0$, $y = -2$, $F = -6i - 2j$.

5. $F_1 = -3j$, $F_2 = 4\cos 45° + 4\sin 45° j = 2\sqrt{2}\,i + 2\sqrt{2}\,j$, $F = -2\sqrt{2}\,i + (3 - 2\sqrt{2})j$.

9. $F_1 = -100j$, $F_2 = T_2(\cos 60° i + \sin 60° j)$

$$= \frac{1}{2}T_2 i + \frac{\sqrt{3}}{2}T_2 j,$$

$F_3 = T_3(\cos 210° i + \sin 210° j) = -\frac{\sqrt{3}}{2}T_3 i - \frac{1}{2}T_3 j.$

$(\frac{1}{2}T_2 - \frac{\sqrt{3}}{2}T_3)i + (-100 + \frac{\sqrt{3}}{2}T_2 - \frac{1}{2}T_3)j = 0,$

$T_2 - \sqrt{3}T_3 = 0$, $\frac{\sqrt{3}}{2}T_2 - \frac{1}{2}T_3 = 100$,

$T_1 = 100$, $T_2 = 100\sqrt{3}$, $T_3 = 100$.

13. $q = \frac{1}{2}u + v + \frac{1}{2}w$, $q = -\frac{1}{2}u + p - \frac{1}{2}w$

Adding, we have $2q = p + v$, $q = \dfrac{p + v}{2}$.

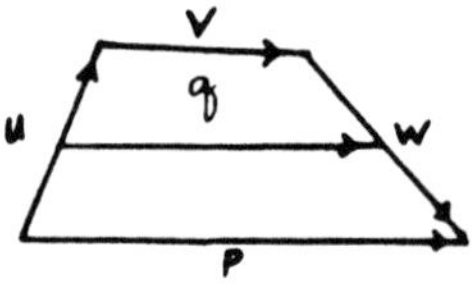

17. $v = a + b$

$|v|^2 = |a + b|^2 = |a|^2 + |b|^2$

$u = a + d = a - b$

$|u| = |a - b|^2 = |a|^2 + |-b|^2 = |a|^2 + |b|^2$

$|v| = |u|$.

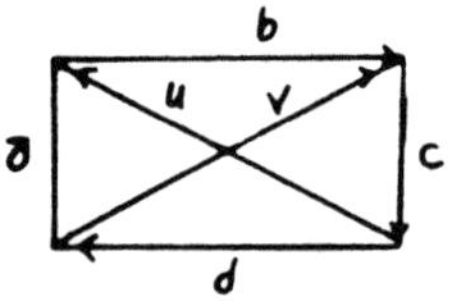

21. $\mu = 30{,}000$, $L = 200$,

$$T_0 = \frac{L^2\mu}{8H} = \frac{4 \times 10^4 \times 3 \times 10^4}{8H} = \frac{3 \times 10^8}{2H}$$

$$T_{max} = \frac{1}{2}\sqrt{4T_0{}^2 + L^2\mu^2}$$

$$4 \times 10^6 = \frac{1}{2}\sqrt{\frac{9 \times 10^{16}}{H^2} + 4 \times 10^4 \times 9 \times 10^8}$$

$$\frac{8}{3} = \sqrt{\frac{10^4}{H^2} + 4}, \quad H^2 = \frac{9}{28} \times 10^4, \quad H = 56.69 \text{ m}.$$

CHAPTER 18, SECTION 6 (p. 875)

1. $$T_0 = \frac{\mu(S^2 - 4H^2)}{8H} = \frac{80(200^2 - 4 \times 10^2)}{8 \times 10} = 39{,}600,$$

$$T_{max} = \sqrt{T_0{}^2 + \mu^2S^2/4} = \sqrt{39{,}600^2 + 80^2 \times 200^2/4}$$

$$= 40{,}400.$$

$$y = \frac{T_0}{\mu}\cosh\frac{\mu}{T_0}x = \frac{39{,}600}{80}\cosh\frac{80x}{39{,}600},$$

$$y = 495\cosh\frac{x}{495}.$$

5. $S = 200$, $\mu = \frac{2400}{200} = 12$

$$T_{max} = \sqrt{T_o{}^2 + \mu^2S^2/4}, \quad 1500 = \sqrt{T_0{}^2 + 12^2 \times 200^2/4},$$

$$2{,}250{,}000 = T_0{}^2 + 1{,}440{,}000, \quad T_0 = 900.$$

$$T_0 = \frac{\mu(S^2 - 4H^2)}{8H}, \quad 900 = \frac{12(200^2 - 4H^2)}{8H},$$

$$H^2 + 150H - 10{,}000 = 0, \quad (H + 200)(H - 50) = 0, \quad H = 50 \text{ m}.$$

CHAPTER 18, REVIEW (pp. 875-877)

1. $\frac{v}{|v|} = (\frac{2}{\sqrt{13}}, -\frac{3}{\sqrt{13}})$.

5. $u \cdot v = 2 \cdot 3 - 1(-2) = 8$.

9. $w = \frac{u \cdot v}{|v|^2} v = \frac{4 \cdot 2 + 1(-2)}{4 + 4}(2i - 2j) = \frac{3}{2} i - \frac{3}{2} j$.

13. $\theta' = 2$, $\theta'' = 0$

When $r = 75$, $\cos\theta = -0.6$, $\sin\theta = 0.8$.

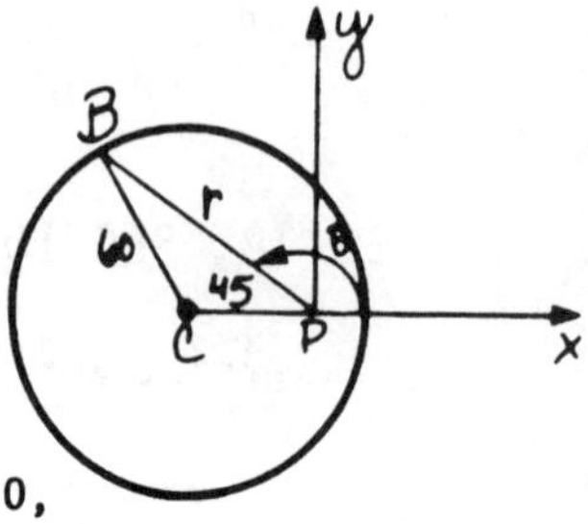

$(x + 45)^2 + y^2 = 3600$,

$r^2 + 90r\cos\theta = 1575$

$2rr' + 90r'\cos\theta - 90r\sin\theta\,\theta' = 0$,

$$r' = \frac{90r\sin\theta}{r + 45\cos\theta} = \frac{90 \times 75(0.8)}{75 + 45(-0.6)} = 112.5$$

$$rr'' + (r')^2 + 45r''\cos\theta - 45r'\sin\theta\,\theta' = 90r'\sin\theta + 90r\cos\theta\,\theta'$$

$$r'' = \frac{180r'\sin\theta + 180r\cos\theta - (r')^2}{r + 45\cos\theta}$$

$$= \frac{180 \times 112.5(0.8) + 180 \times 75(-0.6) - (112.5)^2}{75 + 45(-0.6)}$$

$$= -94.92.$$

$$a = [r'' - r(\theta')^2]i_r + [r\theta'' + 2r'\theta']i_\theta$$

$$= [-94.92 - 75 \times 4]i_r + [75 \times 0 + 2(112.5)(2)]i_\theta$$

$$= -394.92\, i_r + 450\, i_\theta.$$

17. $x = 80\cos 30° = 40\sqrt{3}$ N, $y = 80\sin 30° = 40$ N.

21. $\mu S = 800$

$$\tan 30^\circ = \frac{\mu S}{2T_0}$$

$$\frac{1}{\sqrt{3}} = \frac{400}{T_0}, \quad T_0 = 400\sqrt{3}$$

$$T_{max} = \sqrt{T_0^2 + \mu^2 S^2/4}$$

$$= \sqrt{(400\sqrt{3})^2 + 400^2}$$

$$= 800 \text{ N.}$$

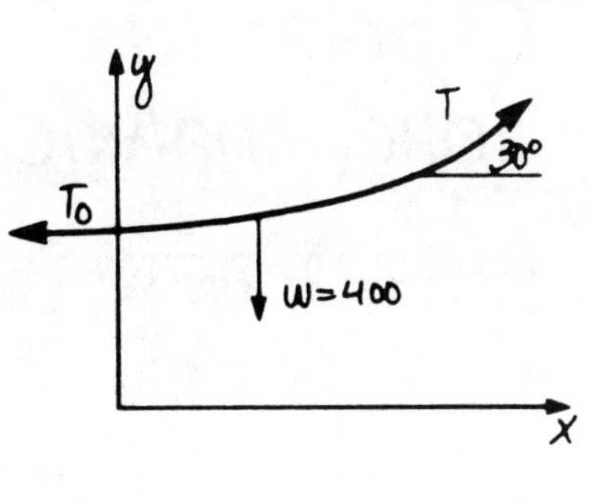

25. $v_1 = u_1$

$$v_1 \cdot v_2 = v_1 \cdot (u_2 - \frac{u_2 \cdot v_1}{v_1 \cdot v_1} v_1)$$

$$= v_1 \cdot u_2 - \frac{u_2 \cdot u_1}{v_1 \cdot v_1} v_1 \cdot v_1$$

$$= u_1 \cdot u_2 - u_2 \cdot u_1$$

$$= 0$$

Thus, v_1 and v_2 are orthogonal.

Chapter 19
Solid Analytic Geometry

CHAPTER 19, SECTION 1 (pp. 889-891)

1. $d = \sqrt{(4-2)^2 + (-2-2)^2 + [1-(-3)]^2} = \sqrt{36} = 6$.

5. $v = \overrightarrow{AB} = (5-3)i + [4-(-2)]j + (-1-4)k$

$= 2i + 6j - 5k$.

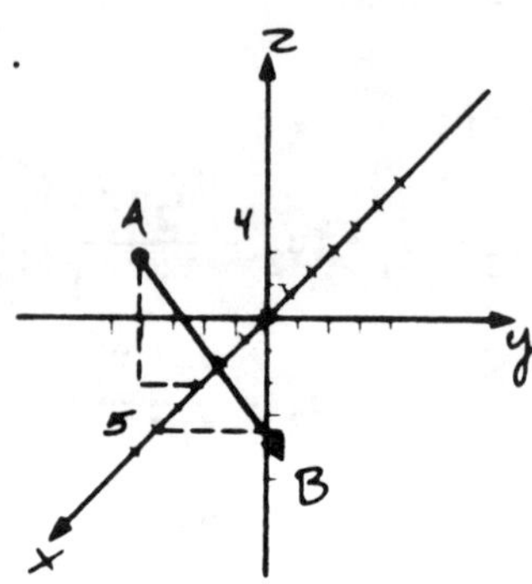

9. $v = \dfrac{i - 2j + 2k}{\sqrt{1 + (-2)^2 + 2^2}} = \frac{1}{3}(i - 2j + 2k)$.

13. $x = \dfrac{x_1 + x_2}{2} = \dfrac{4-2}{2} = 1,\ y = \dfrac{y_1 + y_2}{2} = \dfrac{3-1}{2} = 1,$

$z = \dfrac{z_1 + z_2}{2} = \dfrac{5+2}{2} = \frac{7}{2},\ (1, 1, \frac{7}{2}).$

17. $x = x_1 + r(x_2 - x_1) = 5 + (\frac{2}{5})(-5-5) = 1$,

$y = y_1 + r(y_2 - y_1) = 2 + (\frac{2}{5})(7-2) = 4$,

$z = z_1 + r(z_2 - z_1) = 3 + (\frac{2}{5})(-2-3) = 1$.

21. $u + v = (3+2,\ 1+6,\ -4+3) = (5,7,-1)$;

$u \cdot v = 3\cdot 2 + 1\cdot 6 + (-4)\cdot 3 = 6 + 6 - 12 = 0$;

u and v are orthogonal.

25. $u \cdot v = |u| \cdot |v| \cdot \cos\theta;$

$1 \cdot 2 + 1 \cdot (-1) + 2 \cdot 1 = (1^2 + 1^2 + 2^2)^{1/2} \cdot$

$\cdot [2^2 + (-1)^2 + 1^2]^{1/2} \cos\theta;$

$\cos\theta = \dfrac{3}{\sqrt{6}\,\sqrt{6}} = \dfrac{1}{2}, \quad \theta = \text{Arccos}\,\dfrac{1}{2} = 60° .$

29. $w = \left(\dfrac{u \cdot v}{|v|}\right) \dfrac{v}{|v|} = \dfrac{1 \cdot 0 - 2 \cdot 2 + 4 \cdot 3}{2^2 + 3^2}(2j + 3k) = \dfrac{8}{13}(2j + 3k) .$

33. $x = x_1 + r(x_2 - x_1);\ 6 = 5 + (\frac{1}{3})(x_2 - 5);\ 3 = x_2 - 5;\ x_2 = 8 .$

$y = y_2 + r(y_2 - y_1);\ 0 = -2 + (\frac{1}{3})(y_2 + 2);\ 6 = y_2 + 2;\ y_2 = 4 .$

$z = z_2 + r(z_2 - z_1);\ 0 = 3 + (\frac{1}{3})(z_2 - 3);\ -9 = z_2 - 3;\ z_2 = -6 .$

37. The midpoint of (1,3,2) and (x,-1,6) has coordinates

$(\frac{x + 1}{2}, 1, 4).$

$(\frac{x + 1}{2} + 1)^2 + (5 - 1)^2 + (2 - 4)^2 = 36;\ (\frac{x + 3}{2})^2 + 16 + 4 = 36;$

$(x + 3)^2 = 64;\ x = -3 \pm 8;\ x = 5 \text{ or } x = -11 .$

41. By the distributive property of the dot product,

$(u + v) \cdot w = u \cdot w + v \cdot w$; let $t = u + v$ and project t on w

to get

$$\left(\frac{t \cdot w}{|w|}\right)\left(\frac{w}{|w|}\right) = \frac{t \cdot w}{|w|^2}\, w = \frac{(u + v) \cdot w}{|w|^2}\, w = \frac{u \cdot w + v \cdot w}{|w|^2}\, w$$

$$= \left(\frac{u \cdot w}{|w|}\right)\left(\frac{w}{|w|}\right) + \left(\frac{v \cdot w}{|w|}\right)\left(\frac{w}{|w|}\right) .$$

45. $\overline{AD}^2 = (x_2 - x_1)^2 + (y_2 - y_1)^2$

$\overline{AB}^2 = \overline{AD}^2 + (z_2 - z_1)^2$

$\overline{AB} = \sqrt{(x_2 - x_1)^2 + (y_2 - y_1)^2 + (z_2 - z_1)^2}$.

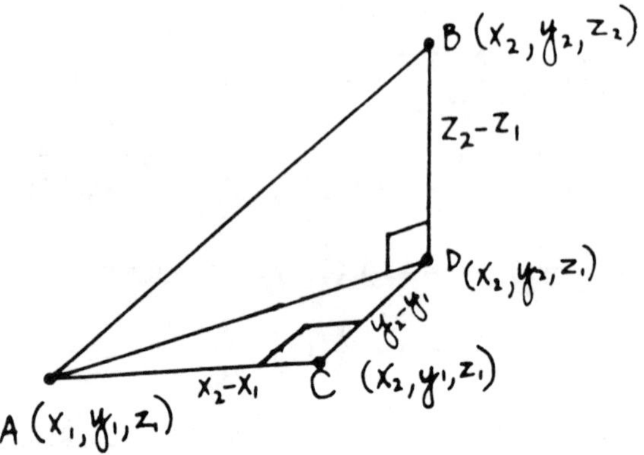

49. Let $r = \frac{1}{2}$ in theorem 4. Then

$$x = x_1 + (\tfrac{1}{2})(x_2 - x_1) = \frac{x_1 + x_2}{2},$$

$$y = y_1 + (\tfrac{1}{2})(y_2 - y_1) = \frac{y_1 + y_2}{2},$$

$$z = z_1 + (\tfrac{1}{2})(z_2 - z_1) = \frac{z_1 + z_2}{2}.$$

53. Substitute $u + v$ for w in problem 50(b).

$(u + v)\cdot w = u\cdot w + v\cdot w = u\cdot(u + v) + v\cdot(u + v)$

$= u\cdot u + u\cdot v + v\cdot u + v\cdot v = |u|^2 + u\cdot v + u\cdot v + |v|^2$

$= |u|^2 + 2u\cdot v + |v|^2$.

CHAPTER 19, SECTION 2 (pp. 896-898)

1. $x = 2 + 2t$, $y = -4 + 3t$, $z = 2 + t$; $\frac{x-2}{2} = \frac{y+4}{3} = \frac{z-2}{1}$.

5. $x = 1 + 3t$, $y = 0 + t$, $z = 5 + 0t$; $\frac{x-1}{3} = \frac{y-0}{1}$ and $z = 5$.

9. $\{3-4, 3-0, 1-2\} = \{-1, 3, -1\}$; $x = 3 - t$, $y = 3 + 3t$, $z = 1 - t$;

   $\frac{x-3}{-1} = \frac{y-3}{3} = \frac{z-1}{-1}$.

13. $\{2-1, 2-2, 4-7\} = \{1, 0, -3\}$; $x = 2 + t$, $y = 2$, $z = 4 - 3t$;

    $\frac{x-2}{1} = \frac{z-4}{-3}$ and $y = 2$.

17. $2 - t = 1 + s$ or $s + t = 1$

    $3 + 2t = -2 + s$ or $s - 2t = 5$

    $4 + t = 5 - 4s$ or $4s + t = 1$

    Solving the first two equations simultaneously, we get $s = 7/3$ and $t = -4/3$. Since this is not a solution of the third, the lines do not intersect.

21. $x = 5 + t$, $y = -2 - 2t$, $z = 3 + 5t$;

    $x = 4 - 2s$, $y = 2 + s$, $z = 4 + 3s$;

    $5 + t = 4 - 2s$ or $2s + t = -1$

    $-2 - 2t = 2 + s$ or $s + 2t = -4$

    $3 + 5t = 4 + 3s$ or $3s - 5t = -1$

    Solving the first two equations simultaneously, we have $s = 2/3$ and $t = -7/3$. Since this is not a solution of the third equation, the lines do not intersect.

25. $\{-1, 2, 1\}$, $\{2, -4, -2\}$

    Parallel or coincident.
    (4,3,1) is on the first line but not the second.
    Parallel.

29. $\{2,0,4\}$, $\{6,3,2\}$

$2\cdot 6 + 0\cdot 3 + 4\cdot 2 = 20 \neq 0$, none.

33.

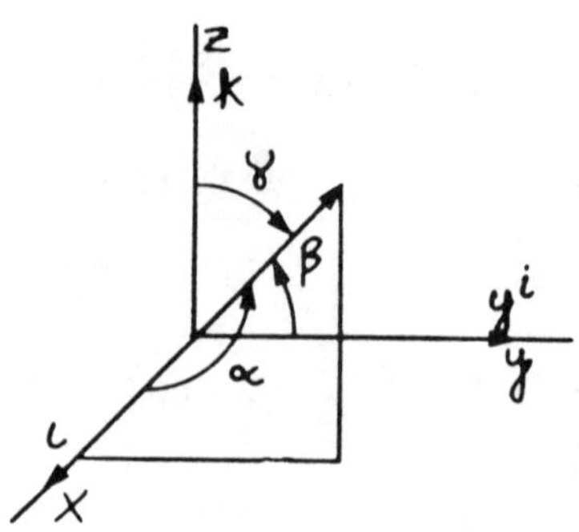

$\frac{v}{|v|} = \ell i + mj + nk$ is a unit vector.

$i \cdot \frac{v}{|v|} = \ell i \cdot i + mi \cdot j + ni \cdot k = \ell + 0m + 0n = \ell.$

Similarly, $j \cdot \frac{v}{|v|} = m$ and $k \cdot \frac{v}{|v|} = n$.

$\cos\alpha = i \cdot \frac{v}{|v|} = \ell$, $\cos\beta = j \cdot \frac{v}{|v|} = m$ and

$\cos\gamma = k \cdot \frac{v}{|v|} = n.$

$\left|\frac{v}{|v|}\right|^2 = 1 = \ell^2 + m^2 + n^2.$

CHAPTER 19, SECTION 3 (pp. 906-908)

1. $u \times v = [1(-4) - 1(-1)]i + [1\cdot 2 - 1(-4)]j + [1(-1) - 1\cdot 2]k$

$= -3i + 6j - 3k.$

5. $u \times v = [1\cdot 3 - (-1)(-1)]i + [(-1)(-1) - 2\cdot 3]j + [2(-1) - 1(-1)]k$

$= 2i - 5j - k.$

9. $u = 4i - j, \quad v = -2i + 2j - k,$

$u \times v = i + 4j + 6k,$ direction numbers $\{1,4,6\}$.

13. $u = i - j + 2k, \quad v = j + k,$

$u \times v = -3i - j + k,$ direction numbers $\{3, 1,-1\}$.

$x = 4 + 3t, \quad y = -1 + t, \quad z = -t$.

17. $u = 2i + j + 3k,$

$(-1,0,5),\ (2,1,-2): \quad v = 3i + j - 7k$

$u \times v = -10i + 23j - k$

$$d = \frac{|u \times v|}{|u|} = \frac{\sqrt{100 + 529 + 1}}{\sqrt{4 + 1 + 9}} = \sqrt{\frac{630}{14}} = 3\sqrt{5}\ .$$

21. $u = i - j + 4k, \quad v = i - 2j + 3k,$

$u \times v = 5i + j - k,$

$w = (2 - 2)i + (1 - 4)j + (0 - 1)k = -3j - k$

$$|\text{proj of } w \text{ on } u \times v| = \frac{|w \cdot (u \times v)|}{|u \times v|} = \frac{|0 \cdot 5 - 3 \cdot 1 - 1(-1)|}{\sqrt{25 + 1 + 1}}$$

$$= \frac{2}{3\sqrt{3}}\ .$$

25. $P(0,4,-2)$, $Q(-2,0,-1)$ (on the given line).
Vectors along the given line and represented by $\overrightarrow{QP}$ are:

$u = 2i + 8j + k, \quad v = 2i + 4j - k,$

$u \times v = -12i + 4j - 8k$. Direction numbers $\{3,-1,2\}$.

$x = 3t, \quad y = 4 - t, \quad z = -2 + 2t$.

29. The area of ΔABC is $A = \frac{1}{2} AC \cdot AB \cdot \sin \measuredangle BAC$. If u is represented by $\overrightarrow{AB}$, v by $\overrightarrow{AC}$ and $\theta = \measuredangle BAC$, then

$A = \frac{1}{2}|u| \cdot |v| \sin\theta = \frac{1}{2}|u \times v|$.

33. $u = (3-4)i + (-1-2)j = -i - 3j,$

$v = (-1-4)i + (0-2)j = -5i - 2j,$

$u \times v = -13k, \quad A = \frac{1}{2}|u \times v| = \frac{13}{2}.$

37. $u = 2i + j + 3k, \quad v = 5i + 3j + k, \quad w = 2i - j + 4k,$

$v \times w = 13i - 18j - 11k$

$$|u \cdot (v \times w)| = |(2i + j + 3k) \cdot (13i - 18j - 11k)|$$

$$= |26 - 18 - 33| = |-25| = 25.$$

41. $u - v$ and $u - w$ are distinct vectors in the plane containing P, Q, and R.

$$(u \times v + v \times w + w \times u) \cdot (u - v) = u \times v \cdot u - u \times v \cdot v + v \times w \cdot u$$

$$- v \times w \cdot v + w \times u \cdot u - w \times u \cdot v = v \times w \cdot u - w \times u \cdot v:$$

the other 4 terms are 0; for instance $u \times v$ is perpendicular to u so that $u \times v \cdot u = 0$. Now

$$v \times w \cdot u = \begin{vmatrix} v_1 & v_2 & v_3 \\ w_1 & w_2 & w_3 \\ u_1 & u_2 & u_3 \end{vmatrix} = -\begin{vmatrix} w_1 & w_2 & w_3 \\ v_1 & v_2 & v_3 \\ u_1 & u_2 & u_3 \end{vmatrix} = \begin{vmatrix} w_1 & w_2 & w_3 \\ u_1 & u_2 & u_3 \\ v_1 & v_2 & v_3 \end{vmatrix}$$

$$= w \times u \cdot v;$$

thus $v \times w \cdot u - w \times u \cdot v = 0$ and $u \times v + v \times w + w \times u$ is perpendicular to $u - v$. Similarly it is perpendicular to $u - w$, and therefore to the plane P,Q,R.

45. Let $w = u$ and $t = v$ in Problem 44.

$$(u \times v) \cdot (u \times v) = (u \cdot u)(v \cdot v) - (u \cdot v)(v \cdot u).$$

Thus $|u \times v|^2 = \begin{vmatrix} u \cdot u & u \cdot v \\ u \cdot v & v \cdot v \end{vmatrix}.$

CHAPTER 19, SECTION 4 (pp. 915-917)

1.

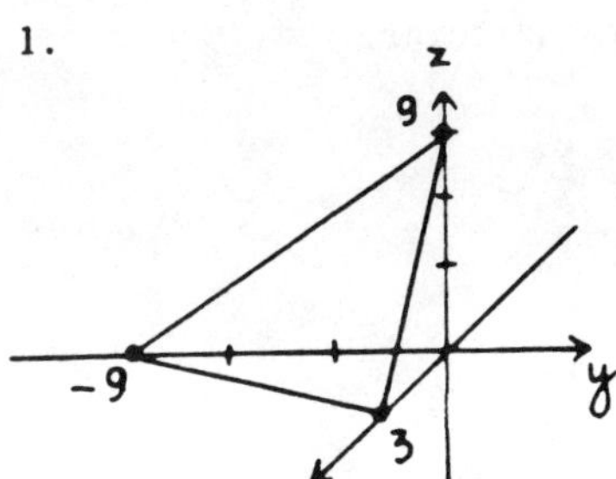

5.

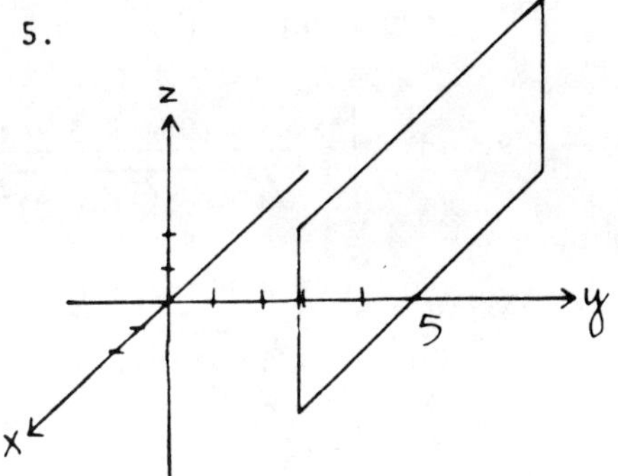

9. $(x - 3) + 3(y - 2) + (z - 5) = 0$, $x + 3y + z - 14 = 0$.

13. u is a vector from (2,-2,-2) to (1,-3,5), v from (1,-3,5) to (-1,4,1).

$u = -i - j + 7k$, $v = -2i + 7j - 4k$,

$u \times v = -45i - 18j - 9k$.

Direction numbers {5,2,1}.

$5(x - 2) + 2(y + 2) + 1(z + 2) = 0$

$5x + 2y + z - 4 = 0$.

17. The lines clearly intersect at (2,-1,4). If {a,b,c} are direction numbers for a line perpendicular to both of the given lines, then

$2a + b - c = 0$ $-a - 2b + 3c = 0$, $5a + b = 0$.

$a = 1$, $b = -5$, $c = -3$.

$(x - 2) - 5(y + 1) - 3(z - 4) = 0$, $x - 5y - 3z + 5 = 0$.

21. (4,-2,3), {3,2,-1}

$x = 4 + 3t$, $y = -2 + 2t$, $z = 3 - t$.

25. $\{4,-2,1\}, \quad \{1,1,-2\}$

$4\cdot 1 - 2\cdot 1 + 1(-2) = 0$, perpendicular.

29. $$d = \frac{|4\cdot 2 + 1\cdot 0 - 8\cdot 3 + 1|}{\sqrt{4^2 + 1^2 + (-8)^2}} = \frac{5}{3}.$$

33. $$d = \frac{|0\cdot 1 + 1\cdot 3 + 0\cdot 1 + 7|}{\sqrt{0^2 + 1^2 + 0^2}} = 10.$$

37. $$\frac{x - 4y - 2z - 5}{\sqrt{1 + 16 + 4}} = 0, \qquad \frac{x - 4y - 2z - 5}{\sqrt{21}} = 0,$$

$$\frac{x - 4y - 3z - 10}{\sqrt{1 + 16 + 4}} = 0, \qquad \frac{x - 4y - 2z - 10}{\sqrt{21}} = 0,$$

$$d = \frac{10}{\sqrt{21}} - \frac{5}{\sqrt{21}} = \frac{5}{\sqrt{21}}.$$

41. If the desired plane is perpendicular to the given planes, it is perpendicular to their line of intersection. Let us find two points on the line of intersection.

$2x - 5y + z = 1, \quad x - 2y - z = 3$

$3x - 7y = 4, \quad y = \dfrac{3x - 4}{7}$

$(-1,-1,-2), \ (6,2,-1).$

Direction numbers for the line of intersection are $\{7,3,1\}$. The desired plane is

$7(x - 3) + 3y + (z + 4) = 0, \quad 7x + 3y + z - 17 = 0.$

45. Two points on the line are $(1,5,2)$ and $(1,9,3)$. Since both are in the given plane, the line lies in the plane.

49. $t = \dfrac{x - 1}{2} = \dfrac{y + 1}{-1} = \dfrac{z}{3}$; $x = 1 + 2t$, $y = -1 - t$, $z = 3t$;

$3(1 + 2t) + 2(-1 - t) - 3t = 5$; $t = 4$, $x = 9$, $y = -5$, $z = 12$.

53. Let $v = ai + bj + ck$ be perpendicular to the plane. The plane is perpendicular to $x + y + z = 0$, so $a \cdot 1 + b \cdot 1 + c \cdot 1 = 0$ and $c = -a - b$.

Also $\cos 45° = \frac{v \cdot k}{|v|} = \frac{c}{\sqrt{a^2 + b^2 + c^2}} = \frac{1}{\sqrt{2}}$,

$2c^2 = a^2 + b^2 + c^2$, $a^2 + b^2 = c^2 = (-a - b)^2$,

$a^2 + b^2 = a^2 + 2ab + b^2$, $ab = 0$, so

$a = 0$ or $b = 0$

$a = 0$	$b = 0$
$c = -b = -1$	$c = -a = -1$
$y - z = 0$	$x - z = 0.$

CHAPTER 19, SECTION 5 (pp. 922-923)

1.

5.

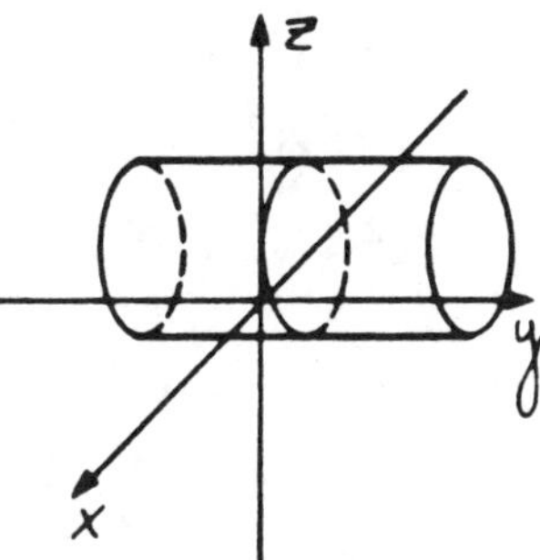

9.

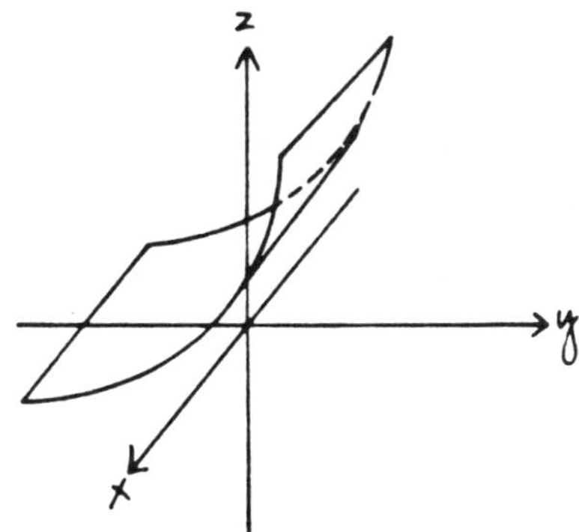

13. $x^2 + y^2 + z^2 + 6x - 8y - 2z + 22 = 0$

$x^2 + 6x + 9 + y^2 - 8y + 16 + z^2 - 2z + 1 = -22 + 9 + 16 + 1$

$(x + 3)^2 + (y - 4)^2 + (z - 1)^2 = 4$

Sphere: center $(-3,4,1)$, $r = 2$.

17. $3x^2 + 3y^2 + 3z^2 + 4x - 2y - 8z + 7 = 0$

$x^2 + y^2 + z^2 + \frac{4}{3}x - \frac{2}{3}y - \frac{8}{3}z + \frac{7}{3} = 0$

$x^2 + \frac{4}{3}x + \frac{4}{9} + y^2 - \frac{2}{3}y + \frac{1}{9} + z^2 - \frac{8}{3}z + \frac{16}{9} = -\frac{7}{3} + \frac{4}{9} + \frac{1}{9} + \frac{16}{9}$

$(x + \frac{2}{3})^2 + (y - \frac{1}{3})^2 + (z - \frac{4}{3})^2 = 0$

Point: $(-\frac{2}{3}, \frac{1}{3}, \frac{4}{3})$.

21. $(x - 3)^2 + (y - 1)^2 + (z - 1)^2 = r^2$

$(0 - 3)^2 + (0 - 1)^2 + (0 - 1)^2 = r^2$, $r^2 = 11$

$(x - 3)^2 + (y - 1)^2 + (z - 1)^2 = 11$

$x^2 + y^2 + z^2 - 6x - 2y - 2z = 0$.

25. $y^2 - z^2 = 1$; about z axis

$x^2 + y^2 - z^2 = 1$.

29. $(x - 1)^2 + y^2 = 1$

$x^2 - 2x + 1 + y^2 = 1$

$x^2 + y^2 = 2x$

$(x^2 + z^2) + y^2 = 2\sqrt{x^2 + z^2}$

$(x^2 + y^2 + z^2)^2 = 4(x^2 + z^2)$.

33. The line through (1,4,4) and perpendicular to the given plane is:

$x = 1 + t, \quad y = 4 + 2t, \quad z = 4 + 2t.$

The sphere with center (1,4,4) and $r = 3$ is

$$(x - 1)^2 + (y - 4)^2 + (z - 4)^2 = 9$$

$$t^2 + 4t^2 + 4t^2 = 9, \quad t^2 = 1, \quad t = \pm 1.$$

The points of intersection of the line and sphere are (2,6,6) and (0,2,2). These are the centers of the desired spheres.

$$(x - 2)^2 + (y - 6)^2 + (z - 6)^2 = 9$$

$$x^2 + y^2 + z^2 - 4x - 12y - 12z + 67 = 0$$

$$x^2 + (y - 2)^2 + (z - 2)^2 = 9$$

$$x^2 + y^2 + z^2 - 4y - 4z - 1 = 0.$$

37. If (x,y,z) is on the sphere, its distance from the center is r.

$$\sqrt{(x - h)^2 + (y - k)^2 + (z - \ell)^2} = r$$

$$(x - h)^2 + (y - k)^2 + (z - \ell)^2 = r^2$$

The above argument is reversible since r is positive.

41.

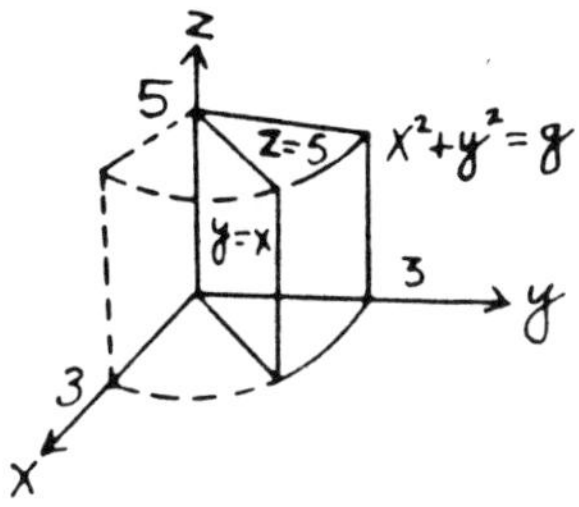

45.

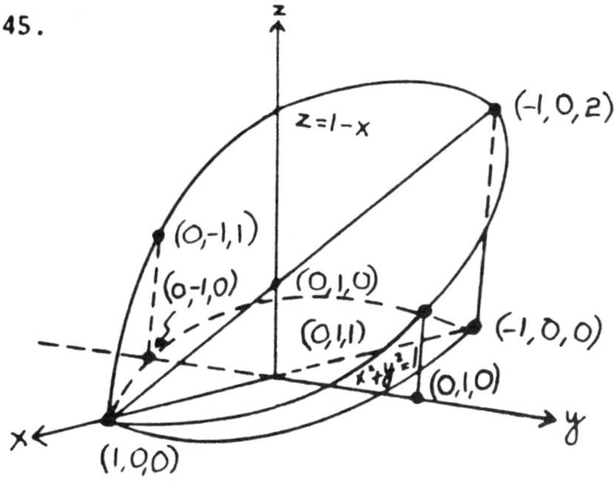

CHAPTER 19, SECTION 6 (pp. 931-932)

1. Ellipsoid (oblate spheroid).

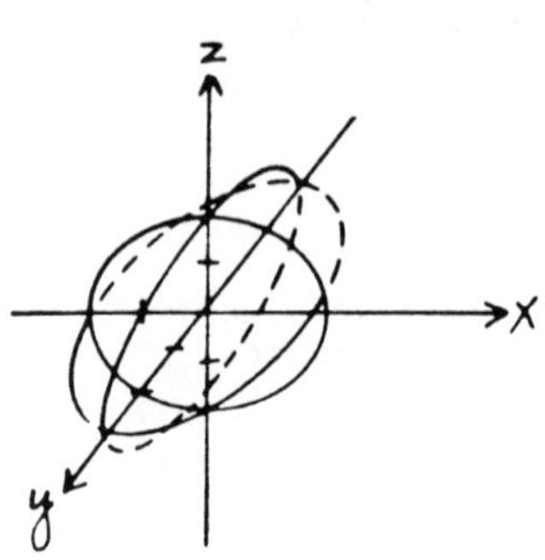

5. Hyperbolic paraboloid.

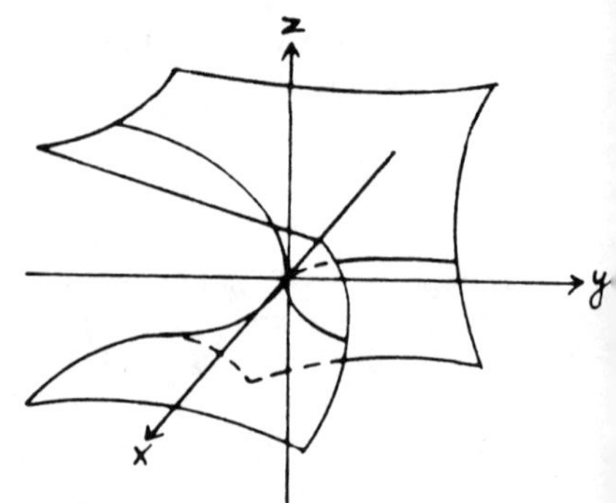

9. Hyperboloid of one sheet.

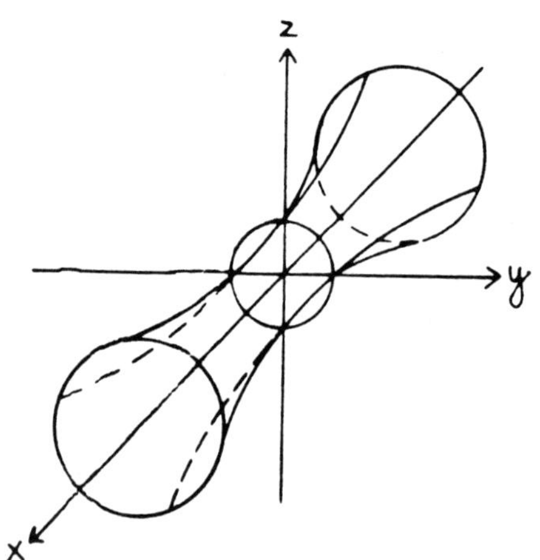

13. Circular paraboloid.

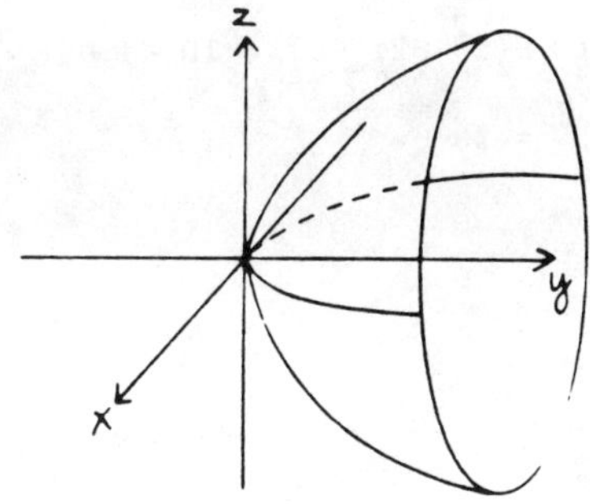

17. Circular paraboloid.

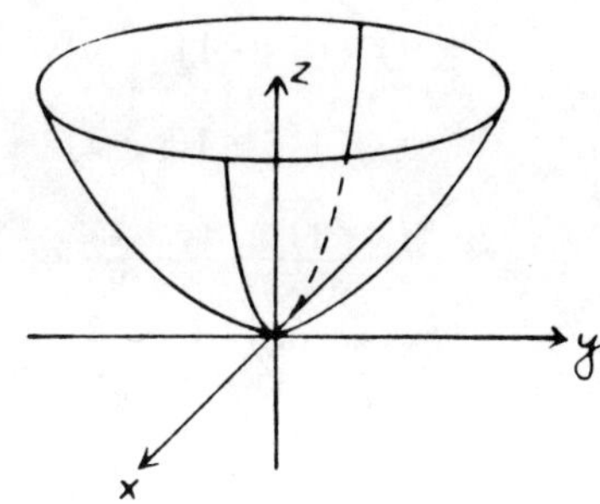

21. Hyperboloid of two sheets.

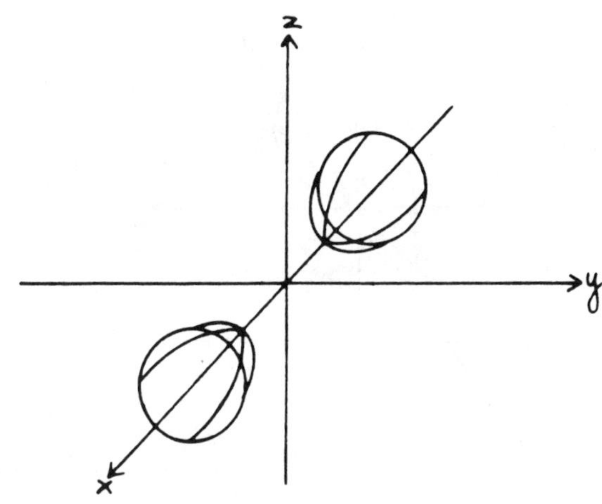

25. Hyperboloid of one sheet.

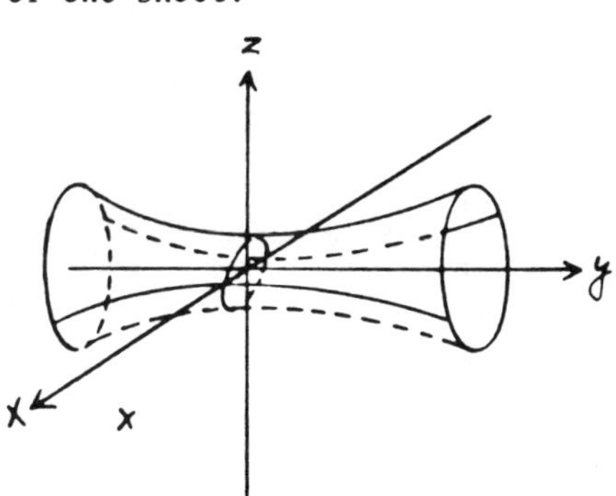

29. $x^2 + 4y^2 + 9z^2 + 2x + 16y - 18z - 10 = 0$

$$(x^2 + 2x + 1) + 4(y^2 + 4y + 4) + 9(z^2 - 2z + 1) = 10 + 1 + 16 + 9$$

$$(x + 1)^2 + 4(y + 2)^2 + 9(z - 1)^2 = 36$$

$$\frac{(x + 1)^2}{36} + \frac{(y + 2)^2}{9} + \frac{(z - 1)^2}{4} = 1$$

$$\frac{x'^2}{36} + \frac{y'^2}{9} + \frac{z'^2}{4} = 1.$$

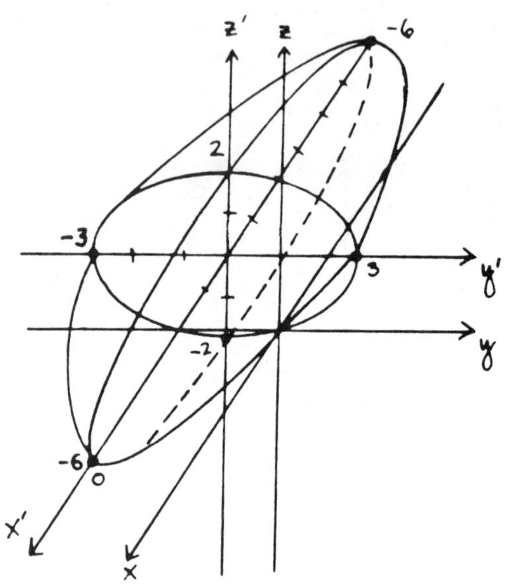

33.

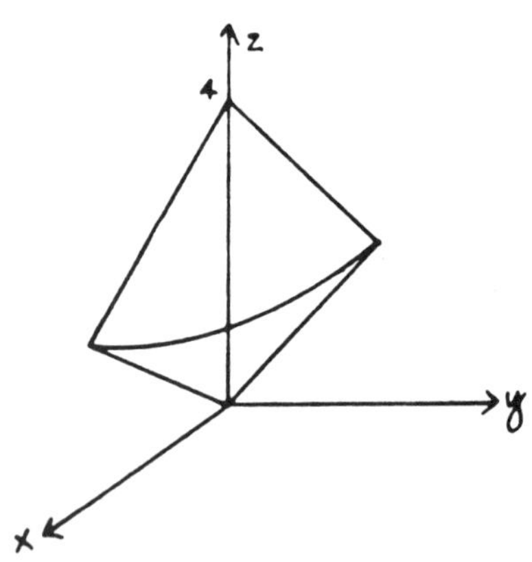

37. a) $x^2 + y^2 + (z - 4)^2 = (z + 4)^2$; $x^2 + y^2 = 16z$.

b) $x^2 + y^2 + (z - 4)^2 = 4(z + 4)^2$;

$x^2 + y^2 + z^2 - 8z + 16 = 4z^2 + 32z + 64$;

$x^2 + y^2 - 3z^2 - 40z - 48 = 0$.

c) $4x^2 + 4y^2 + 4(z - 4)^2 = (z + 4)^2$

$4x^2 + 4y^2 + 3z^2 - 40z + 48 = 0$.

CHAPTER 19, SECTION 7 (pp. 939-940)

1. $f'(t) = -\sin t i + \cos t j + e^t k$,

$f'(0) = j + k$.

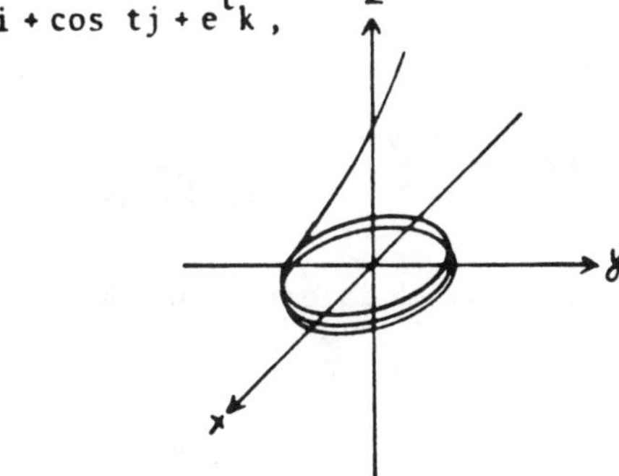

5. $r = \cos t i + \sin t j + t k$,

$r' = -\sin t i + \cos t j + k$, $|r'|$ $\sqrt{\sin^2 t + \cos^2 t + 1} = \sqrt{2}$,

$T = \frac{1}{\sqrt{2}} (- \sin t i + \cos t j + k)$,

$\frac{dT}{dt} = \frac{1}{\sqrt{2}} (- \cos t i - \sin t j)$,

$N = \frac{dT/dt}{|dT/dt|} = -(\cos t i + \sin t j)$,

$B = T \times N = \frac{1}{\sqrt{2}} (-\sin t i + \cos t j + k) \times (-\cos t i - \sin t j)$

$= \frac{1}{\sqrt{2}} (\sin t i - \cos t j + k)$.

9. $\mathbf{r} = \cos t\mathbf{i} + \sin t\mathbf{j}$,

$\mathbf{r}' = -\sin t\mathbf{i} + \cos t\mathbf{j}$, $\quad \mathbf{r}'' = -(\cos t\mathbf{i} + \sin t\mathbf{j})$,

$$K = \frac{|\mathbf{r}' \times \mathbf{r}''|}{|\mathbf{r}'|^3} = \frac{|(-\sin t\mathbf{i} + \cos t\mathbf{j}) \times (\cos t\mathbf{i} + \sin t\mathbf{j})|}{(\sin^2 t + \cos^2 t)^{3/2}} = 1 .$$

13. $\mathbf{r} = t^2\mathbf{i} + t^3\mathbf{j}$, $\quad \mathbf{r}' = 2t\mathbf{i} + 3t^2\mathbf{j}$, $\quad \mathbf{r}'' = 2\mathbf{i} + 6t\mathbf{j}$,

$$v = |\mathbf{r}'| = \sqrt{4t^2 + 9t^4} = |t|\sqrt{4 + 9t^2} .$$

The tangential component of acceleration $a_T = a$

$$= \frac{4t + 18t^3}{\sqrt{4t^2 + 9t^4}} = \frac{2 + 9t^2}{\sqrt{4 + 9t^2}} \cdot \frac{2t}{|t|} .$$

The normal component of acceleration a_N

$$= Kv^2 = \frac{|\mathbf{r}' \times \mathbf{r}''|}{v} = \frac{6t^2}{\sqrt{4t^2 + 9t^4}} = \frac{6|t|}{\sqrt{4 + 9t^2}} .$$

17. $\mathbf{r}' = \mathbf{c}\times\mathbf{r}$, so $\mathbf{r}'$ is orthogonal to $\mathbf{c}$ and $\mathbf{r}$, or

$\mathbf{r}'\cdot\mathbf{c} = 0$, $\quad \mathbf{r}'\cdot\mathbf{r} = 0$.

$$\mathbf{a}\cdot\mathbf{c} = \mathbf{r}''\cdot\mathbf{c} = (\mathbf{c}\times\mathbf{r})'\cdot\mathbf{c} = (\mathbf{c}\times\mathbf{r}' + \mathbf{c}'\times\mathbf{r})\cdot\mathbf{c}$$

$$= (\mathbf{c}\times\mathbf{r}')\cdot\mathbf{c} + (\mathbf{c}'\times\mathbf{r})\cdot\mathbf{c} = 0$$

because $\mathbf{c}' = 0$, as $\mathbf{c}$ is a constant vector, and $\mathbf{c}$ is perpendicular to $\mathbf{c}\times\mathbf{r}'$. Thus $\mathbf{a}$ is perpendicular to $\mathbf{c}$.

To show that v is a constant, show that $\frac{dv}{dt} = 0$.

$$\frac{dv}{dt} = \frac{d}{dt}|\mathbf{c}\times\mathbf{r}| = \frac{d}{dt}\sqrt{c^2r^2 - (\mathbf{c}\cdot\mathbf{r})^2}$$

$$= \frac{c^2(2\mathbf{r}\cdot\mathbf{r}') - 2(\mathbf{c}\cdot\mathbf{r})(\mathbf{c}\cdot\mathbf{r}' + \mathbf{c}'\cdot\mathbf{r})}{2\sqrt{c^2r^2 - (\mathbf{c}\cdot\mathbf{r})^2}} = 0$$

because $\mathbf{r}\perp\mathbf{r}'$, $\mathbf{r}'\perp\mathbf{c}$, $\mathbf{c}' = 0$, and

$$|\mathbf{c}\times\mathbf{r}|^2 = |\mathbf{c}|^2|\mathbf{r}^2|\sin^2\theta = |\mathbf{c}|^2|\mathbf{r}|^2(1 - \cos^2\theta)$$

$$= (\mathbf{c}\cdot\mathbf{c})(\mathbf{r}\cdot\mathbf{r}) - (|\mathbf{c}||\mathbf{r}|\cos\theta)^2 = c^2r^2 - (\mathbf{c}\cdot\mathbf{r})^2 .$$

CHAPTER 19, SECTION 8 (pp. 944-945)

1. a) $x = r \cos\theta = 2 \cos 45° = \frac{2}{\sqrt{2}} = \sqrt{2}$

$y = r \sin\theta = 2 \sin 45° = \frac{2}{\sqrt{2}} = \sqrt{2}$

$(\sqrt{2}, \sqrt{2}, 2)$.

b) $x = r \cos\theta = 3 \cos \frac{2\pi}{3} = 3(-\frac{1}{2}) = -\frac{3}{2}$

$y = r \sin\theta = 3 \sin \frac{2\pi}{3} = 3 \cdot \frac{\sqrt{3}}{2} = \frac{3\sqrt{3}}{2}$

$(-\frac{3}{2}, \frac{3\sqrt{3}}{2}, -2)$.

c) $x = r \cos\theta = 2 \cos 0° = 2 \cdot 1 = 2$

$y = r \sin\theta = 2 \sin 0° = 2 \cdot 0 = 0$

$(2,0,1)$.

d) $x = r \cos\theta = 0 \cos \frac{\pi}{4} = 0$

$y = r \sin\theta = 0 \sin \frac{\pi}{4} = 0$

$(0,0,-3)$.

5. a) $x = r \cos\theta = 2 \cos \frac{\pi}{4} = 2 \cdot \frac{1}{\sqrt{2}} = \sqrt{2}$

$y = r \sin\theta = 2 \sin \frac{\pi}{4} = 2 \cdot \frac{1}{\sqrt{2}} = \sqrt{2}$

$(\sqrt{2}, \sqrt{2}, -2)$

$\rho^2 = x^2 + y^2 + z^2 = 2 + 2 + 4 = 8$, $\quad \rho = 2\sqrt{2}$

$z = \rho \cos\phi$, $\quad -2 = 2\sqrt{2} \cos\phi$,

$\cos\phi = -\frac{1}{\sqrt{2}}$, $\quad \phi = \frac{3\pi}{4}$

$(2\sqrt{2}, \frac{\pi}{4}, \frac{3\pi}{4})$.

b) $x = r\cos\theta = 3\cos 30° = 3\cdot\frac{\sqrt{3}}{2} = \frac{3\sqrt{3}}{2}$

$y = r\sin\theta = 3\sin 30° = 3\cdot\frac{1}{2} = \frac{3}{2}$

$(\frac{3\sqrt{3}}{2}, \frac{3}{2}, 4)$

$\rho^2 = x^2 + y^2 + z^2 = \frac{27}{4} + \frac{9}{4} + 16 = 25, \quad \rho = 5$

$z = \rho\cos\phi, \quad 4 = 5\cos\phi, \quad \cos\phi = \frac{4}{5},$

$\phi = \text{Arccos}\,\frac{4}{5}$

$(5, 30°, \text{Arccos}\,\frac{4}{5})$.

c) $x = r\cos\theta = 2\cos\frac{\pi}{2} = 2\cdot 0 = 0$

$y = r\sin\theta = 2\sin\frac{\pi}{2} = 2\cdot 1 = 2$

$(0, 2, -4)$

$\rho^2 = x^2 + y^2 + z^2 = 0 + 4 + 16 = 20, \quad \rho = 2\sqrt{5}$

$z = \rho\cos\phi, \quad -4 = 2\sqrt{5}\cos\phi, \quad \cos\phi = -\frac{2}{\sqrt{5}},$

$\phi = \text{Arccos}(-\frac{2}{\sqrt{5}})$

$(2\sqrt{5}, \frac{\pi}{2}, \text{Arccos}(-\frac{2}{\sqrt{5}}))$.

d) $x = r\cos\theta = 0\cos 45° = 0$

$y = r\sin\theta = 0\sin 45° = 0$

$(0, 0, 3)$

$\rho^2 = x^2 + y^2 + z^2 = 0 + 0 + 9 = 9, \quad \rho = 3$

$z = \rho\cos\phi, \quad 3 = 3\cos\phi, \quad \cos\phi = 1, \quad \phi = 0°$

$(3, 45°, 0°)$.

9. $x^2 - y^2 = z$

$r^2\cos^2\theta - r^2\sin^2\theta = z$

$r^2(\cos^2\theta - \sin^2\theta) = z$

$r^2\cos 2\theta = z$

$r^2 = z \sec 2\theta$

$x^2 - y^2 = z$

$\rho^2\sin^2\phi \cos^2\phi - \rho^2\sin^2\phi \sin^2\theta$

$= \rho \cos \phi$

$\rho^2\sin^2\phi(\cos^2\theta - \sin^2\theta) = \rho \cos \phi$

$\rho \sin^2\phi \cos 2\theta = \cos \phi$

$$\rho = \frac{\cos \phi}{\sin^2\phi} \sec 2\theta$$

$\rho = \csc \phi \cot \phi \sec 2\theta$

13. $4x^2 + 9y^2 + 9z^2 = 1$

$4r^2\cos^2\theta + 9r^2\sin^2\theta + 9z^2 = 1$

$r^2(4 \cos^2\theta + 9 \sin^2\theta) = 1 - 9z^2$

$4x^2 + 9y^2 + 9z^2 = 1$

$4\rho^2\sin^2\phi \cos^2\theta + 9\rho^2\sin^2\phi \sin^2\theta + 9\rho^2\cos^2\phi = 1$

$\rho^2(4 \sin^2\phi \cos^2\theta + 9 \sin^2\phi \sin^2\theta + 9 \cos^2\phi) = 1$

17. $z = r + r \sin \theta$

$z - y = r$

$(z - y)^2 = r^2$

$y^2 - 2yz + z^2 = x^2 + y^2$

$x^2 - z^2 + 2yz = 0$.

21. $\rho^2\sin \phi \cos \phi = 1$

$\rho^4\sin^2\phi \cos^2\phi = 1$

$\rho^4(1 - \cos^2\phi)\cos^2\phi = 1$

$(p^2 - \rho^2\cos^2\phi)\rho^2\cos^2\phi = 1$

$(x^2 + y^2 + z^2 - z^2)z^2 = 1$

$(x^2 + y^2)z^2 = 1$.

25. The two equations imply a curve. Eliminating θ, $r = 2z$ or $x^2 + y^2 = 4z^2$, the equation of a cone. The curve is wrapped around the cone. The rise above the xy plane is equal to the angle θ (in radians) while the distance from the z axis is twice that number. The curve is a conical helix (spiral, spring).

29. The region is approximately that of one-eighth of an ice-cream cone bounded by the sphere $\rho = 4$ above, the cone $x^2 + y^2 = z^2$ or $r = z$ below, lying between the plane $y = 0$ and the plane $y = x$ (which makes an angle of 45° with the xz plane).

CHAPTER 19, REVIEW (pp. 945-948)

1. $x = x_1 + r(x_2 - x_1) = -2 + \frac{1}{3}(1 + 2) = -1$,

$y = y_1 + r(y_2 - y_1) = 1 + \frac{1}{3}(4 - 1) = 2$,

$z = z_1 + r(z_2 - z_1) = 5 + \frac{1}{3}(-4 - 5) = 2$.

5. ℓ_1: $\{3,4,-3\}$

ℓ_2: $\{6,8,-6\}$ Parallel or coincident.

ℓ_3: $(5,1,-2)$, $(-3,0,7)$; $\{8,1,-9\}$. Parallel.

9.

$2 + t = -3 + 2s$	$2s - t = 5$
$3 - 2t = -1 + 3s$	$3s + 2t = 4$
$3 + 5t = -s$	$s + 5t = -3$

The solution of the first pair of equations is $s = 2$, $t = -1$. Since this also satisfies the third, there is a point of intersection at $(1,5,-2)$.

13. $$d = \frac{|2\cdot 4 - 2\cdot 3 + 1\cdot 1 - 4|}{\sqrt{4 + 4 + 1}} = \frac{1}{3}.$$

17. $u = 2i - 5j + k$, $v = i + j - 2k$

$$\cos\theta = \frac{|u \cdot v|}{|u|\,|v|} = \frac{|2\cdot 1 - 5\cdot 1 + 1(-2)|}{\sqrt{4 + 25 + 1}\,\sqrt{1 + 1 + 4}} = \frac{5}{\sqrt{30}\,\sqrt{6}} = \frac{\sqrt{5}}{6}$$

$$= 0.373,$$

$\theta = 68°$.

21. $z^2 - x^2 = 4$.

Hyperbolic cylinder.

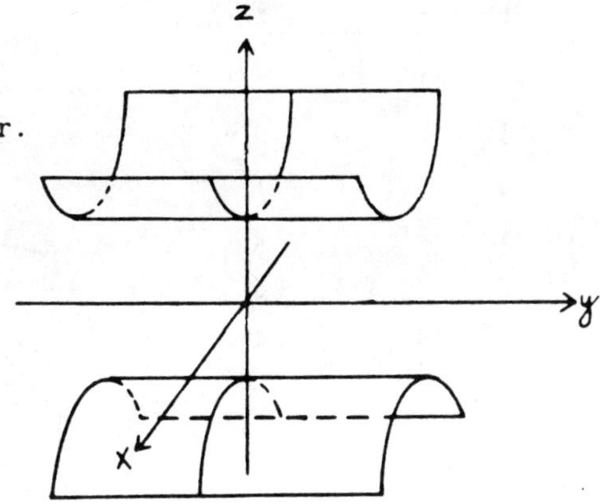

25. $z = r \cos \theta$,

$z = x$.

29. $(5,7,2)$, $(1,1,z)$: midpoint $(3,4,\frac{z+2}{2})$

$$3 = \sqrt{(5-3)^2 + (3-4)^2 + (\frac{z+2}{2} + 2)^2}$$

$$9 = 4 + 1 + \frac{(z+6)^2}{4}, \quad (z+6)^2 = 16,$$

$$z + 6 = \pm 4, \quad z = -6 \pm 4 = -2, -10.$$

33. $(-1,3,1)$, $(4,2,3)$: $u = 5i - j + 2k$

$(2,5,-3)$, $(4,2,3)$: $v = 2i - 3j + 6k$

$u \times v = -26j - 13k$, direction numbers $\{0,2,1\}$.

$x = 4$, $y = 2 + 2t$, $z = 3 + t$.

37.

t	x	y	z
-2	-1	3	-1
-1	0	0	0
0	1	-1	1
1	2	0	2
2	3	3	3
3	4	8	4

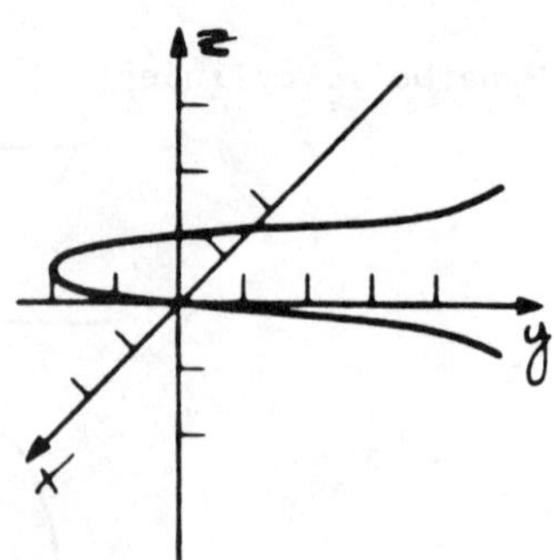

41. Let $u = i$, $v = j$, $w = j + k$

$v \times w = i$, $u \times (v \times w) = 0$,

$u \times v = k$, $(u \times v) \times w = -i$,

$u \times (v \times w) \neq (u \times v) \times w$.

45. $v_1 \cdot v_2 = v_1 \cdot u_2 - \dfrac{u_2 \cdot v_1}{v_1 \cdot v_1} v_1 \cdot v_1 = v_1 \cdot u_2 - u_2 \cdot v_1 = 0$

$$v_1 \cdot v_3 = u_3 \cdot v_1 - \frac{u_3 \cdot v_1}{v_1 \cdot v_1} v_1 \cdot v_1 - \frac{u_2 \cdot v_2}{v_2 \cdot v_2} v_2 \cdot v_1$$

$$= u_3 \cdot v_1 - u_3 \cdot v_1 - 0 = 0.$$

$$v_2 \cdot v_3 = u_3 \cdot v_2 - \frac{u_3 \cdot v_1}{v_1 \cdot v_1} v_1 \cdot v_2 - \frac{u_3 \cdot v_2}{v_2 \cdot v_2} v_2 \cdot v_2$$

$$= u_3 \cdot v_2 - 0 - u_3 \cdot v_2 = 0$$

Chapter 20
Partial Derivatives

CHAPTER 20, SECTION 1 (pp. 956-957)

1. $f(1,2) = 1^2 + 3 \cdot 1 \cdot 2 - 2^2 = 3.$

5. $f(2,1) = \dfrac{2^2 + 1^2}{2^2 + 1^2 + 1} = \dfrac{5}{6}.$

9.

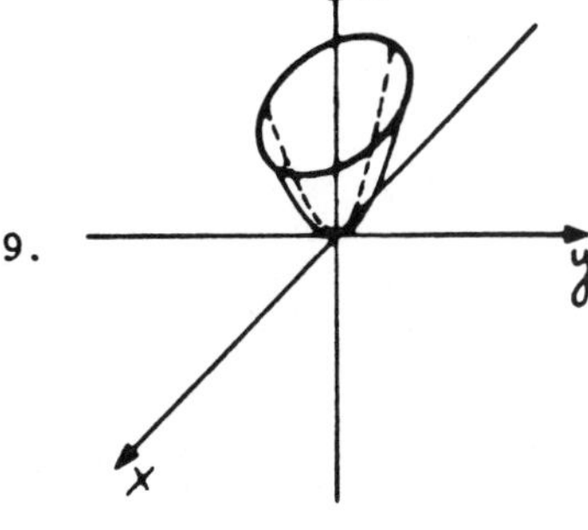

Domain: $\{(x,y) \mid x,y \text{ real}\}$

Range: $\{z \mid z \geq 0\}$.

13.

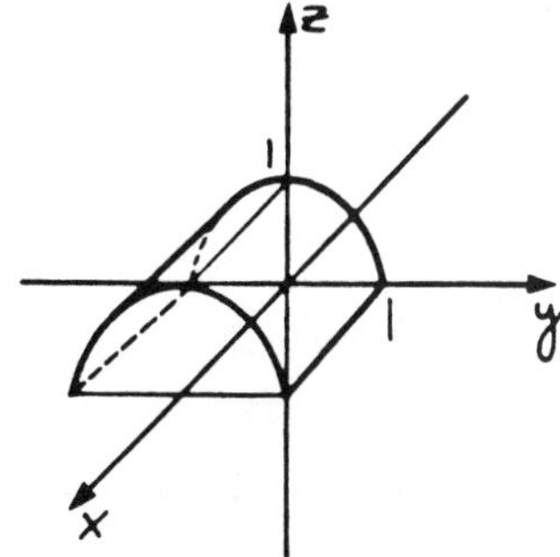

Domain: $\{(x,y) \mid x,y \text{ real}\}$

Range: $\{z \mid z \leq 1\}$.

17.

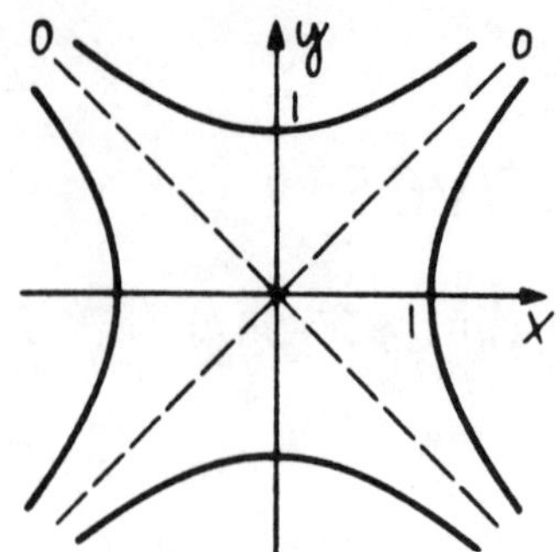

21. $z = \frac{2}{y^2 - 1}$, $x = 1$; $z = \frac{x^2 + 1}{3}$, $y = 2$.

25. $\frac{2 + 6 - 1}{1 - 2 + 2} = 7$.

29.

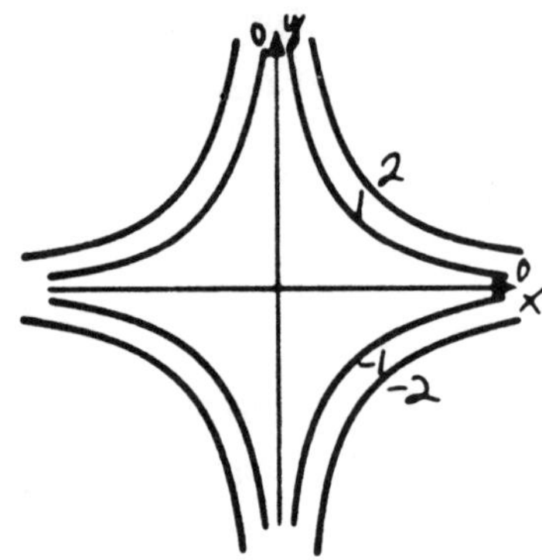

33. $\lim\limits_{\substack{y=mx \\ x\to 0}} \frac{x \cdot mx}{x^2 + (mx)^2} = \frac{m}{1 + m^2}$.

Different limits approached along different lines $y = mx$.

CHAPTER 20, SECTION 2 (pp. 962-964)

1. $\frac{\partial z}{\partial x} = 2x, \frac{\partial z}{\partial y} = 2y$.

5. $\frac{\partial z}{\partial x} = \frac{y}{2\sqrt{xy}}, \quad \frac{\partial z}{\partial y} = \frac{x}{2\sqrt{xy}}$.

9. $\frac{\partial z}{\partial x} = \cos x \cos y , \ \frac{\partial z}{\partial y} = - \sin x \sin y$.

13. $\frac{\partial z}{\partial x} = y = -1 , \quad \frac{\partial z}{\partial y} = x - 3y^2 = 2 - 3(-1)^2 = -1$.

17. $\frac{\partial w}{\partial x} = \sin yz , \ \frac{\partial w}{\partial y} = xz \cos yz , \quad \frac{\partial w}{\partial z} = xy \cos yz$.

21. $\frac{\partial w}{\partial x} = e^{y/z}, \quad \frac{\partial w}{\partial y} = \frac{x}{z} e^{y/z}, \quad \frac{\partial w}{\partial z} = - \frac{xy}{z^2} e^{y/z}$.

25. $\frac{\partial z}{\partial x} = \cos x \cos y , \quad \frac{\partial z}{\partial y} = - \sin x \sin y , \quad \frac{\partial^2 z}{\partial x^2} = - \sin x \cos y,$
$\frac{\partial^2 z}{\partial y \partial x} = - \cos x \sin y , \quad \frac{\partial^2 z}{\partial x \partial y} = -\cos x \sin y ,$
$\frac{\partial^2 z}{\partial y^2} = -\sin x \cos y$.

29. $z = 1 + y^2, \ x = 1; \ \frac{\partial z}{\partial y} = 2y = 4.$

$z = x^2 + 4, \ y = 2; \ \frac{\partial z}{\partial x} = 2x = 2.$

33. $z = \pi \sin y - y, \ x = \pi; \ \frac{\partial z}{\partial y} = \pi \cos y - 1 = -\pi - 1.$

$z = -\pi \cos x, \ y = -\pi; \ \frac{\partial z}{\partial x} = \pi \sin x = 0$

37. $z^2 = x^2 + y^2, \ z = \sqrt{x^2 + y^2}, \ \frac{\partial z}{\partial x} = \frac{x}{\sqrt{x^2 + y^2}}$.

41. $\{(x,y,z) \mid x^2 + y^2 + z^2 \leq 4\}$, the solid sphere bounded by $x^2 + y^2 + z^2 = 4$.

CHAPTER 20, SECTION 3 (pp. 972-975)

1. $\frac{dz}{dt} = \frac{\partial z}{\partial x}\frac{dx}{dt} + \frac{\partial z}{\partial y}\frac{dy}{dt} = 2xy \cdot 2t + x^2 e^t = 4xyt + x^2 e^t$.

5. $$\frac{dw}{dt} = \frac{\partial w}{\partial x}\frac{dx}{dt} + \frac{\partial w}{\partial y}\frac{dy}{dt} + \frac{\partial w}{\partial z}\frac{dz}{dt}$$
$$= \frac{2x}{z} \cdot 1 + \frac{2y}{z} \cdot 2t - \frac{x^2 + y^2}{z^2} \cdot 2t$$
$$= \frac{2xz + 4t\,yz - 2tx^2 - 2ty^2}{z^2} .$$

9. $$\frac{\partial z}{\partial u} = \frac{\partial z}{\partial x}\frac{\partial x}{\partial u} + \frac{\partial z}{\partial y}\frac{\partial y}{\partial u} = \frac{1}{y}\sin v - \frac{x}{y^2}(-v \sin u)$$
$$= \frac{y \sin v + xv \sin u}{y^2} ,$$
$$\frac{\partial z}{\partial v} = \frac{\partial z}{\partial x}\frac{\partial x}{\partial v} + \frac{\partial z}{\partial y}\frac{\partial y}{\partial v} = \frac{1}{y} u \cos v - \frac{x}{y^2} \cos u$$
$$= \frac{yu \cos v - x \cos u}{y^2} .$$

13. $$\frac{\partial w}{\partial u} = \frac{\partial w}{\partial x}\frac{\partial x}{\partial u} + \frac{\partial w}{\partial y}\frac{\partial y}{\partial u} + \frac{\partial w}{\partial z}\frac{\partial z}{\partial u} = yz \cdot 1 + xz \cdot 1 + xy \cdot 2$$
$$= yz + xz + 2xy ,$$
$$\frac{\partial w}{\partial v} = \frac{\partial w}{\partial x}\frac{\partial x}{\partial v} + \frac{\partial w}{\partial y}\frac{\partial y}{\partial v} + \frac{\partial w}{\partial z}\frac{\partial z}{\partial v} = yz \cdot 1 + xz(-1) + xy \cdot 3$$
$$= yz - xz + 3xy .$$

17. $$\frac{\partial w}{\partial t} = \frac{\partial w}{\partial x}\frac{\partial x}{\partial t} + \frac{\partial w}{\partial y}\frac{\partial y}{\partial t} + \frac{\partial w}{\partial z}\frac{\partial z}{\partial t} = 2 \cdot 2t + 1 \cdot 0 - 1 \cdot 2t = 2t ,$$
$$\frac{\partial w}{\partial u} = \frac{\partial w}{\partial x}\frac{\partial x}{\partial u} + \frac{\partial w}{\partial y}\frac{\partial y}{\partial u} + \frac{\partial w}{\partial z}\frac{\partial z}{\partial u} = 2 \cdot 2u + 1 \cdot 2u - 1 \cdot 0 = 6u ,$$
$$\frac{\partial w}{\partial v} = \frac{\partial w}{\partial x}\frac{\partial x}{\partial v} + \frac{\partial w}{\partial y}\frac{\partial y}{\partial v} + \frac{\partial w}{\partial z}\frac{\partial z}{\partial v} = 2 \cdot 0 + 1 \cdot 2v - 1 \cdot 2v = 0 .$$

21. $$\frac{dz}{dx} = \frac{\partial z}{\partial x} + \frac{\partial z}{\partial y}\frac{dy}{dx} = y + x(e^x \operatorname{cox} x + e^x \sin x)$$
$$= y + xe^x(\cos x + \sin x).$$

25. $\frac{dy}{dx} = -\frac{\partial f/\partial x}{\partial f/\partial y} = -\frac{2x - 2y}{-2x - 2y} = \frac{x - y}{x + y}$.

29. $\frac{\partial y}{\partial x} = -\frac{\partial f/\partial x}{\partial f/\partial y} = -\frac{2x - yz}{-xz} = \frac{2x - yz}{xz}$,

$\frac{\partial y}{\partial z} = -\frac{\partial f/\partial z}{\partial f/\partial y} = -\frac{-xy + 2z}{-xz} = \frac{2z - xy}{xz}$.

33. $\frac{\partial z}{\partial r} = \frac{\partial z}{\partial x}\frac{\partial x}{\partial r} + \frac{\partial z}{\partial y}\frac{\partial y}{\partial r} = \frac{\partial z}{\partial x}\cos\theta + \frac{\partial z}{\partial y}\sin\theta$,

$\frac{\partial z}{\partial \theta} = \frac{\partial z}{\partial x}\frac{\partial x}{\partial \theta} + \frac{\partial z}{\partial y}\frac{\partial y}{\partial \theta} = \frac{\partial z}{\partial x}(-r\sin\theta) + \frac{\partial z}{\partial y}\cdot r\cos\theta$,

$\frac{1}{r}\frac{\partial z}{\partial \theta} = -\frac{\partial z}{\partial x}\sin\theta + \frac{\partial z}{\partial y}\cos\theta$

$$\left(\frac{\partial z}{\partial r}\right)^2 + \frac{1}{r^2}\left(\frac{\partial z}{\partial \theta}\right)^2 = \left(\frac{\partial z}{\partial x}\right)^2\cos^2\theta + 2\frac{\partial z}{\partial x}\frac{\partial z}{\partial y}\sin\theta\cos\theta$$

$$+ \left(\frac{\partial z}{\partial y}\right)^2\sin^2\theta + \left(\frac{\partial z}{\partial x}\right)^2\sin^2\theta - 2\frac{\partial z}{\partial x}\frac{\partial z}{\partial y}\sin\theta\cos\theta$$

$$+ \left(\frac{\partial z}{\partial y}\right)^2\cos^2\theta$$

$$= \left(\frac{\partial z}{\partial x}\right)^2(\cos^2\theta + \sin^2\theta) + \left(\frac{\partial z}{\partial y}\right)^2(\sin^2\theta + \cos^2\theta)$$

$$= \left(\frac{\partial z}{\partial x}\right)^2 + \left(\frac{\partial z}{\partial y}\right)^2 .$$

37. $z = f(u)$, where $u = ax + by$.

$\frac{\partial z}{\partial x} = f'(u)\frac{\partial u}{\partial x} = af'(u)$, $\frac{\partial z}{\partial y} = f'(u)\frac{\partial u}{\partial y} = bf'(u)$,

$b\frac{\partial z}{\partial x} = abf'(u) = a\frac{\partial z}{\partial y}$.

41. $A = xy$,

$\frac{dA}{dt} = \frac{\partial A}{\partial x}\frac{dx}{dt} + \frac{\partial A}{\partial y}\frac{dy}{dt} = y\frac{dx}{dt} + x\frac{dy}{dt}$

$= 4\cdot 2 + 5\cdot 3 = 23\ \text{cm}^2/\text{min}$.

45. $\frac{dz}{dt} = \frac{\partial z}{\partial x}\frac{dx}{dt} + \frac{\partial z}{\partial y}\frac{dy}{dt} = 2x\cdot 2t - 2y(-2) = 4tx + 4y$,

$$\frac{d^2z}{dt^2} = \frac{\partial z'}{\partial t} + \frac{\partial z'}{\partial x}\frac{dx}{dt} + \frac{\partial z'}{\partial y}\frac{dy}{dt}$$

$$= 4x + 4t\cdot 2t + 4(-2) = 4x + 8t^2 - 8 = 12t^2 - 8 .$$

49. $f(x + \Delta x,\ y + \Delta y) - f(x,y) = (x + \Delta x)^2 - 2(x + \Delta x)(y + \Delta y)$

$$- (y + \Delta y)^2 - x^2 + 2xy + y^2$$

$$= x^2 + 2x\ \Delta x + (\Delta x)^2 - 2xy - 2x\ \Delta y - 2y\ \Delta x - 2\Delta x\ \Delta y$$

$$- y^2 - 2y\ \Delta y - (\Delta y)^2 - x^2 + 2xy + y^2$$

$$= (2x - 2y)\Delta x + (-2x - 2y)\Delta y + (\Delta x)^2 - (\Delta y)^2 - 2\Delta x\ \Delta y$$

$$= \frac{\partial f}{\partial x}\Delta x + \frac{\partial f}{\partial y}\Delta y + (\Delta x - \Delta y)\Delta x + (-\Delta x - \Delta y)\Delta y\ .$$

53. $y + 3y\frac{\partial z}{\partial x} - 2z\frac{\partial z}{\partial x} = 0$, $\frac{\partial z}{\partial x} = \frac{y}{2z - 3y}$

$$x + 3z + 3y\frac{\partial z}{\partial y} - 2z\frac{\partial z}{\partial y} = 0\ ,\quad \frac{\partial z}{\partial y} = \frac{x + 3z}{2z - 3y}$$

$$3y\frac{\partial^2 z}{\partial x^2} - 2\left(\frac{\partial z}{\partial x}\right)^2 - 2z\frac{\partial^2 z}{\partial x^2} = 0\ ,\quad \frac{\partial^2 z}{\partial x^2} = -\frac{2\left(\frac{\partial z}{\partial x}\right)^2}{2z - 3y}$$

$$= -\frac{2y^2}{(2z - 3y)^3}$$

$$3\frac{\partial z}{\partial y} + 3\frac{\partial z}{\partial y} + 3y\frac{\partial^2 z}{\partial y^2} - 2\left(\frac{\partial z}{\partial y}\right)^2 - 2z\frac{\partial^2 z}{\partial y^2} = 0$$

$$\frac{\partial^2 z}{\partial y^2} = \frac{6\frac{\partial z}{\partial y} - 2\left(\frac{\partial z}{\partial y}\right)^2}{2z - 3y} = \frac{6(x + 3z)(2z - 3y) - 2(x + 3z)^2}{(2z - 3y)^3}$$

$$= \frac{(x + 3z)(12z - 18y - 2x - 6z)}{(2z - 3y)^3} = \frac{2(x + 3z)(3z - x - 9y)}{(2z - 3y)^3}$$

$$1 + 3\frac{\partial z}{\partial x} + 3y\frac{\partial^2 z}{\partial y\partial x} - 2\frac{\partial z}{\partial y}\frac{\partial z}{\partial x} - 2z\frac{\partial^2 z}{\partial y\partial x} = 0$$

$$\frac{\partial^2 z}{\partial x \partial y} = \frac{\partial^2 z}{\partial y \partial x} = \frac{1 + 3\ \partial z/\partial x - 2\ \partial z/\partial x\ \partial z/\partial y}{2z - 3y}$$

$$= \frac{(2z - 3y)^2 + 3y(2z - 3y) - 2y(x + 3z)}{(2z - 3y)^3}$$

$$= \frac{4z^2 - 12yz + 9y^2 + 6yz - 9y^2 - 2xy - 6yz}{(2z - 3y)^3}$$

$$= \frac{4z^2 - 12yz - 2xy}{(2z - 3y)^3} .$$

CHAPTER 20, SECTION 4 (pp. 981-983)

1. $\frac{\partial z}{\partial x} = y = 1 , \ \frac{\partial z}{\partial y} = x = 2 ,$

   $1(x - 2) + 2(y - 1) - (z - 2) = 0 ,$

   $x + 2y - z - 2 = 0$: tangent plane,

   $\frac{x - 2}{1} = \frac{y - 1}{2} = \frac{z - 2}{-1}$: normal line.

5. $\frac{\partial z}{\partial x} = y - 1 = 1 , \ \frac{\partial z}{\partial y} = x - 2 = -1 ,$

   $1(x - 1) - (y - 2) - (z + 1) = 0 ,$

   $x - y - z = 0$: tangent plane,

   $\frac{x - 1}{1} = \frac{y - 2}{-1} = \frac{z + 1}{-1}$: normal line.

9. $F = x^2 + y^2 - z^2 = 0 ,$

   $\frac{\partial F}{\partial x} = 2x = 6 , \quad \frac{\partial F}{\partial y} = 2y = 8 , \quad \frac{\partial F}{\partial z} = -2z = -10 ,$

   $6(x - 3) + 8(y - 4) - 10(z - 5) = 0 ,$

   $3x + 4y - 5z = 0$: tangent plane,

   $\frac{x - 3}{3} = \frac{y - 4}{4} = \frac{z - 5}{-5}$: normal line.

13. $F = xyz + x + y + z + 3 = 0$,

$\frac{\partial F}{\partial x} = yz + 1 = -3$, $\frac{\partial F}{\partial y} = xz + 1 = 3$, $\frac{\partial F}{\partial z} = xy + 1 = -1$,

$-3(x - 1) + 3(y + 2) - (z - 2) = 0$,

$3x - 3y + z - 11 = 0$: tangent plane,

$\frac{x - 1}{-3} = \frac{y + 2}{3} = \frac{z - 2}{-1}$: normal line.

17. $\frac{dx}{dt} = -4 \sin t = -4$, $\frac{dy}{dt} = 3 \cos t = 0$, $\frac{dz}{dt} = 1$, $(0,3,\frac{\pi}{2})$,

$x = -4s$, $y = 3$, $z = \frac{\pi}{2} + s$: tangent line,

$-4(x - 0) + 0(y - 3) + (z - \frac{\pi}{2}) = 0$, $8x - 2z + \pi = 0$: normal plane.

21. $\frac{dx}{dt} = 2t = 6$, $\frac{dy}{dt} = 3t^2 = 27$, $\frac{dz}{dt} = 2$, $(8,28,6)$,

$x = 8 + 6s$, $y = 28 + 27s$, $z = 6 + 2s$: tangent line,

$6(x - 8) + 27(y - 28) + 2(z - 6) = 0$,

$6x + 27y + 2z - 816 = 0$: normal plane.

25. **By Theorem 9, the tangent plane to $f(x,y,z) = 0$ has coefficients $A_1 = \frac{\partial f}{\partial x}$, $B_1 = \frac{\partial f}{\partial y}$, $C_1 = \frac{\partial f}{\partial z}$ and the tangent plane to $g(x,y,z) = 0$ has coefficients $A_2 = \frac{\partial g}{\partial x}$, $B_2 = \frac{\partial g}{\partial y}$, $C_2 = \frac{\partial g}{\partial z}$. The tangent planes are orthogonal if and only if**

$A_1A_2 + B_1B_2 + C_1C_2 = 0$ or $\frac{\partial f}{\partial x}\frac{\partial g}{\partial x} + \frac{\partial f}{\partial y}\frac{\partial g}{\partial y} + \frac{\partial f}{\partial z}\frac{\partial g}{\partial z} = 0$,

i.e., their normal vectors are orthogonal.

29. $f = x^2 + y^2 + z^2 - 12 = 0, \quad g = 2x - 3y - 5z = 0,$

$\frac{\partial f}{\partial x} = 2x = 4, \quad \frac{\partial f}{\partial y} = 2y = -4, \quad \frac{\partial f}{\partial z} = 2z = 4,$

$\frac{\partial g}{\partial x} = 2, \quad \frac{\partial g}{\partial y} = -3, \quad \frac{\partial g}{\partial z} = -5,$

$u = i - j + k, \quad v = 2i - 3j - 5k,$

$u \times v = 8i + 7j - k,$

$x = 2 + 8t, \quad y = -2 + 7t, \quad z = 2 - t$: tangent line.

33. Let $F(x,y,z) = (yz)^{xz}$. $f(2,1,2) = 2^4 = 16,$

$\ln f = xz\ln(yz),$

$\frac{1}{f}\frac{\partial f}{\partial x} = z\ln(yz) = 2\ln z, \frac{\partial f}{\partial x} = 32\ln z,$

$\frac{1}{f}\frac{\partial f}{\partial y} = \frac{xz^2}{yz} = 4, \frac{\partial f}{\partial y} = 64,$

$\frac{1}{f}\frac{\partial f}{\partial z} = x\ln(yz) + \frac{xyz}{yz} = 2\ln 2 + 2, \frac{\partial f}{\partial z} = 32(\ln 2 + 1).$

Direction numbers $\{\ln 2, 2, \ln 2 + 1\}$,

$\ln 2(x-2) + 2(y-1) + (\ln 2 + 1)(z-2) = 0,$

$(\ln 2)x + 2y + (\ln 2 + 1)z = 4(\ln 2 + 1).$

$x = 2 + t\ln 2, \ y = 1 + 2t, \ z = 2 + t(1 + \ln 2).$

CHAPTER 20, SECTION 5 (pp. 991-993)

1. $D_v z = \frac{\partial z}{\partial x}\cos\theta + \frac{\partial z}{\partial y}\sin\theta = (3x^2 + y)\frac{\sqrt{3}}{2} + (x + 3y^2)(-\frac{1}{2})$

$= \frac{3\sqrt{3}\,x^2 - 3y^2 - x + \sqrt{3}\,y}{2}.$

5. $D_v z = \frac{\partial z}{\partial x}\cos\theta + \frac{\partial z}{\partial y}\sin\theta = ye^{xy}\cdot\frac{1}{2} + xe^{xy}(-\frac{\sqrt{3}}{2})$

$= \frac{e^{xy}}{2}(y - \sqrt{3}\,x).$

9. $w = \frac{1}{y} + \frac{2}{x} - \frac{z}{xy}$, $\frac{v}{|v|} = -\frac{1}{2\sqrt{2}} i + \frac{1}{2} j - \frac{\sqrt{5}}{2\sqrt{2}} k$.

$$D_v w = \frac{\partial w}{\partial x} \cos \alpha + \frac{\partial w}{\partial y} \cos \beta + \frac{\partial w}{\partial z} \cos \gamma = \frac{z - 2y}{x^2 y} \cdot \frac{1}{2\sqrt{2}}$$

$$+ \frac{z - x}{xy^2} \cdot \frac{1}{2} - \frac{1}{xy} (- \frac{\sqrt{5}}{2\sqrt{2}})$$

$$= \frac{yz - 2y^2 + \sqrt{2}\, xz - \sqrt{2}\, x^2 + \sqrt{5}\, xy}{2\sqrt{2}\, x^2 y^2} .$$

13. $D_v z = \frac{\partial z}{\partial x} \cos \theta + \frac{\partial z}{\partial y} \sin \theta = 2(x + y) \cdot \frac{1}{\sqrt{5}} + 2(x + y)(- \frac{2}{\sqrt{5}})$

$$= - \frac{2}{\sqrt{5}} (x + y) .$$

At $(1,-1,0)$, $D_v z = 0$.

17. $\frac{v}{|v|} = \frac{1}{\sqrt{6}} i - \frac{2}{\sqrt{6}} j + \frac{1}{\sqrt{6}} k$.

$$D_v w = \frac{\partial w}{\partial x} \cos \alpha + \frac{\partial w}{\partial y} \cos \beta + \frac{\partial w}{\partial z} \cos \gamma = 2x \cdot \frac{1}{\sqrt{6}}$$

$$- 2(y + 2)(- \frac{2}{\sqrt{6}}) - 2(y + z) \frac{1}{\sqrt{6}} = \frac{2x + 2y + 2z}{\sqrt{6}} .$$

At $(2,1,1,0)$, $D_v w = \frac{8}{\sqrt{6}} = \frac{4}{3} \sqrt{6}$.

21. $\nabla z = \frac{\partial z}{\partial x} i + \frac{\partial z}{\partial y} j = 3x^2 yi + x^3 j$,

$$\frac{df}{dn} = |\nabla z| = \sqrt{9x^4 y^2 + x^6} = x^2 \sqrt{9y^2 + x^2} ,$$

$$v = \frac{3x^2 yi + x^3 j}{x^2 \sqrt{9y^2 + x^2}} = \frac{3yi + xj}{\sqrt{9y^2 + x^2}} .$$

25. $\nabla w = \frac{\partial w}{\partial x} i + \frac{\partial w}{\partial y} j + \frac{\partial w}{\partial z} k = yzi + xzj + xyk,$

$$\frac{dw}{dn} = |\nabla w| = \sqrt{y^2z^2 + x^2z^2 + x^2y^2},$$

$$v = \frac{yzi + xzj + xyk}{\sqrt{y^2z^2 + x^2z^2 + x^2y^2}}.$$

29. $\nabla w = \frac{\partial w}{\partial x} i + \frac{\partial w}{\partial y} j + \frac{\partial w}{\partial z} k = yi + (x - z)j - yk$

$$= i + 2j - k \quad \text{at} \quad (1,1,-1,2),$$

$$\frac{dw}{dn} = |\nabla w| = \sqrt{1 + 4 + 1} = \sqrt{6}, \quad v = \frac{i + 2j - k}{\sqrt{6}}.$$

33. $P_o = (2,-1), \quad P_1 = (0,1), \quad u_1 = -2i + 2j,$

$$v_1 = \frac{u_1}{|u_1|} = \frac{-2i + 2j}{\sqrt{4 + 4}} = \frac{-i + j}{\sqrt{2}},$$

$$P_o = (2,-1), \quad P_2 = (-1,1), \quad u_2 = -3i + 2j,$$

$$v_2 = \frac{u_2}{|u_2|} = \frac{-3i + 2j}{\sqrt{9 + 4}} = \frac{-3i + 2j}{\sqrt{13}},$$

$$\frac{\partial f}{\partial x}\left(-\frac{1}{\sqrt{2}}\right) + \frac{\partial f}{\partial y}\left(\frac{1}{\sqrt{2}}\right) = 7, \quad \frac{\partial f}{\partial x}\left(-\frac{3}{\sqrt{13}}\right) + \frac{\partial f}{\partial y}\left(\frac{2}{\sqrt{13}}\right) = 3$$

$$\frac{\partial f}{\partial x} = 14\sqrt{2} - 3\sqrt{13}, \quad \frac{\partial f}{\partial y} = 21\sqrt{2} - 3\sqrt{13},$$

$$P_o = (2,-1), \ P_3 = (6,2), \quad u_3 = 4i + 3j,$$

$$v_3 = \frac{u_3}{|u_3|} = \frac{4i + 3j}{5},$$

$$D_{v_3} f = \frac{\partial f}{\partial x} \cdot \frac{4}{5} + \frac{\partial f}{\partial y} \cdot \frac{3}{5} = \frac{4}{5}(14\sqrt{2} - 3\sqrt{13}) + \frac{3}{5}(21\sqrt{2} - 3\sqrt{13})$$

$$= \frac{119\sqrt{2} - 21\sqrt{13}}{5}.$$

37. $g(s) = f(x + s\cos\theta,\ y + s\sin\theta) = f(u,v)$, where

$u = x + s\cos\theta$, $\quad v = y + s\sin\theta$.

$$\frac{dg}{ds} = g'(s) = \frac{\partial f}{\partial u}\frac{du}{ds} + \frac{\partial f}{\partial v}\frac{dv}{ds} = \frac{\partial f}{\partial u}\cos\theta + \frac{\partial f}{\partial v}\sin\theta .$$

At $s = 0$, $u = x$ and $v = y$, so

$$g'(0) = \frac{\partial f}{\partial x}\cos\theta + \frac{\partial f}{\partial y}\sin\theta .$$

$$D_v f(x,y) = \lim_{s\to 0}\frac{f(x + s\cos\theta,\ y + s\sin\theta) - f(x,y)}{s} = g'(0)$$

Thus $D_v f = \dfrac{\partial f}{\partial x}\cos\theta + \dfrac{\partial f}{\partial y}\sin\theta$.

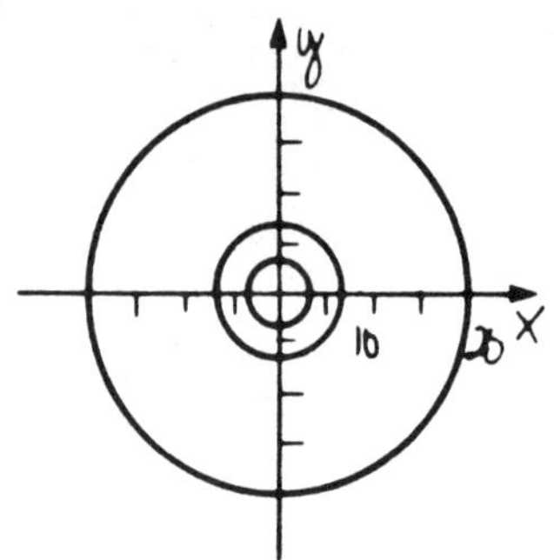

CHAPTER 20, SECTION 6 (pp. 997-998)

1. $df = \dfrac{\partial f}{\partial x}dx + \dfrac{\partial f}{\partial y}dy = (2xy + y^2)dx + (x^2 + 2xy)dy$.

5. $df = \dfrac{\partial f}{\partial x}dx + \dfrac{\partial f}{\partial y}dy = 3(x + \tan xy)^2(1 + y\sec^2 xy)dx$

   $+ 3(x + \tan xy)^2 x\sec^2 xy\, dy$.

9. $df = \dfrac{\partial f}{\partial x}dx + \dfrac{\partial f}{\partial y}dy + \dfrac{\partial f}{\partial z}dz = 2xy\, dx + (\dfrac{z}{\sqrt{1 - y^2}} + x^2)dy$

   $+ \text{Arcsin } y\, dz$.

13. $z = \dfrac{\sqrt[3]{x}}{y}$,

$$dz = \frac{\partial z}{\partial x}\,dx + \frac{\partial z}{\partial y}\,dy = \frac{dx}{3yx^{2/3}} - \frac{\sqrt[3]{x}}{y^2}\,dy$$

$$= \frac{0.02}{3\cdot 4\cdot 4} - \frac{2}{16}\,(0.11) = -0.13$$

$x = 8 \qquad y = 4 \qquad z = 0.5$

$dx = 0.02 \qquad dy = 0.11 \qquad dz = -0.13$

$$\frac{\sqrt[3]{8.02}}{4.11} \approx 0.5 - 0.13 = 0.487\ .$$

17. $V = xyz\ , \quad dV = yz\,dx + xz\,dy + xy\,dz$

$$|dV| \le yz|dx| + xz|dy| + xy|dz|$$

$$\le 3\cdot 6(0.01) + 4\cdot 6(0.01) + 4\cdot 3(0.01) = 0.54\ \text{in}^3\ .$$

21. $\sin\theta = \dfrac{x}{y}, \quad x = y\sin\theta$

$$dx = \sin\theta\,dy + y\cos\theta\,d\theta$$

$$|dx| = |\sin\theta\,dy + y\cos\theta\,d\theta|$$

$$\le \sin\theta|dy| + y\cos\theta|d\theta|$$

$$\le (\sin 21^\circ)(0.3) + 500(\cos 21^\circ)(30'')$$

$$= (0.36)(0.3) + 500(0.93)(.00015) = 0.175.$$

25. $f(x,y) = (2x+y)^{x-2y}, \ D = \{(x,y)\,|\,2x+y>0\}.$

$$\ln f = (x-2y)\ln(2x+y)$$

$$\frac{1}{f}\frac{\partial f}{\partial x} = \ln(2x+y) + \frac{2x-4y}{2x+y}\ ,$$

$$\frac{\partial f}{\partial x} = (2x+y)^{x-2y}[\ln(2x+y) + \frac{2x-4y}{2x+y}],$$

$$\frac{1}{f}\frac{\partial f}{\partial y} = -2\ln(2x+y) + \frac{x-2y}{2x+y},$$

$$\frac{\partial f}{\partial y} = (2x+y)^{x-2y}[-2\ln(2x+y) + \frac{x-2y}{2x+y}],$$

$$df = \frac{\partial f}{\partial x}dx + \frac{\partial f}{\partial y}dy$$

$$= (2x+y)^{x-2y}\{[\ln(2x+y) + \frac{2x-4y}{2x+y}]dx$$

$$+ [-2\ln(2x+y) + \frac{x-2y}{2x+y}]dy\}.$$

CHAPTER 20, SECTION 7 (pp. 1004-1005)

1. $\frac{\partial z}{\partial x} = 2x+1$, $\frac{\partial z}{\partial y} = 8y+8$

$2x+1 = 0$, $8y+8 = 0$ gives $x = -\frac{1}{2}$, $y = -1$,

$z = -\frac{13}{4}$.

$\frac{\partial^2 z}{\partial x^2} = 2$, $\frac{\partial^2 z}{\partial x^2} = 8$, $\frac{\partial^2 z}{\partial y\partial x} = 0$, $D = 2\cdot 8 - 0^2 = 16$.

$(-\frac{1}{2}, -1, -\frac{13}{4})$ is a relative minimum.

5. $\frac{\partial z}{\partial x} = 4x+y+2$, $\frac{\partial z}{\partial y} = x+2y-3$

$4x+y+2 = 0$, $x+2y-3 = 0$ gives $x = -1$, $y = 2$,

$z = -2$.

$\frac{\partial^2 z}{\partial x^2} = 4$, $\frac{\partial^2 z}{\partial y^2} = 2$, $\frac{\partial^2 z}{\partial y\partial x} = 1$, $D = 4\cdot 2 - 1^2 = 7$.

$(-1,2,-2)$ is a relative minimum.

9. $\frac{\partial z}{\partial x} = 2x - 4$, $\frac{\partial z}{\partial y} = 6y^2 - 2y - 4$

$x - 2 = 0$ gives $x = 2$; $3y^2 - y - 2 = 0$ gives $y = 1$,

$y = -\frac{2}{3}$.

Critical points: $(2,1,-6)$., $(2,-\frac{2}{3},-\frac{37}{27})$

$\frac{\partial^2 z}{\partial x^2} = 2$, $\frac{\partial^2 z}{\partial y^2} = 12y-2$, $\frac{\partial^2 z}{\partial y\partial x} = 0$, $D = 24y - 4$

$(2,1,-6)$: $D = 20$, $\frac{\partial^2 z}{\partial x^2} = 2$, $\frac{\partial^2 z}{\partial y} = 10$;

relative minimum.

$(2,-\frac{2}{3},-\frac{37}{27})$: $D = -20$; neither maximum nor minimum.

13. $\frac{\partial z}{\partial x} = 9x^2 - y^2$, $\qquad \frac{\partial z}{\partial y} = -2xy + 1$

$9x^2 - y^2 = 0$ gives $y = \pm 3x$.

$y = 3x$ and $-2xy + 1 = 0$ gives $x = \pm 1/\sqrt{6}$.

$y = -3x$ and $-2xy + 1 = 0$ gives $6x^2 = -1$.

Critical points: $(1/\sqrt{6}, 3/\sqrt{6}, 2/\sqrt{6}), (-1/\sqrt{6}, -3/\sqrt{6}, -2/\sqrt{6})$.

$\frac{\partial^2 z}{\partial x^2} = 18x$, $\qquad \frac{\partial^2 z}{\partial y^2} = -2x$.

Since these two derivatives must have opposite signs, neither critical point is a maximum or minimum.

17. $x + y + z = 25$, $\quad z = 25 - x - y$

$w = x^2y^2z = x^2y^2(25 - x - y) = 25x^2y^2 - x^3y^2 - x^2y^3$

$\frac{\partial w}{\partial x} = 50xy^2 - 3x^2y^2 - 2xy^3$, $\quad \frac{\partial w}{\partial y} = 50x^2y - 2x^3y - 3x^2y^2$

$xy^2(50 - 3x - 2y) = 0$ gives $3x + 2y = 50$ (x and y must be positive); $x^2y(50 - 2x - 3y) = 0$ gives $2x + 3y = 50$;

thus $x = 10$, $y = 10$, $z = 5$.

21. $v = xyz$, $\quad 2x + y + z = 4$,

$V = xy(4 - 2x - y) = 4xy - 2x^2y - xy^2$

$\frac{\partial V}{\partial x} = 4y - 4xy - y^2 = y(4 - 4x - y)$

$\frac{\partial V}{\partial y} = 4x - 2x^2 - 2xy = 2x(2 - x - y)$

$4x + y = 4$, $\quad x + y = 2$

$x = \frac{2}{3}$, $\quad y = \frac{4}{3}$, $\quad z = \frac{4}{3}$

$V = xyz = \frac{2}{3} \cdot \frac{4}{3} \cdot \frac{4}{3} = \frac{32}{27}$.

25. $P = (x - 65)(1200 - 40x + 20y) + (y - 80)(2650 + 10x - 30y)$

$$= -40x^2 + 30xy - 30y^2 + 3000x + 3750y - 290{,}000$$

$$\frac{\partial P}{\partial x} = -80x + 30y + 3000 = 0\ , \quad \frac{\partial P}{\partial y} = 30x - 60y + 3750 = 0$$

$$8x - 3y = 300\ , \quad -x + 2y = 125$$

$$x = 75\ , \quad y = 100$$

$$\frac{\partial^2 P}{\partial x^2} = -80\ , \quad \frac{\partial^2 P}{\partial y^2} = -60\ , \quad \frac{\partial^2 P}{\partial y \partial x} = 30$$

$$D = \begin{vmatrix} -80 & 30 \\ 30 & -60 \end{vmatrix} = 4800 - 900 > 0\ .$$

Hence $x = 75$, $y = 100$ gives a maximum.

CHAPTER 20, SECTION 8 (pp. 1011-1012)

1. $u = x^2 + y^2 + z^2 + \lambda(xyz - 1)$

$$\frac{\partial u}{\partial x} = 2x + \lambda yz = 0, \quad \frac{\partial u}{\partial y} = 2y + \lambda xz = 0, \quad \frac{\partial u}{\partial z} = 2z + \lambda xy = 0$$

$$\frac{\partial u}{\partial \lambda} = xyz - 1 = 0$$

$$2x^2 = -\lambda xyz, \quad 2y^2 = -\lambda xyz, \quad 2z^2 = -\lambda xyz$$

$$x^2 = y^2 = z^2\ , \quad |x| = |y| = |z|$$

$$\pm x^3 = 1, \quad x = \pm 1, \quad y = \pm 1, \quad z = \pm 1$$

(one or three positive)

$$w = x^2 + y^2 + z^2 = 3\ .$$

5. $u = x^2 + 3y^2 + 2z^2 + \lambda(2x - 3y + 5z - 1)$

$$\frac{\partial u}{\partial x} = 2x + 2\lambda = 0, \quad \frac{\partial u}{\partial y} = 6y - 3\lambda = 0, \quad \frac{\partial u}{\partial z} = 4z + 5\lambda = 0$$

$$\frac{\partial u}{\partial \lambda} = 2x - 3y + 5z - 1 = 0, \quad -\lambda = x = -2y = \frac{4}{5} z$$

$$2x + \frac{3}{2}x + \frac{25}{4}x = 1, \quad x = \frac{4}{39}, \quad y = -\frac{2}{39}, \quad z = \frac{5}{39}$$

$$w = \frac{16}{1521} + \frac{12}{1521} + \frac{50}{1521} = \frac{2}{39}\ .$$

9. $u = xyz + \lambda(2x^2 + y^2 + 4z^2 - 4)$

$\frac{\partial u}{\partial x} = yz + 4\lambda x = 0, \quad \frac{\partial u}{\partial y} = xz + 2\lambda y = 0, \quad \frac{\partial u}{\partial z} = xy + 8\lambda z = 0$

$\frac{\partial u}{\partial \lambda} = 2x^2 + y^2 + 4z^2 - 4 = 0$

$-8\lambda xyz = 2y^2z^2 = 4x^2z^2 = x^2y^2; \quad y^2 = 2x^2, \quad z^2 = \frac{x^2}{2}$

$2x^2 + 2x^2 + 2x^2 = 4, \quad x^2 = \frac{2}{3}, \quad x = \pm\sqrt{\frac{2}{3}}, \quad y = \pm\frac{2}{\sqrt{3}}$

$z = \pm\frac{1}{\sqrt{3}}$.

We have a maximum when one or three are positive.

$w = \sqrt{\frac{2}{3}} \cdot \frac{2}{\sqrt{3}} \cdot \frac{1}{\sqrt{3}} = \frac{2\sqrt{2}}{3\sqrt{3}}$.

13. $u = x^2 + y^2 + z^2 + \lambda(x - y + z - 1) + \mu(2x + y + 3z - 6)$

$\frac{\partial u}{\partial x} = 2x + \lambda + 2\mu = 0, \quad \frac{\partial u}{\partial y} = 2y - \lambda + \mu = 0$

$\frac{\partial u}{\partial z} = 2z + \lambda + 3\mu = 0, \quad \frac{\partial u}{\partial \lambda} = x - y + z - 1 = 0$

$\frac{\partial u}{\partial \mu} = 2x + y + 3z - 6 = 0$

$2x + \lambda + 2\mu = 0$, $2y - \lambda + \mu = 0$, and $2z + \lambda + 3\mu = 0$

gives $2x + 2y + 3\mu = 0$, $2y + 2z + 4\mu = 0$, and

$8x + 2y - 6z = 0$ or $4x + y - 3z = 0$.

Solving this together with $x - y + z = 1$ and

$2x + y + 3z = 6$, we have $x = \frac{9}{13}$, $y = \frac{12}{13}$, $z = \frac{16}{13}$.

Thus $w = x^2 + y^2 + z^2 = \frac{81}{169} + \frac{144}{169} + \frac{256}{169} = \frac{37}{13}$.

17. $d = \sqrt{x^2 + y^2 + z^2}$ is to be a minimum. d is a minimum whenever $w = d^2$ is a minimum. Thus

$w = x^2 + y^2 + z^2, \quad -x^2 + y - z^2 = 2$

$u = x^2 + y^2 + z^2 + \lambda(-x^2 + y - z^2 - 2)$

$\frac{\partial u}{\partial x} = 2x - 2x\lambda = 0, \quad \frac{\partial u}{\partial y} = 2y + \lambda = 0, \quad \frac{\partial u}{\partial z} = 2z - 2z\lambda = 0$

$\frac{\partial u}{\partial \lambda} = -x^2 + y - z^2 - 2 = 0$

$2x = 2x\lambda$ gives $x = 0$ or $\lambda = 1$. Assume $\lambda = 1$.

Then $y = -\frac{1}{2}$ and $x^2 + z^2 = -\frac{5}{2}$, which is impossible.

Thus $x = 0$. Similarly $2z = 2z\lambda$ gives $z = 0$.

Since $-x^2 + y - z^2 = 2$, $y = 2$. The nearest point is $(0,2,0)$.

21. $V = xyz$, $\quad u = xyz + \lambda(x + 2y + z - 2)$

$\frac{\partial u}{\partial x} = yz + \lambda = 0, \quad \frac{\partial u}{\partial y} = xz + 2\lambda = 0, \quad \frac{\partial u}{\partial z} = xy + \lambda = 0$

$\frac{\partial u}{\partial \lambda} = x + 2y + z - 2 = 0$

$-2\lambda = 2yz = xz = 2xy$. Thus $x = z = 2y$ (since none of them can be zero at a maximum) and $x + 2y + z = 2$ becomes $3x = 2$ or $x = \frac{2}{3}$, $y = \frac{1}{3}$, $z = \frac{2}{3}$, and

$V = xyz = \frac{2}{3} \cdot \frac{1}{3} \cdot \frac{2}{3} = \frac{4}{27}$.

CHAPTER 20, REVIEW (pp. 1012-1014)

1. $\frac{\partial w}{\partial x} = 2x + y, \quad \frac{\partial w}{\partial y} = x - z, \quad \frac{\partial w}{\partial z} = -y.$

5. $\frac{\partial w}{\partial u} = \frac{\partial w}{\partial x}\frac{\partial x}{\partial u} + \frac{\partial w}{\partial y}\frac{\partial y}{\partial u} + \frac{\partial w}{\partial z}\frac{\partial z}{\partial u} = (y + 3z)2u + (x + 2z)2u + (2y+3x)$

$= 2u(x + y + 5z) + 4y + 6x$

$\frac{\partial w}{\partial v} = \frac{\partial w}{\partial x}\frac{\partial x}{\partial v} + \frac{\partial w}{\partial y}\frac{\partial y}{\partial v} + \frac{\partial w}{\partial z}\frac{\partial z}{\partial v} = (y + 3z)2v + (x + 2z)(-2v)$

$+ (2y + 3x)3$

$= 2v(y - x + z) + 6y + 9x.$

9. $df = \frac{\partial f}{\partial x} dx + \frac{\partial f}{\partial y} dy = (y \cos xy + \cos y)\, dx$

$+ (x \cos xy - x \sin y) dy.$

13. $\frac{\partial z}{\partial x} = 6x^2 = 6, \quad \frac{\partial z}{\partial y} = -12y^2 = -12$

$6(x - 1) - 12(y - 1) - (z + 2) = 0,$

$6x - 12y - z + 4 = 0$; tangent plane,

$x = 1 + 6t, \quad y = 1 - 12t, \quad z = -2 - t$: normal line.

17.

$x = \sin t$	$y = \cos 2t$	$z = t$
$x' = \cos t = \frac{\sqrt{3}}{2}$	$y' = -2 \sin 2t = -\sqrt{3}$	$z' = 1$
$x = \frac{1}{2}$	$y = \frac{1}{2}$	$z = \frac{\pi}{6}$
$x = \frac{1}{2} + \frac{\sqrt{3}}{2} s,$	$y = \frac{1}{2} - \sqrt{3}\, s,$	$z = \frac{\pi}{6} + s.$

21. $\frac{\partial z}{\partial x} = y^2 - 2y = 0, \quad \frac{\partial z}{\partial y} = 3y^2 + 2xy - 2x = 0$

$y(y - 2) = 0, \quad y = 0$ or $y = 2.$

If $y = 0$, then $3y^2 + 2xy - 2x = -2x = 0$; $x = 0.$

If $y = 2$, then $3y^2 + 2xy - 2x = 12 + 2x = 0$; $x = -6.$

	(0,0,0)	(-6,2,8)
$\frac{\partial^2 z}{\partial x^2} = 0$	0	0
$\frac{\partial^2 z}{\partial y^2} = 6y + 2x$	0	0
$\frac{\partial^2 z}{\partial y \partial x} = 2y - 2$	-2	2
D	-4	-4

(0,0,0) neither; (-6,2,8) neither.

25. $V = 4xyz = 4xy(4 - x^2 - 4y^2) = 16xy - 4x^3y - 16xy^3$

$$\frac{\partial V}{\partial x} = 16y - 12x^2y - 16y^3 = 0, \quad 4y(4 - 3x^2 - 4y^2) = 0$$

$$\frac{\partial V}{\partial y} = 16x - 4x^3 - 48xy^2 = 0, \quad 4x(4 - x^2 - 12y^2) = 0$$

$$3x^2 + 4y^2 = 4, \qquad x^2 + 12y^2 = 4$$

$$x^2 = 1, \quad y^2 = 1/4; \quad x = 1, \quad y = 1/2$$

$$V = 4xy(4 - x^2 - 4y^2) = 4.$$

29. $|f(x,y) - f(0,0)| \le \dfrac{|x|y^2}{x^2 + y^2} \le |x|$, so for

$|x| < \varepsilon$, $|f(x,y) - f(0,0)| < \varepsilon$; choose $\delta = \varepsilon$.

$$\frac{\partial f}{\partial x} = \lim_{x \to 0} \frac{f(x,0) - f(0,0)}{x} = \lim_{x \to 0} \frac{0 - 0}{x} = 0.$$

Similarly, $\dfrac{\partial f}{\partial y} = 0$.

Suppose $f(\Delta x, \Delta y) = \dfrac{\partial f}{\partial x}\Delta x + \dfrac{\partial f}{\partial y}\Delta y + \varepsilon\Delta x + \eta\Delta y$

$= 0\Delta x + 0\Delta y + \varepsilon\Delta x + \eta\Delta y.$

Then

$$\frac{\Delta x \Delta y^2}{\Delta x^2 + \Delta y^2} = \varepsilon\Delta x + \eta\Delta y,$$

$$\frac{\Delta y^2}{\Delta x^2 + \Delta y^2} = \varepsilon + \eta\frac{\Delta y}{\Delta x}.$$

For $\Delta x = \Delta y$, $\frac{1}{2} = \varepsilon + \eta$, and not both $\varepsilon \to 0$ and $\eta \to 0$ as $\Delta x = \Delta y \to 0$. Thus f is not differentiable at (0,0) (see Problem 28). It follows that not both $\dfrac{\partial f}{\partial x}$ and $\dfrac{\partial f}{\partial y}$ are continuous at (0,0).

33. $\frac{\partial z}{\partial x} = a^3y - \frac{1}{x^2}$, $\frac{\partial z}{\partial y} = a^3x - \frac{1}{y^2}$

$a^3y - \frac{1}{x^2} = 0$, $a^3x^2y = 1$

$a^3x - \frac{1}{y^2} = 0$, $a^3xy^2 = 1$

$x = \frac{1}{a}$, $y = \frac{1}{a}$

$\frac{\partial^2 z}{\partial x^2} = \frac{2}{x^3} = 2a^3$, $\frac{\partial^2 z}{\partial y^2} = \frac{2}{y^3} = 2a^3$,

$\frac{\partial^2 z}{\partial y \partial x} = a^3$

$D = 4a^6 - a^6 = 3a^6 > 0$

$a > 0$, $\frac{\partial^2 z}{\partial x^2} = 2a^3 > 0$, relative minimum

$a < 0$, $\frac{\partial^2 z}{\partial x^2} = 2a^3 < 0$, relative maximum.

No. For $y = 1$, $z \to \pm\infty$ as $x \to 0^{\pm}$.

Chapter 21
Multiple Integrals

CHAPTER 21, SECTION 1 (pp. 1027-1029)

1. $\int_0^2 \int_0^3 (xy + x - y)\,dy\,dx$

$= \int_0^2 \left(x\,\frac{y^2}{2} + xy - \frac{y^2}{2}\Big|_0^3\right) dx$

$= \int_0^2 \left(x \cdot \frac{9}{2} + 3x - \frac{9}{2}\right) dx$

$= \frac{15x^2}{4} - \frac{9x}{2}\Big|_0^2 = 15 - 9 = 6\ .$

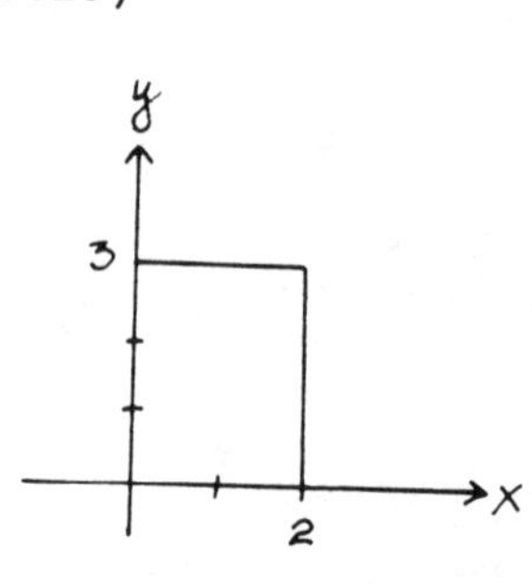

5. $\int_0^2 \int_0^{x^2} (x^2 - y^2)\,dy\,dx$

$= \int_0^2 \left(x^2 y - \frac{y^3}{3}\right)\Big|_0^{x^2} dx$

$= \int_0^2 \left(x^4 - \frac{x^6}{3}\right) dx$

$= \left(\frac{x^5}{5} - \frac{x^7}{21}\right)\Big|_0^2$

$= \frac{32}{5} - \frac{128}{21} = \frac{32}{105}\ .$

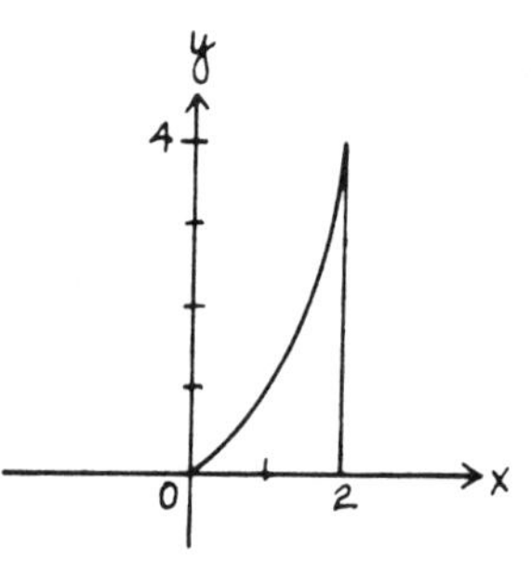

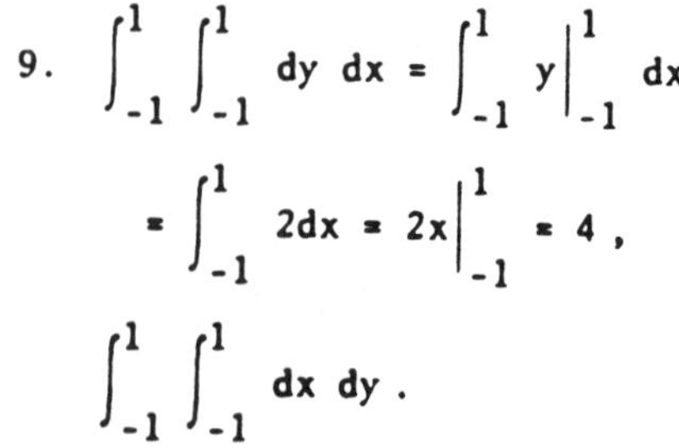

9. $\int_{-1}^1 \int_{-1}^1 dy\,dx = \int_{-1}^1 y\Big|_{-1}^1 dx$

$= \int_{-1}^1 2dx = 2x\Big|_{-1}^1 = 4\ ,$

$\int_{-1}^1 \int_{-1}^1 dx\,dy\ .$

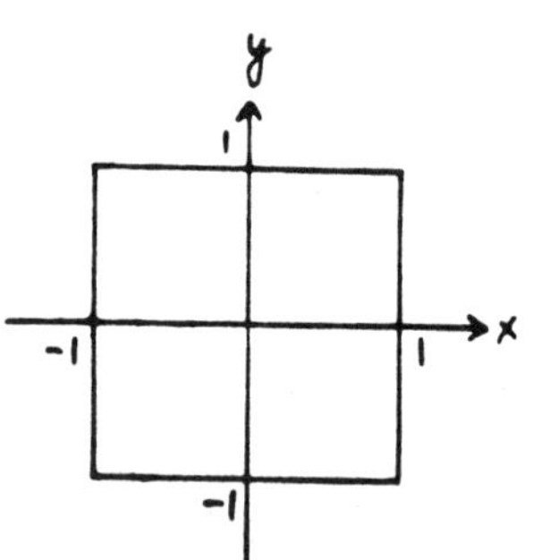

13. $\int_0^1 \int_0^y e^{x+y} dx\, dy = \int_0^1 e^{x+y}\Big|_0^y dy$

$= \int_0^1 (e^{2y} - e^y) dy = (\frac{1}{2} e^{2y} - e^y)\Big|_0^1$

$= \frac{1}{2} e^2 - e - \frac{1}{2} + 1 = \frac{1}{2} e^2 - e + \frac{1}{2},$

$\int_0^1 \int_x^1 e^{x+y}\, dy\, dx .$

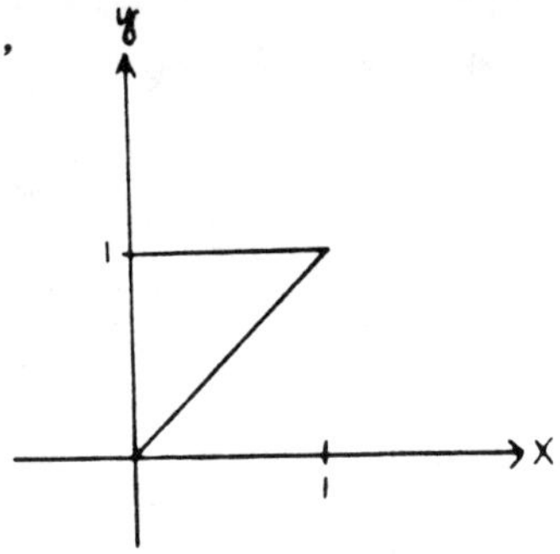

17. $\int_0^1 \int_0^{\sqrt[3]{y}} (x^3 - y^3) dx\, dy = \int_0^1 (\frac{x^4}{4} - xy^3)\Big|_0^{y^{1/3}} dy$

$= \int_0^1 (\frac{y^{4/3}}{4} - y^{10/3}) dy = (\frac{3y^{7/3}}{28} - \frac{3y^{13/3}}{13})\Big|_0^1 = \frac{3}{28} - \frac{3}{13}$

$= - \frac{45}{364} .$

21. $\int_0^1 \int_{x^3}^x (x + y)^2 dy\, dx = \int_0^1 \frac{(x + y)^3}{3}\Big|_{x^3}^x dx$

$= \int_0^1 (\frac{8x^3}{3} - \frac{(x + x^3)^3}{3}) dx = \int_0^1 (\frac{7}{3}x^3 - x^5 - x^7 - \frac{1}{3} x^9) dx$

$= (\frac{7x^4}{12} - \frac{x^6}{6} - \frac{x^8}{8} - \frac{x^{10}}{30})\Big|_0^1 = \frac{7}{12} - \frac{1}{6} - \frac{1}{8} - \frac{1}{30} = \frac{31}{120}.$

25. $\int_0^1 \int_y^{9y} \sqrt{xy}\, dx\, dy = \int_0^1 y^{1/2}(\frac{2}{3} x^{3/2}\Big|_y^{9y})dy$

$= \frac{2}{3}\int_0^1 y^{1/2}(27y^{3/2} - y^{3/2})dy = \frac{52}{3}\int_0^1 y^2 dy = \frac{52}{3} \cdot \frac{y^3}{3}\Big|_0^1 = \frac{52}{9}$

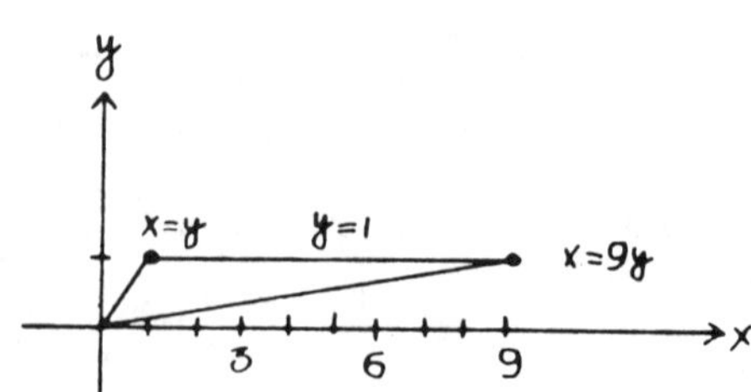

29. $\int_a^b \int_a^y f(x)\, dx\, dy$

$= \int_a^b \int_x^b f(x) dy\, dx$

$= \int_a^b f(x)(y\Big|_x^b)dx$

$= \int_a^b (b - x)f(x)dx .$

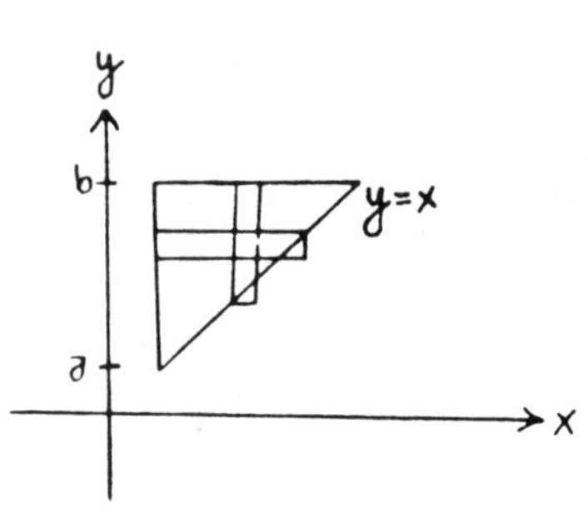

CHAPTER 21, SECTION 2 (pp. 1034-1035)

1. $V = \int_0^1 \int_0^{1-x} (1 - x - y)dy\, dx = \int_0^1 (y - xy - \frac{y^2}{2})\Big|_0^{1-x} dx$

$= \int_0^1 (1 - x - x + x^2 - \frac{1 - 2x + x^2}{2})dx = \int_0^1 (\frac{1}{2} - x + \frac{x^2}{2})dx$

$= (\frac{1}{2}x - \frac{x^2}{2} + \frac{x^3}{6})\Big|_0^1 = \frac{1}{2} - \frac{1}{2} + \frac{1}{6} = \frac{1}{6} .$

5. $V = \int_0^4 \int_0^{4-x} xydydx = \int_0^4 \frac{xy^2}{2}\Big|_0^{4-x} dx = \frac{1}{2}\int_0^4 x(4-x)^2 dx$

$= \frac{1}{2}\int_0^4 (16x - 8x^2 + x^3)dx = \frac{1}{2}(8x^2 - \frac{8x^3}{3} + \frac{x^4}{4})\Big|_0^4$

$= \frac{1}{2}(128 - \frac{512}{3} + 64) = \frac{32}{3}$.

9. $A = \int_0^2 \int_{x^2-2x}^0 dydx = \int_0^2 (2x - x^2)dx = (x^2 - \frac{x^3}{3})\Big|_0^2$

$= 4 - \frac{8}{3} = \frac{4}{3}$.

13. $A = \int_{-\infty}^0 \int_0^{e^x} dydx = \int_{-\infty}^0 e^x dx = \lim_{k\to-\infty} \int_k^0 e^x dx = \lim_{k\to-\infty} e^x\Big|_k^0$

$= \lim_{k\to-\infty} (1 - e^k) = 1$.

17. $m = \int_0^1 \int_{x^2}^x (x^2 + y^2)dydx = \int_0^1 (x^2y + \frac{y^3}{3})\Big|_{x^2}^x dx$

$= \int_0^1 (x^3 + \frac{x^3}{3} - x^4 - \frac{x^6}{3})dx = (\frac{x^4}{3} - \frac{x^5}{5} - \frac{x^7}{21})\Big|_0^1$

$= \frac{1}{3} - \frac{1}{5} - \frac{1}{21} = \frac{3}{35}$.

21. $V = \int_0^2 \int_0^{\sqrt{4-x^2}} y^2 dydx = \int_0^2 \frac{1}{3}(4 - x^2)^{3/2} dx$

(Formula 63, $a = 2$, $u = x$)

$= \frac{1}{3}[\frac{x}{4}(4 - x^2)^{3/2} + \frac{12x}{8}\sqrt{4 - x^2} + \frac{3\cdot16}{8}\text{Sin}^{-1}\frac{x}{2}\Big|_0^2]$

$= \frac{1}{3}\cdot 6 \cdot \frac{\pi}{2} = \pi$.

25. $V = 4\int_0^{\sqrt{2}}\int_0^{\sqrt{2-x^2}} (2 - x^2 - y^2)dydx$

$$= 4\int_0^{\sqrt{2}} [(2 - x^2)y - \frac{y^3}{3}]\Big|_0^{\sqrt{2-x^2}} dx$$

$$= 4\int_0^{\sqrt{2}} [(2 - x^2)^{3/2} - \frac{1}{3}(2 - x^2)^{3/2}]dx = \frac{8}{3}\int_0^{\sqrt{2}} (2-x^2)^{3/2}dx$$

$x = \sqrt{2}\sin\theta, \quad dx = \sqrt{2}\cos\theta\, d\theta$

$$= \frac{8}{3}\int_0^{\pi/2} (2 - 2\sin^2\theta)^{3/2}\sqrt{2}\cos\theta\, d\theta = \frac{32}{3}\int_0^{\pi/2} \cos^4\theta\, d\theta$$

$$= \frac{8}{3}\int_0^{\pi/2} (1 + \cos 2\theta)^2 d\theta = \frac{8}{3}\int_0^{\pi/2} (1 + 2\cos 2\theta + \cos^2 2\theta)d\theta$$

$$= \frac{8}{3}\int_0^{\pi/2} (1 + 2\cos 2\theta + \frac{1 + \cos 4\theta}{2})d\theta$$

$$= \frac{8}{3}(\frac{3\theta}{2} + \sin 2\theta + \frac{1}{8}\sin 4\theta)\Big|_0^{\pi/2} = 2\pi .$$

29. $A = \text{area } R = n\Delta A,\ \Delta A = \frac{A}{n}$

$$M(S) = \frac{1}{n}\sum_{i=1}^{n} f(x_i,y_i) = \frac{1}{A}\sum_{i=1}^{n} f(x_i,y_i)\frac{A}{n}$$

$$= \frac{1}{A}\sum_{i=1}^{n} f(x_i,y_i)\Delta A$$

$$\lim_{\|S\|\to 0} M(S) = \frac{1}{A}\lim_{\|S\|\to 0}\sum_{i=1}^{n} f(x_i,y_i)\Delta A$$

$$= \frac{1}{A}\iint_R f(x,y)dA.$$

CHAPTER 21, SECTION 3 (pp. 1040-1042)

1. $$V = \int_0^{\pi}\int_0^{\sin\theta} 3r\,drd\theta = \int_0^{\pi} \frac{3}{2} r^2\Big|_0^{\sin\theta} d\theta = \frac{3}{2}\int_0^{\pi} \sin^2\theta\, d\theta$$

$$= \frac{3}{4}\int_0^{\pi}(1 - \cos 2\theta)d\theta = \frac{3}{4}(\theta - \frac{1}{2}\sin 2\theta)\Big|_0^{\pi} = \frac{3\pi}{4}.$$

5. $$V = 4\int_0^{\pi/2}\int_0^{\sin 2\theta}(1 + r^2)rdrd\theta = 4\int_0^{\pi/2}\int_0^{\sin 2\theta}(r+r^3)drd\theta$$

$$= 4\int_0^{\pi/2}(\frac{r^2}{2} + \frac{r^4}{4}\Big|_0^{\sin 2\theta} d\theta = 4\int_0^{\pi/2}(\frac{\sin^2 2\theta}{2} + \frac{\sin^4 2\theta}{4})d\theta$$

$$= 4\int_0^{\pi/2}(\frac{1 - \cos 4\theta}{4} + \frac{(1 - \cos 4\theta)^2}{16})d\theta$$

$$= \frac{1}{4}\int_0^{\pi/2}(5 - 6\cos 4\theta + \frac{1 + \cos 8\theta}{2})d\theta$$

$$= \frac{1}{4}(\frac{11\theta}{2} - \frac{3}{2}\sin 4\theta + \frac{1}{16}\sin 8\theta)\Big|_0^{\pi/2} = \frac{11\pi}{16}.$$

9. $$m = \int_0^{2\pi}\int_0^{1-\sin\theta} r\cdot rdr\theta$$

$$= \int_0^{2\pi}(\frac{r^3}{3}\Big|_0^{1-\sin\theta})d\theta$$

$$= \frac{1}{3}\int_0^{2\pi}(1 - 3\sin\theta + 3\sin^2\theta - \sin^3\theta)d\theta$$

$$= \frac{1}{3}\int_0^{2\pi}[1 - 3\sin\theta + \frac{3}{2} - \frac{3}{2}\cos 2\theta - (1 - \cos^2\theta)\sin\theta]d\theta$$

$$= \frac{1}{3}\int_0^{2\pi}(\frac{5}{2} - 4\sin\theta - \frac{3}{2}\cos 2\theta + \cos^2\theta\sin\theta)d\theta$$

$$= \frac{1}{3}(\frac{5\theta}{2} + 4\cos\theta - \frac{3}{4}\sin 2\theta - \frac{1}{3}\cos^3\theta)\Big|_0^{2\pi} = \frac{5\pi}{3}.$$

13. $m = 8 \int_0^{\pi/4} \int_0^{\cos 2\theta} r^2 dr d\theta = \frac{8}{3} \int_0^{\pi/4} r^3 \Big|_0^{\cos 2\theta} d\theta$

$= \frac{8}{3} \int_0^{\pi/4} \cos^3 2\theta = \frac{4}{3} \int_0^{\pi/4} (1 - \sin^2 2\theta) 2 \cos 2\theta d\theta$

$= \frac{4}{3}(\sin 2\theta - \frac{\sin^3 2\theta}{3}) \Big|_0^{\pi/4} = \frac{4}{3}(1 - \frac{1}{3}) = \frac{8}{9}$.

17. $V = 4 \int_0^{\pi/2} \int_0^1 r^3 dr d\theta = \int_0^{\pi/2} r^4 \Big|_0^1 d\theta = \int_0^{\pi/2} d\theta$

$= \theta \Big|_0^{\pi/2} = \frac{\pi}{2}$.

21. $V = 2 \int_0^{\pi/2} \int_0^{2 \cos \theta} 2r^2 dr d\theta = \frac{4}{3} \int_0^{\pi/2} r^3 \Big|_0^{2 \cos \theta} d\theta$

$= \frac{4}{3} \int_0^{\pi/2} 8 \cos^3\theta \, d\theta = \frac{32}{3} \int_0^{\pi/2} (1 - \sin^2\theta) \cos\theta \, d\theta$

$= \frac{32}{3}(\sin \theta - \frac{\sin^3\theta}{3}) \Big|_0^{\pi/2} = \frac{32}{3}(1 - \frac{1}{3}) = \frac{64}{9}$.

25.

$$\frac{\partial(x,y)}{\partial(r,\theta)} = \begin{vmatrix} \frac{\partial x}{\partial r} & \frac{\partial x}{\partial \theta} \\ \frac{\partial y}{\partial r} & \frac{\partial y}{\partial \theta} \end{vmatrix} = \begin{vmatrix} \cos \theta & -r \sin \theta \\ \sin \theta & r \cos \theta \end{vmatrix}$$

$= r \cos^2\theta + r \sin^2\theta = r$.

29. $m = \int_{1/2}^{1} \int_{0}^{1/r} kr^2 \cdot r d\theta dr$

$= k \int_{1/2}^{1} r^3\theta \Big|_{0}^{1/r} dr$

$= k \int_{1/2}^{1} r^2 dr$

$= \frac{kr^3}{3}\Big|_{1/2}^{1} = k(\frac{1}{3} - \frac{1}{24}) = \frac{7k}{24}$.

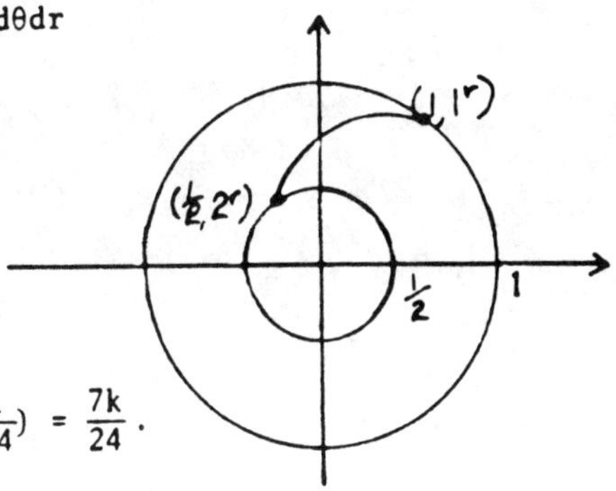

$(\frac{1}{2}, 2^r)$ $(1, 1^r)$ $\frac{1}{2}$ 1

33. $(\int_{-\infty}^{+\infty} e^{-x^2/2} dx)^2 = \int_{-\infty}^{+\infty} e^{-x^2/2} dx \int_{-\infty}^{+\infty} e^{-y^2/2} dy$

$= \int_{-\infty}^{+\infty}\int_{-\infty}^{+\infty} e^{-x^2/2} e^{-y^2/2} dydx$

$= \int_{-\infty}^{+\infty}\int_{-\infty}^{+\infty} e^{-(x^2+y^2)/2} dy\, dx$.

Using $x = r\cos\theta$, $y = r\sin\theta$ and Problem 25,

$= \int_{0}^{2\pi} \int_{0}^{+\infty} e^{-r^2/2} r dr d\theta$

$= \int_{0}^{+\infty} \int_{0}^{2\pi} e^{-r^2/2} r d\theta dr = \int_{0}^{+\infty} 2\pi e^{-r^2/2} r dr$

$= \lim_{k\to+\infty} \int_{0}^{k} -2\pi e^{-r^2/2}(-r)dr = \lim_{k\to+\infty} -2\pi e^{-r^2/2}\Big|_{0}^{k}$

$= \lim_{k\to+\infty} (2\pi - 2\pi e^{-k^2/2}) = 2\pi$.

Therefore $\int_{-\infty}^{+\infty} e^{-x^2/2} dx = \sqrt{2\pi}$.

CHAPTER 21, SECTION 4 (pp. 1048-1050)

1. $m = \int_0^1 \int_0^{\sqrt{y}} xydxdy = \int_0^1 \frac{x^2y}{2}\Big|_0^{\sqrt{y}} dy = \int_0^1 \frac{y^2}{2} dy = \frac{y^3}{6}\Big|_0^1 = \frac{1}{6}$

$$M_x = \int_0^1 \int_0^{\sqrt{y}} xy^2dxdy = \int_0^1 \frac{x^2y^2}{2}\Big|_0^{\sqrt{y}} dy = \int_0^1 \frac{y^3}{2} dy$$

$$= \frac{y^4}{8}\Big|_0^1 = \frac{1}{8}$$

$$M_y = \int_0^1 \int_0^{\sqrt{y}} x^2ydxdy = \int_0^1 \frac{x^3y}{3}\Big|_0^{\sqrt{y}} dy = \int_0^1 \frac{y^{5/2}}{3} dy$$

$$= \frac{2y^{7/2}}{21}\Big|_0^1 = \frac{2}{21}$$

$$\bar{x} = \frac{M_y}{m} = \frac{2/21}{1/6} = \frac{4}{7}, \quad \bar{y} = \frac{M_x}{m} = \frac{1/8}{1/6} = \frac{3}{4}.$$

5. $m = \int_0^1 \int_{x^2}^{x} x^2ydydx = \int_0^1 \frac{x^2y^2}{2}\Big|_{x^2}^{x} dx = \int_0^1 (\frac{x^4}{2} - \frac{x^6}{2})dx$

$$= (\frac{x^5}{10} - \frac{x^7}{14})\Big|_0^1 = \frac{1}{10} - \frac{1}{14} = \frac{1}{35}$$

$$M_x = \int_0^1 \int_{x^2}^{x} x^2y^2dydx = \int_0^1 \frac{x^2y^3}{3}\Big|_{x^2}^{x} dx = \int_0^1 (\frac{x^5}{3} - \frac{x^8}{3})dx$$

$$= (\frac{x^6}{18} - \frac{x^9}{27}\Big|_0^1 = \frac{1}{18} - \frac{1}{27} = \frac{1}{54}$$

$$M_y = \int_0^1 \int_{x^2}^{x} x^3ydydx = \int_0^1 \frac{x^3y^2}{2}\Big|_{x^2}^{x} dx = \int_0^1 (\frac{x^5}{2} - \frac{x^7}{2})dx$$

$$= (\frac{x^6}{12} - \frac{x^8}{16})\Big|_0^1 = \frac{1}{12} - \frac{1}{16} = \frac{1}{48}$$

$$\bar{x} = \frac{M_y}{m} = \frac{1/48}{1/35} = \frac{35}{48}, \quad \bar{y} = \frac{M_x}{m} = \frac{1/54}{1/35} = \frac{35}{54}.$$

9. $m = \int_0^1 \int_0^{\sqrt{y}} (x + y)dx\ dy = \int_0^1 (\frac{x^2}{2} + xy)\Big|_0^{\sqrt{y}} dy$

$= \int_0^1 (\frac{y}{2} + y^{3/2})dy = \frac{y^2}{4} + \frac{2}{5} y^{5/2}\Big|_0^1 = \frac{1}{4} + \frac{2}{5} = \frac{13}{20}$

$I_x = \int_0^1 \int_0^{\sqrt{y}} (xy^2 + y^3)dx\ dy = \int_0^1 (\frac{x^2y^2}{2} + xy^3)\Big|_0^{\sqrt{y}} dy$

$= \int_0^1 (\frac{y^3}{2} + y^{7/2})dy = \frac{y^4}{8} + \frac{2y^{9/2}}{9}\Big|_0^1 = \frac{1}{8} + \frac{2}{9} = \frac{25}{72}$

$R = \sqrt{\frac{I}{m}} = \sqrt{\frac{25/72}{13/20}} = \frac{5\sqrt{5}}{3\sqrt{26}}$.

13. $m = \int_0^1 \int_0^{x^3} (x + y)dy\ dx = \int_0^1 (xy + \frac{y^2}{2})\Big|_0^{x^3} dx$

$= \int_0^1 (x^4 + \frac{x^6}{2})dx = (\frac{x^5}{5} + \frac{x^7}{14})\Big|_0^1 = \frac{1}{5} + \frac{1}{14} = \frac{19}{70}$

$I_y = \int_0^1 \int_0^{x^3} x^2(x + y)dydx = \int_0^1 (x^3y + \frac{x^2y^2}{2})\Big|_0^{x^3} dx$

$= \int_0^1 (x^6 + \frac{x^8}{2})dx = (\frac{x^7}{7} + \frac{x^9}{18})\Big|_0^1 = \frac{1}{7} + \frac{1}{18} = \frac{25}{126}$

$R = \sqrt{\frac{I}{m}} = \sqrt{\frac{25/126}{19/70}} = \frac{5\sqrt{5}}{3\sqrt{19}}$.

17. $V = \int_0^1 \int_0^{1-x} (1 - x - y)dy\ dx = \int_0^1 [(1 - x)y - \frac{y^2}{2}]\Big|_0^{1-x} dx$

$= \int_0^1 [(1-x)^2 - \frac{(1-x)^2}{2}]dx = \frac{1}{2}\int_0^1 (1 - x)^2 = -\frac{1}{6}(1 - x)^3\Big|_0^1 = \frac{1}{6}$

$I_z = \int_0^1 \int_0^{1-x} (x^2 + y^2)(1 - x - y)dy\ dx$

$$= \int_0^1 \int_0^{1-x} (x^2 - x^3 - x^2y + y^2 - xy^2 - y^3)\,dy\,dx$$

$$= \int_0^1 (x^2y - x^3y - \frac{x^2y^2}{2} + \frac{y^3}{3} - \frac{xy^3}{3} - \frac{y^4}{4})\Big|_0^{1-x} dx$$

$$= \int_0^1 [x^2(1 - x) - x^3(1 - x) - \frac{x^2(1-x)^2}{2} + \frac{(1-x)^3}{3} - \frac{x(1-x)^3}{3}$$

$$- \frac{(1 - x)^4}{4}]\,dx$$

$$= \int_0^1 [\frac{1}{2}x^2(1 - x)^2 + \frac{(1 - x)^4}{3} - \frac{(1 - x)^4}{4}]dx$$

$$= \int_0^1 [\frac{x^2 - 2x^3 + x^4}{2} + \frac{(1 - x)^4}{12}]dx$$

$$= [\frac{1}{2}(\frac{x^3}{3} - \frac{x^4}{2} + \frac{x^5}{5}) - \frac{(1 - x)^5}{60}]\Big|_0^1 = \frac{1}{2}(\frac{1}{3} - \frac{1}{2} + \frac{1}{5}) + \frac{1}{60} = \frac{1}{30}$$

$$R = \sqrt{\frac{I}{V}} = \sqrt{\frac{1/30}{1/6}} = \frac{1}{\sqrt{5}}\ .$$

21. $I_y = 2\int_{-1}^1 \int_0^{1-x} (x^2 + z^2)\sqrt{1 - x^2}\,dz\,dx$

$$= 2\int_{-1}^1 (x^2z + \frac{z^3}{3})\sqrt{1 - x^2}\Big|_0^{1-x} dx$$

$$= 2\int_{-1}^1 [x^2(1 - x) + \frac{(1 - x)^3}{3}]\sqrt{1 - x^2}\,dx$$

$$= \frac{2}{3}\int_{-1}^1 (1 - 3x + 6x^2 - 4x^3)\sqrt{1 - x^2}\,dx$$

$x = \sin\theta$, $dx = \cos\theta\,d\theta$

$$= \frac{2}{3}\int_{-\pi/2}^{\pi/2} (1 - 3\sin\theta + 6\sin^2\theta - 4\sin^3\theta)\cos^2\theta\,d\theta$$

$$= \frac{2}{3}\int_{-\pi/2}^{\pi/2} [\frac{1 + \cos 2\theta}{2} - 3\sin\theta\cos^2\theta$$

$$+ 6\,\frac{1 - \cos\, 2\theta}{2}\,\frac{1 + \cos\, 2\theta}{2} - 4\,\sin\theta\cos^2\theta(1 - \cos^2\theta)]d\theta$$

$$= \frac{2}{3}\int_{-\pi/2}^{\pi/2} (\frac{1}{2} + \frac{1}{2}\cos\, 2\theta - 7\,\sin\theta\cos^2\theta + \frac{3}{2} - \frac{3}{2}\,\frac{1 + \cos\, 4\theta}{2}$$

$$+ 4\,\sin\,\theta\,\cos^4\theta)d\theta$$

$$= \frac{2}{3}\int_{-\pi/2}^{\pi/2} (\frac{5}{4} + \frac{1}{2}\cos\, 2\theta - 7\,\sin\theta\cos^2\theta - \frac{3}{4}\cos\, 4\theta$$

$$+ 4\,\sin\,\theta\,\cos^4\theta)d\theta$$

$$= \frac{2}{3}(\frac{5\theta}{4} + \frac{1}{4}\sin\, 2\theta + \frac{7}{3}\cos^3\theta - \frac{3}{16}\sin\, 4\theta - \frac{4}{5}\cos^5\theta)\Big|_{-\pi/2}^{\pi/2}$$

$$= \frac{5\pi}{6}$$

$$R = \sqrt{\frac{I}{V}} = \sqrt{\frac{5\pi/6}{\pi}} = \sqrt{\frac{5}{6}}\ .$$

25. $$m = \int_0^{2\pi}\int_0^{1-\cos\,\theta} \frac{k}{r}\cdot r\ dr\ d\theta = k\int_0^{2\pi} (1 - \cos\,\theta)d\theta$$

$$= k(\theta - \sin\,\theta)\Big|_0^{2\pi} = 2k\pi$$

$$M_y = \int_0^{2\pi}\int_0^{1-\cos\,\theta} r\,\cos\,\theta\cdot\frac{k}{r}\cdot r\ dr\ d\theta$$

$$= k\int_0^{2\pi} (\frac{r^2}{2}\Big|_0^{1-\cos\,\theta})\cos\theta\ d\theta$$

$$= \frac{k}{2}\int_0^{2\pi} (\cos\,\theta - 2\,\cos^2\theta + \cos^3\theta)d\theta$$

$$= \frac{k}{2}\int_0^{2\pi} [\cos\theta - 1 - \cos\, 2\theta + (1 - \sin^2\theta)\cos\,\theta]d\theta$$

$$= \frac{k}{2}\,(2\,\sin\,\theta - \theta - \frac{1}{2}\sin\, 2\theta - \frac{1}{3}\sin^3\theta)\Big|_0^{2\pi} = -\,k\pi$$

$$\bar{x} = \frac{M_y}{m} = \frac{-k\pi}{2k\pi} = -\,\frac{1}{2}\,, \quad \bar{y} = 0 \quad \text{(by symmetry)}\ .$$

If the density is $\frac{k}{r^2}$, then the inner integral for the mass is

$$\int_0^{1-\cos\theta} \frac{k}{r}\,dr = \lim_{\varepsilon\to 0^+} k\int_\varepsilon^{1-\cos\theta} \frac{1}{r}\,dr$$

$$= \lim_{\varepsilon\to 0^+} k \ln r\Big|_\varepsilon^{1-\cos\theta}$$

$$= \lim_{\varepsilon\to 0^+} k[\ln(1-\cos\theta) - \ln\varepsilon],$$

which diverges.

29. $I_x = \int_0^{2\pi}\int_a^b (r\sin\theta)^2 r\,dr\,d\theta$

$$= \int_0^{2\pi} \left(\frac{r^4}{4}\Big|_a^b\right)\sin^2\theta\,d\theta$$

$$= \frac{b^4 - a^4}{4}\int_0^{2\pi}\left(\frac{1}{2} - \frac{1}{2}\cos 2\theta\right)d\theta$$

$$= \frac{b^4 - a^4}{4}\left(\frac{\theta}{2} - \frac{1}{4}\sin 2\theta\right)\Big|_0^{2\pi} = \frac{b^4 - a^4}{4}\pi$$

$I_y = I_x$ (by symmetry).

33. $y = 1 - x^2$, $y' = -2x = -2x_0$ at (x_0, y_0).

$y - y_0 = -2x_0(x - x_0)$

$$y = y_0 + 2x_0^2 - 2x_0x = 1 - x_0^2 + 2x_0^2 - 2x_0x$$

$$= 1 + x_0^2 - 2x_0x$$

$$y = 0 : x = \frac{1 + x_0^2}{2x_0}$$

$$I_z(x_0) = \int_0^{(1+x_0^2)/2x_0}\int_0^{1+x_0^2-2x_0x}(x^2 + y^2)\,dy\,dx$$

$$= \int_0^{(1+x_0^2)/2x_0} (x^2y + \frac{y^3}{3})\Big|_0^{1+x_0^2-2x_0x} dx$$

$$= \int_0^{(1+x_0^2)/2x_0} [(1+x_0^2)x^2 - 2x_0x^3 + \frac{(1+x_0^2-2x_0x)^3}{3}]dx$$

$$= \frac{x^3}{3}(1+x_0^2) - \frac{x_0}{2}x^4 - \frac{(1+x_0^2-2x_0x)^4}{24x_0}\Big|_0^{(1+x_0^2)/2x_0}$$

$$= \frac{(1+x_0^2)^4}{24x_0^3} - \frac{x_0}{2}\cdot\frac{(1+x_0)^4}{16x_0^4} + \frac{(1+x_0^2)^4}{24x_0}$$

$$= \frac{(1+x_0^2)^4}{24}(\frac{1}{4x_0^3} + \frac{1}{x_0})$$

$$\frac{d}{dx_0}I_z(x_0) = \frac{1}{24}[4(1+x_0^2)^3 2x_0(\frac{1}{4x_0^3} + \frac{1}{x_0})$$

$$+ (1+x_0^2)^4(-\frac{3}{4x_0^4} - \frac{1}{x_0^2})]$$

$$= \frac{(1+x_0^2)^3}{24}[\frac{2}{x_0^2} + 8 - (1+x_0^2)(\frac{3}{4x_0^4} + \frac{1}{x_0^2})] = 0$$

$$\frac{2}{x_0^2} + 8 - \frac{3}{4x_0^4} - \frac{1}{x_0^2} - \frac{3}{4x_0^2} - 1 = 0$$

$$28x_0^4 + x_0^2 - 3 = 0$$

$$x_0^2 = \frac{-1+\sqrt{1+336}}{56} = 0.306$$

$$x_0 = \sqrt{0.306} = 0.557$$

$$y_0 = 1 - x_0^2 = 0.690.$$

CHAPTER 21, SECTION 5 (pp. 1057-1058)

1. $$\int_0^2\int_0^x\int_0^{x+y} x\,dz\,dy\,dx = \int_0^2\int_0^x xz\Big|_0^{x+y} dy\,dx$$

$$= \int_0^2\int_0^x (x^2 + xy)dy\,dx = \int_0^2 (x^2y + \frac{xy^2}{2})\Big|_0^x dx$$

$$= \int_0^2 (x^3 + \frac{x^3}{2})dx = \int_0^2 \frac{3x^3}{2}dx = \frac{3x^4}{8}\Big|_0^2 = 6 .$$

5. $$V = \int_0^3\int_{2x-6}^0\int_0^{(6-2x+y)/3} dz\,dy\,dx$$

$$= \int_0^3\int_{2x-6}^0 \frac{6 - 2x + y}{3} dy\,dx$$

$$= \frac{1}{3}\int_0^3 [(6 - 2x)y + \frac{y^2}{2}]\Big|_{2x-6}^0 dx$$

$$= \frac{1}{3}\int_0^3 [(2x - 6)^2 - \frac{(2x - 6)^2}{2}]dx$$

$$= \frac{1}{6}\int_0^3 (2x - 6)^2dx = \frac{1}{12}\frac{(2x - 6)^3}{3}\Big|_0^3 = 6 .$$

9. $$m = \int_0^1\int_0^1\int_0^1 (x + y + z)dz\,dy\,dx = \int_0^1\int_0^1 [(x+y)z + \frac{z^2}{2}]\Big|_0^1 dy\,dx$$

$$= \int_0^1\int_0^1 (x + y + \frac{1}{2})dy\,dx = \int_0^1 [(x + \frac{1}{2})y + \frac{y^2}{2}]\Big|_0^1 dx$$

$$= \int_0^1 (x + \frac{1}{2} + \frac{1}{2})dx = (\frac{x^2}{2} + x)\Big|_0^1 = \frac{1}{2} + 1 = \frac{3}{2} .$$

13. $m = \int_0^1 \int_0^{1-x} \int_0^{1-x-y} (1 - x - y)dz\ dy\ dx$

$= \int_0^1 \int_0^{1-x} (1 - x - y)^2 dy\ dx = \int_0^1 -\frac{1}{3}(1 - x - y)^3 \Big|_0^{1-x} dx$

$= \int_0^1 \frac{1}{3}(1 - x)^3 dx = -\frac{1}{12}(1 - x)^4 \Big|_0^1 = \frac{1}{12}$.

17. $\int_0^3 \int_0^y \int_0^x dz\ dx\ dy = \int_0^3 \int_0^y (z\Big|_0^x)dx\ dy$

$= \int_0^3 \int_0^y x\ dx\ dy = \int_0^3 (\frac{x^2}{2}\Big|_0^y)dy$

$= \int_0^3 \frac{y^2}{2} dy = \frac{y^3}{6}\Big|_0^3 = \frac{9}{2}$

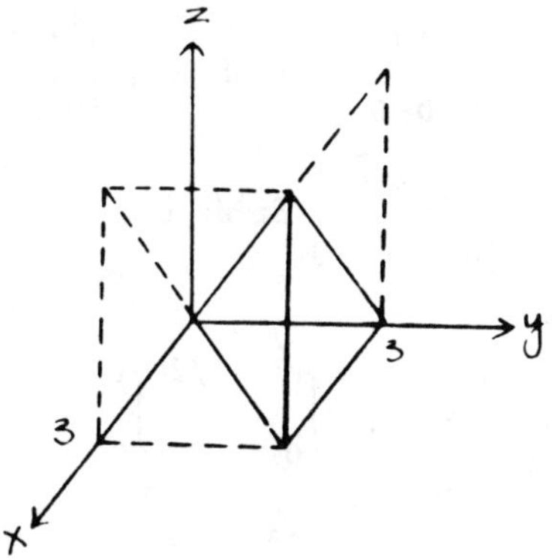

$\int_0^3 \int_x^3 \int_0^x dz\ dy\ dx$,

$\int_0^3 \int_0^x \int_x^3 dy\ dz\ dx$,

$\int_0^3 \int_z^3 \int_x^3 dy\ dx\ dz$,

$\int_0^3 \int_z^3 \int_z^y dx\ dy\ dz$, $\int_0^3 \int_0^y \int_z^y dx\ dz\ dy$.

z
y
3
3
x

21. $\int_0^1 \int_0^{1-z^2} \int_z^{\sqrt{1-y}} dx\ dy\ dz$

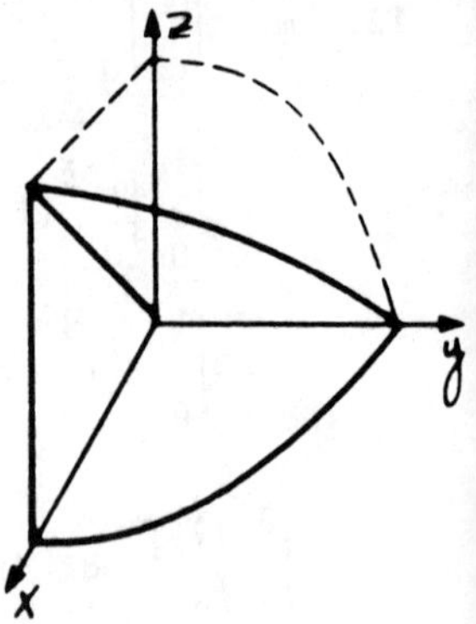

$$= \int_0^1 \int_0^{1-z^2} (\sqrt{1-y} - z)dy\ dz$$

$$= \int_0^1 [-\frac{2}{3}(1-y)^{3/2} - yz]\Big|_0^{1-z^2} dz$$

$$= \int_0^1 (\frac{2}{3} - \frac{2}{3}z^3 - z + z^3)dz$$

$$= \frac{2}{3}z - \frac{z^2}{z} + \frac{z^4}{12}\Big|_0^1 = \frac{2}{3} - \frac{1}{2} + \frac{1}{12} = \frac{1}{4}$$

$\int_0^1 \int_0^{\sqrt{1-y}} \int_z^{\sqrt{1-y}} dx\ dz\ dy$, $\int_0^1 \int_z^1 \int_0^{1-x^2} dy\ dx\ dz$,

$\int_0^1 \int_0^x \int_0^{1-x^2} dy\ dz\ dx$, $\int_0^1 \int_0^{1-x^2} \int_0^x dz\ dy\ dx$,

$\int_0^1 \int_0^{\sqrt{1-y}} \int_0^x dz\ dx\ dy$.

25. $V = 4 \int_0^2 \int_0^{\sqrt{4-x^2}/2} \int_0^{1-y^2} dz\ dy\ dx$

$$= 4 \int_0^2 \int_0^{\sqrt{4-x^2}/2} (1-y^2)dy\ dx$$

$$= 4 \int_0^2 (y - \frac{y^3}{3})\Big|_0^{\sqrt{4-x^2}/2} dx$$

$$= \frac{4}{3} \int_0^2 (3 \frac{\sqrt{4-x^2}}{2} - \frac{(4-x^2)\sqrt{4-x^2}}{8})dx$$

$$= \frac{1}{6} \int_0^2 (8 + x^2) \sqrt{4-x^2}\ dx$$

$x = 2\sin\theta\ , \quad dx = 2\cos\theta\, d\theta$

$$= \frac{1}{6}\int_0^{\pi/2} (8 + 4\sin^2\theta)2\cos\theta \cdot 2\cos\theta\, d\theta$$

$$= \frac{8}{3}\int_0^{\pi/2} (2\cos^2\theta + \sin^2\theta\cos^2\theta)d\theta$$

$$= \frac{8}{3}\int_0^{\pi/2} (1 + \cos 2\theta + \frac{1 - \cos 2\theta}{2}\ \frac{1 + \cos 2\theta}{2})d\theta$$

$$= \frac{8}{3}\int_0^{\pi/2} (1 + \cos 2\theta + \frac{1 - \cos^2 2\theta}{4})d\theta$$

$$= \frac{8}{3}\int_0^{\pi/2} [\frac{5}{4} + \cos 2\theta - \frac{1}{8}\ (1 + \cos 4\theta]d\theta$$

$$= \frac{8}{3}(\frac{9\theta}{8} + \frac{1}{2}\sin 2\theta - \frac{1}{32}\sin 4\theta)\Big|_0^{\pi/2} = \frac{3\pi}{2}\ .$$

29. $\dfrac{x^2}{a^2} + \dfrac{y^2}{b^2} + \dfrac{z^2}{c^2} = 1, \quad z = c\sqrt{1 - \dfrac{x^2}{a^2} - \dfrac{y^2}{b^2}}$

$$V = 8\int_0^a \int_0^{b\sqrt{1-x^2/a^2}} \int_0^{c\sqrt{1-x^2/a^2-y^2/b^2}} dz\ dy\ dx$$

$$= 8bc\int_0^a \int_0^{b\sqrt{a^2-x^2}/a} \sqrt{(1 - \frac{x^2}{a^2}) - \frac{y^2}{b^2}}\ (\frac{1}{b}\ dy)dx$$

(Formula 56, $a^2 = 1 - x^2/a^2$, $u = y/b$)

$$= 8bc\int_0^a (\frac{y}{2b}\sqrt{1 - \frac{x^2}{a^2} - \frac{y^2}{b^2}}$$

$$+ \frac{1 - \frac{x^2}{a^2}}{2}\ \mathrm{Sin}^{-1}\ \frac{\frac{y}{b}}{\sqrt{1 - \frac{x^2}{a^2}}}\Bigg|_0^{b\sqrt{1 - \frac{x^2}{a^2}}}\ )dx$$

$$= 8bc \int_0^a \left(\frac{a^2 - x^2}{2a^2} \text{Sin}^{-1} 1\right)dx = \frac{2\pi bc}{a^2} \int_0^a (a^2 - x^2)dx$$

$$= \frac{2\pi bc}{a^2} \left(a^3 - \frac{x^3}{3}\right)\Big|_0^a = \frac{4}{3} \pi abc .$$

CHAPTER 21, SECTION 6 (pp. 1062-1063)

1. $m = \frac{1}{8}$ (by Problem 8, Section 5)

$$M_{yz} = \int_0^1 \int_0^1 \int_0^1 x^2yz \, dz \, dy \, dx = \int_0^1 \int_0^1 \frac{x^2yz^2}{2}\Big|_0^1 dy \, dx$$

$$= \int_0^1 \int_0^1 \frac{x^2y}{2} dy \, dx = \int_0^1 \frac{x^2y^2}{4}\Big|_0^1 dx$$

$$= \int_0^1 \frac{x^2}{4} dx = \frac{x^3}{12}\Big|_0^1 = \frac{1}{12}$$

$$M_{xz} = \int_0^1 \int_0^1 \int_0^1 xy^2z \, dz \, dy \, dx = \int_0^1 \int_0^1 \frac{xy^2z^2}{2}\Big|_0^1 dy \, dx$$

$$= \int_0^1 \int_0^1 \frac{xy^2}{2} dy \, dx = \int_0^1 \frac{xy^3}{6}\Big|_0^1 dx$$

$$= \int_0^1 \frac{x}{6} dx = \frac{x^2}{12}\Big|_0^1 = \frac{1}{12}$$

$$M_{xy} = \int_0^1 \int_0^1 \int_0^1 xyz^2 dz \, dy \, dx = \int_0^1 \int_0^1 xy \frac{z^3}{3}\Big|_0^1 dy \, dx$$

$$= \int_0^1 \int_0^1 \frac{xy}{3} dy \, dx = \int_0^1 \frac{xy^2}{6}\Big|_0^1 dx = \int_0^1 \frac{x}{6} dx$$

$$= \frac{x^2}{12}\Big|_0^1 = \frac{1}{12}$$

$$\bar{x} = \frac{M_{yz}}{m} = \frac{1/12}{1/8} = \frac{2}{3}, \quad \bar{y} = \frac{M_{xz}}{m} = \frac{1/12}{1/8} = \frac{2}{3},$$

$$\bar{z} = \frac{M_{xy}}{m} = \frac{1/12}{1/8} = \frac{2}{3}.$$

5. $m = 4\int_0^1 \int_0^{\sqrt{1-x^2}} \int_0^1 xyz\,dz\,dy\,dx = 4\int_0^1 \int_0^{\sqrt{1-x^2}} \frac{xyz^2}{2}\Big|_0^1 dy\,dx$

$$= 2\int_0^1 \int_0^{\sqrt{1-x^2}} xy\,dy\,dx = \int_0^1 xy^2\Big|_0^{\sqrt{1-x^2}} dx = \int_0^1 x(1-x^2)dx$$

$$= \left(\frac{x^2}{2} - \frac{x^4}{4}\right)\Big|_0^1 = \frac{1}{2} - \frac{1}{4} = \frac{1}{4}$$

$$M_{xy} = 4\int_0^1 \int_0^{\sqrt{1-x^2}} \int_0^1 xyz^2 dz\,dy\,dx$$

$$= 4\int_0^1 \int_0^{\sqrt{1-x^2}} \frac{xyz^3}{3}\Big|_0^1 dy\,dx$$

$$= \frac{4}{3}\int_0^1 \int_0^{\sqrt{1-x^2}} xy\,dy\,dx = \frac{2}{3}\int_0^1 xy^2\Big|_0^{\sqrt{1-x^2}} dx$$

$$= \frac{2}{3}\int_0^1 x(1-x^2)dx = \frac{2}{3}\left(\frac{x^2}{2} - \frac{x^4}{4}\right)\Big|_0^1 = \frac{2}{3}\left(\frac{1}{2} - \frac{1}{4}\right) = \frac{1}{6}$$

$\bar{x} = \bar{y} = 0$ (by symmetry), $\bar{z} = \frac{M_{xy}}{m} = \frac{1/6}{1/4} = \frac{2}{3}$.

9. $m = \frac{1}{8}$ (from Problem 1)

$$I_z = \int_0^1 \int_0^1 \int_0^1 xyz(x^2 + y^2)dz\,dy\,dx$$

$$= \frac{1}{2}\int_0^1 \int_0^1 xyz^2(x^2 + y^2)\Big|_0^1 dy\,dx$$

$$= \frac{1}{2}\int_0^1 \int_0^1 xy(x^2 + y^2)dy\,dx = \frac{1}{2}\int_0^1 \left(\frac{x^3y^2}{2} + \frac{xy^4}{4}\right)\Big|_0^1 dx$$

$$= \frac{1}{8}\int_0^1 (2x^3 + x)dx = \frac{1}{8}(\frac{x^4}{2} + \frac{x^2}{2})\Big|_0^1 = \frac{1}{8}$$

$$R = \sqrt{\frac{I}{m}} = \sqrt{\frac{1/8}{1/8}} = 1 .$$

13. $$m = \int_0^1 \int_0^{x^2} \int_0^x 4 \; dz \; dy \; dx = \int_0^1 \int_0^{x^2} 4z\Big|_0^x dy \; dx$$

$$= \int_0^1 \int_0^{x^2} 4x \; dy \; dx = \int_0^1 4xy\Big|_0^{x^2} dx = \int_0^1 4x^3 dx = x^4\Big|_0^1 = 1$$

$$I_x = \int_0^1 \int_0^{x^2} \int_0^x 4(y^2 + z^2)dz \; dy \; dx$$

$$= \int_0^1 \int_0^{x^2} 4(y^2 z + \frac{z^3}{3})\Big|_0^x dy \; dx = \int_0^1 \int_0^{x^2} 4(xy^2 + \frac{x^3}{3})dy \; dx$$

$$= \int_0^1 4(\frac{xy^3}{3} + \frac{x^3 y}{3})\Big|_0^{x^2} dx = \frac{4}{3}\int_0^1 (x^7 + x^5)dx$$

$$= \frac{4}{3}(\frac{x^8}{8} + \frac{x^6}{6})\Big|_0^1 = \frac{4}{3}(\frac{1}{8} + \frac{1}{6}) = \frac{7}{18}$$

$$R = \sqrt{\frac{I}{m}} = \sqrt{\frac{7/18}{1}} = \frac{\sqrt{7}}{3\sqrt{2}} = \frac{\sqrt{14}}{6} .$$

17. Let M be the total mass of the body and dM be the differential of mass. Then the moment of the body about a plane through the center of mass is

$$\iiint_V [\ell(x - \bar{x}) + m(y - \bar{y}) + n(z - \bar{z})]dM$$

$$= \ell \iiint_V x \; dM + m \iiint_V ydm + n \iiint_V zdM - \ell\bar{x}M - m\bar{y}M - n\bar{z}M$$

$$= \ell(\bar{x}M) + m(\bar{y}M) + n(\bar{z}M) - \ell\bar{x}M - m\bar{y}M - n\bar{z}M$$

$$= 0, \text{ since } \bar{x} = \frac{\iiint_V xdM}{M}, \; \bar{y} = \frac{\iiint_V ydM}{M}, \text{ and}$$

$$\bar{z} = \frac{\iiint_V zdM}{M} .$$

CHAPTER 21, SECTION 7 (pp. 1071-1073)

1. $m = 4k \int_0^{\pi/2} \int_0^1 \int_0^4 r^2 dz\, dr\, d\theta = 4k \int_0^{\pi/2} \int_0^1 4r^2 dr\, d\theta$

$= 4k \int_0^{\pi/2} \left. \frac{4r^3}{3} \right|_0^1 d\theta = 4k \int_0^{\pi/2} \frac{4}{3}\, d\theta = \frac{16k}{3} \cdot \frac{\pi}{2} = \frac{8\pi k}{3}$.

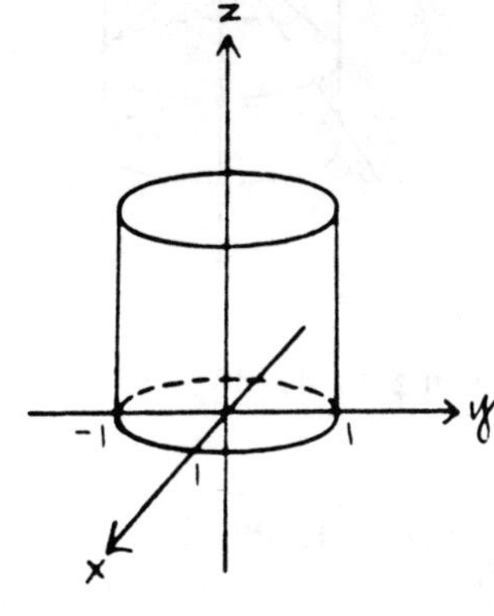

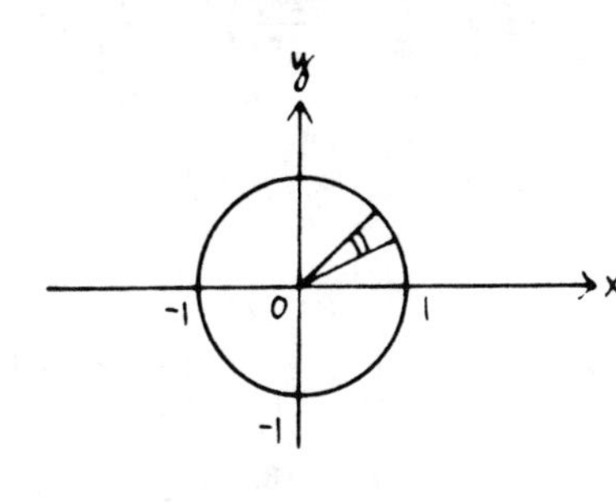

5. Cylinder: $r = 1$.

Cone: $z = r$.

$m = 4 \int_0^{\pi/2} \int_0^1 \int_0^r zr\, dz\, dr\, d\theta = 4 \int_0^{\pi/2} \int_0^1 \left. \frac{z^2 r}{2} \right|_0^r dr\, d\theta$

$= 2 \int_0^{\pi/2} \int_0^1 r^3 dr\, d\theta = 2 \int_0^{\pi/2} \left. \frac{r^4}{4} \right|_0^1 d\theta = \frac{1}{2} \int_0^{\pi/2} d\theta = \frac{\pi}{4}$,

$M_{xy} = 4 \int_0^{\pi/2} \int_0^1 \int_0^r z^2 r\, dz\, dr\, d\theta$

$= 4 \int_0^{\pi/2} \int_0^1 \left. \frac{z^3 r}{3} \right|_0^r dr\, d\theta$

$= \frac{4}{3} \int_0^{\pi/2} \int_0^1 r^4 dr\, d\theta$

$$= \frac{4}{3}\int_0^{\pi/2} \frac{1}{5}\, d\theta$$

$$= \frac{4}{15}\cdot\frac{\pi}{2} = \frac{2\pi}{15}$$

$\bar{x} = \bar{y} = 0$ (by symmetry),

$$\bar{z} = \frac{M_{xy}}{m} = \frac{2\pi/15}{\pi/4} = \frac{8}{15}\ .$$

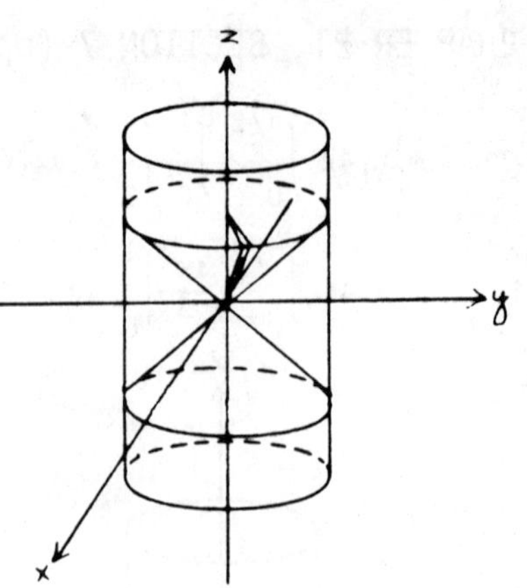

9. $$m = 2\int_0^{2\pi}\int_0^{\pi/2}\int_0^2 (4-\rho)\rho^2 \sin\phi\, d\rho\, d\phi\, d\theta$$

$$= 2\int_0^{2\pi}\int_0^{\pi/2} \left(\frac{4}{3}\rho^3 - \frac{1}{4}\rho^4\Big|_0^2\right)\sin\phi\, d\phi\, d\theta$$

$$= 2\int_0^{2\pi}\int_0^{\pi/2} \frac{20}{3}\sin\phi\, d\phi\, d\theta$$

$$= \frac{40}{3}\int_0^{2\pi}\left(-\cos\phi\Big|_0^{\pi/2}\right)d\theta = \frac{40}{3}\int_0^{2\pi} d\theta = \frac{80\pi}{3}\ .$$

13. Sphere: $\rho = 1$.

Cone: $\phi = \pi/4$.

$$m = 4\int_0^{\pi/4}\int_0^{\pi/2}\int_0^1 (4-2\rho)\rho^2 \sin\phi\, d\rho\, d\theta\, d\phi$$

$$= 4\int_0^{\pi/4}\int_0^{\pi/2} \left(\frac{4\rho^3}{3} - \frac{\rho^4}{2}\right)\sin\phi\Big|_0^1\, d\theta\, d\phi$$

$$= 4\int_0^{\pi/4}\int_0^{\pi/2} \frac{5}{6}\sin\phi\, d\theta\, d\phi = \frac{10}{3}\int_0^{\pi/4}\frac{\pi}{2}\sin\phi\, d\phi$$

$$= -\frac{5\pi}{3}\cos\phi\Big|_0^{\pi/4} = \frac{5\pi}{3}\left(1 - \frac{1}{\sqrt{2}}\right) = \frac{5\pi(\sqrt{2}-1)}{3\sqrt{2}}\ ,$$

$$M_{xy} = 4\int_0^{\pi/4}\int_0^{\pi/2}\int_0^1 z(4-2\rho)\rho^2 \sin\phi \, d\rho \, d\theta \, d\phi$$

$$= 4\int_0^{\pi/4}\int_0^{\pi/2}\int_0^1 \rho\cos\phi(4-2\rho)\rho^2 \sin\phi \, d\rho \, d\theta \, d\phi$$

$$= 4\int_0^{\pi/4}\int_0^{\pi/2} (\rho^4 - \frac{2\rho^5}{5})\Big|_0^1 \sin\phi\cos\phi \, d\theta \, d\phi$$

$$= 4\int_0^{\pi/4}\int_0^{\pi/2} \frac{3}{5}\sin\phi\cos\phi \, d\theta \, d\phi$$

$$= \frac{12}{5}\int_0^{\pi/4} \frac{\pi}{2}\sin\phi\cos\phi \, d\phi$$

$$= \frac{6\pi}{5}\frac{\sin^2\phi}{2}\Big|_0^{\pi/4} = \frac{3\pi}{5}\cdot\frac{1}{2} = \frac{3\pi}{10}$$

$$\bar{x} = 0 , \quad \bar{y} = 0 , \quad \bar{z} = \frac{M_{xy}}{m} = \frac{3\pi}{10}\cdot\frac{3\sqrt{2}}{5\pi(\sqrt{2}-1)} = \frac{9(2+\sqrt{2})}{50} .$$

17. $\frac{z}{h} = \frac{a-r}{a}$, $z = \frac{h}{a}(a-r)$

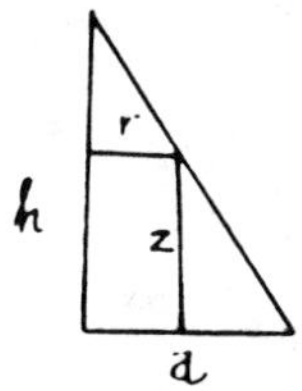

Density k

$m = \frac{1}{3}ka^2h\pi$ by geometry.

$$I_z = \int_0^{2\pi}\int_0^a\int_0^{\frac{h}{a}(a-r)} r^2 \cdot kr \, dz \, dr \, d\theta$$

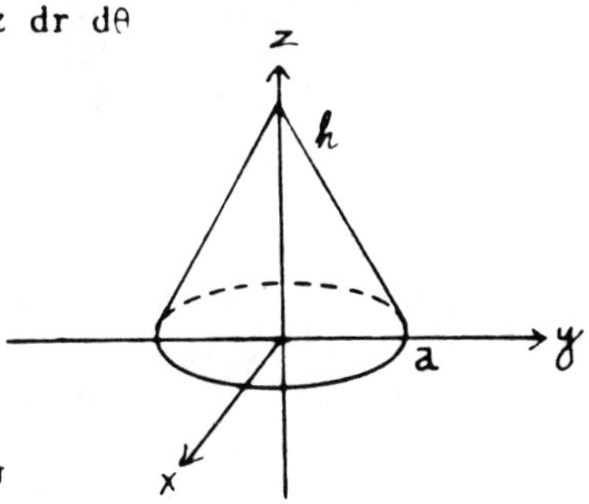

$$= \frac{kh}{a}\int_0^{2\pi}\int_0^a (a-r)r^3 dr \, d\theta$$

$$= \frac{kh}{a}\int_0^{2\pi} (\frac{ar^4}{4} - \frac{r^5}{5}\Big|_0^a) d\theta$$

$$= \frac{kh}{a}\int_0^{2\pi} \frac{a^5}{20} d\theta = \frac{kha^4}{20}\cdot 2\pi$$

$$= \frac{kha^4\pi}{10} ,$$

$$R = \sqrt{\frac{I_z}{m}} = \sqrt{\frac{kha^4\pi/10}{\pi ka^2h/3}} = a\sqrt{\frac{3}{10}} .$$

21. $H = \int_0^{2\pi}\int_0^{\pi}\int_0^{a} Ck\rho\ \delta\rho^2 \sin\phi\ d\rho\ d\phi\ d\theta$

$= Ck\delta \int_0^{2\pi}\int_0^{\pi} \frac{\rho^4}{4} \sin\phi \Big|_0^a d\phi\ d\theta$

$= \frac{Ck\delta a^4}{4}\int_0^{2\pi} - \cos\phi\Big|_0^{\pi} d\theta$

$= \frac{Ck\delta a^4}{2}\int_0^{2\pi} d\theta = \pi Ck\delta a^4.$

CHAPTER 21, REVIEW (pp. 1073-1075)

1. $\int_0^3\int_0^{4x-x^2} (x + y)dy\ dx = \int_0^3 (xy + \frac{y^2}{2})\Big|_0^{4x-x^2} dx$

$= \int_0^3 [x(4x - x^2) + \frac{1}{2}(4x - x^2)^2]dx$

$= \int_0^3 (12x^2 - 5x^3 + \frac{x^4}{2})dx$

$= (4x^3 - \frac{5x^4}{4} + \frac{x^5}{10})\Big|_0^3$

$= 4 \cdot 27 - \frac{5 \cdot 81}{4} + \frac{243}{10} = \frac{621}{20}$.

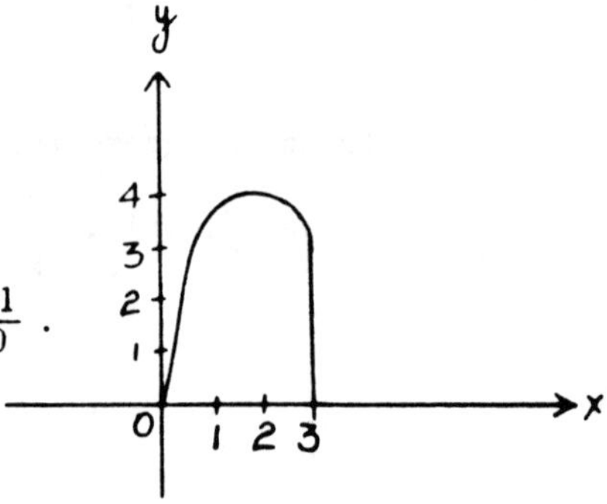

5. $\iint_R (x^2 - y)dA = \int_0^2\int_y^{4-y} (x^2 - y)dx\ dy$

$= \int_0^2 (\frac{x^3}{3} - xy)\Big|_y^{4-y} dy$

$= \int_0^2 [\frac{(4 - y)^3}{3} - (4 - y)y - \frac{y^3}{3} + y^2]dy$

$$= \left(-\frac{(4-y)^4}{12} - 2y^2 + \frac{y^3}{3} - \frac{y^4}{12} + \frac{y^3}{3}\right)\Bigg|_0^2$$

$$= -\frac{16}{12} - 8 + \frac{8}{3} - \frac{4}{3} + \frac{8}{3} + \frac{256}{12}$$

$$= 16 .$$

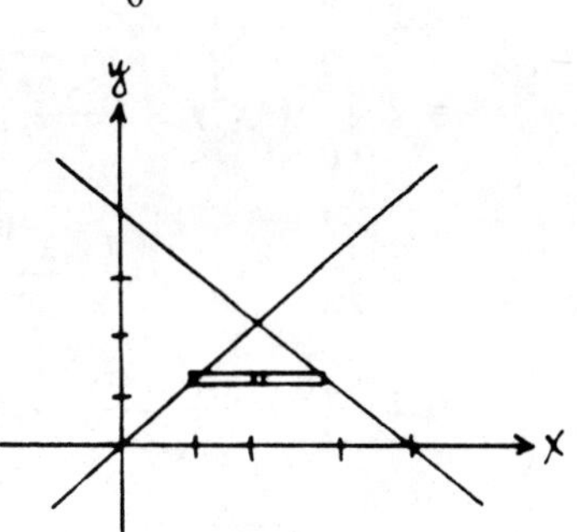

9. $$m = \int_0^{4/3}\int_y^{4-2y} (x+y)\,dx\,dy = \int_0^{4/3} \left(\frac{x^2}{2} + xy\right)\Bigg|_y^{4-2y} dy$$

$$= \int_0^{4/3} \left[\frac{4(2-y)^2}{2} + (4-2y)y - \frac{3y^2}{2}\right]dy$$

$$= \int_0^{4/3} \left(8 - 4y - \frac{3y^2}{2}\right)dy$$

$$= \left(8y - 2y^2 - \frac{y^3}{2}\right)\Bigg|_0^{4/3}$$

$$= \frac{32}{3} - \frac{32}{9} - \frac{32}{27} = \frac{160}{27} .$$

13. $$m = 4\int_0^{\pi/2}\int_a^b r^2\,dr\,d\theta = 4\int_0^{\pi/2} \frac{r^3}{3}\Bigg|_a^b d\theta$$

$$= \frac{4}{3}\int_0^{\pi/2} (b^3 - a^3)\,d\theta = \frac{4}{3}(b^3 - a^3)\frac{\pi}{2} = \frac{2\pi}{3}(b^3 - a^3) .$$

17. $V = \frac{256}{15}$ from Problem 16.

$$I_z = 2\int_0^2\int_{x^2}^4\int_0^{4-y} (x^2 + y^2)\,dz\,dy\,dx$$

$$= 2\int_0^2 \int_{x^2}^4 (x^2 + y^2)(4 - y)dy\ dx$$

$$= 2\int_0^2 (4x^2y - \frac{x^2y^2}{2} + \frac{4y^3}{3} - \frac{y^4}{4}\Big|_{x^2}^4)dx$$

$$= 2\int_0^2 (8x^2 - 4x^4 - \frac{5x^6}{6} + \frac{x^8}{4} + \frac{64}{3})dx$$

$$= 2(\frac{8x^3}{3} + \frac{64x}{3} - \frac{4x^5}{5} - \frac{5x^7}{42} + \frac{x^9}{36})\Big|_0^2 = \frac{23,552}{315},$$

$$R_z = \sqrt{\frac{23,552/315}{256/15}} = \frac{2\sqrt{23}}{\sqrt{21}}.$$

21. $$m = 2\int_0^2 \int_{x^2}^4 \int_{\frac{y-4}{4}}^{\frac{4-y}{2}} (1 + y + z^2)dz\ dy\ dx$$

$$= 2\int_0^2 \int_{x^2}^4 [(1 + y)z + \frac{z^3}{3}]\Big|_{\frac{y-4}{4}}^{\frac{4-y}{2}} dy\ dx$$

$$= 2\int_0^2 \int_{x^2}^4 [\frac{(1 + y)(4-y)}{2} + \frac{(4-y)^3}{24} - \frac{(1+y)(y-4)}{4} - \frac{(y-4)^3}{192}]dy\ dx$$

$$= 2\int_0^2 \int_{x^2}^4 [\frac{3(4 + 3y - y^2)}{4} + \frac{3(4 - y)^3}{64}]dy\ dx$$

$$= 2\int_0^2 (3y + \frac{9y^2}{8} - \frac{y^3}{4} - \frac{3(4 - y)^4}{256})\Big|_{x^2}^4 dx$$

$$= 2\int_0^2 (12 + 18 - 16 - 3x^2 - \frac{9x^4}{8} + \frac{x^6}{4} + \frac{3(4 - x^2)^4}{256})dx$$

$$= 2\int_0^2 (17 - 6x^2 + \frac{x^6}{16} + \frac{3x^8}{256})dx$$

$$= (17x - 2x^3 + \frac{x^7}{112} + \frac{x^9}{768})\Big|_0^2 = 2(34 - 16 + \frac{8}{7} + \frac{2}{3}) = \frac{832}{21}.$$

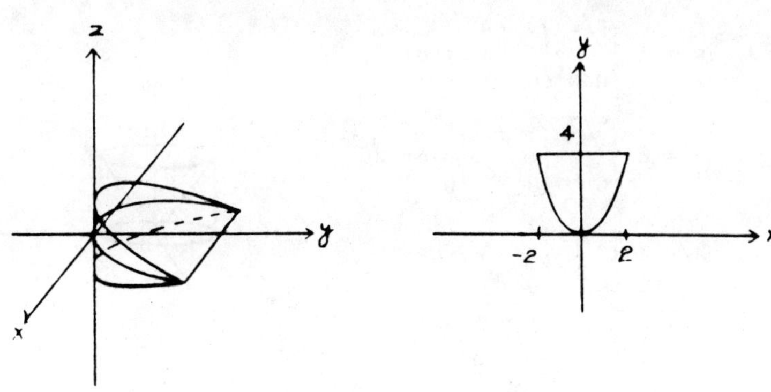

25. Sphere: $\rho = 2$, Cone: $\phi = \pi/4$, Plane: $\phi = \pi/2$.

$$m = 4\int_{\pi/4}^{\pi/2}\int_{0}^{\pi/2}\int_{0}^{2}(1 + \rho\cos\phi)\rho^2\sin\phi\, d\rho\, d\theta\, d\phi$$

$$= 4\int_{\pi/4}^{\pi/2}\int_{0}^{\pi/2}\int_{0}^{2}(\rho^2\sin\phi + \rho^3\sin\phi\cos\phi)d\rho\, d\theta\, d\phi$$

$$= 4\int_{\pi/4}^{\pi/2}\int_{0}^{\pi/2}\left(\frac{\rho^3\sin\phi}{3} + \frac{\rho^4\sin\phi\cos\phi}{4}\right)\Big|_{0}^{2} d\theta\, d\phi$$

$$= 4\int_{\pi/4}^{\pi/2}\int_{0}^{\pi/2}\left(\frac{8}{3}\sin\phi + 4\sin\phi\cos\phi\right)d\theta\, d\phi$$

$$= 2\pi\int_{\pi/4}^{\pi/2}\left(\frac{8}{3}\sin\phi + 4\sin\phi\cos\phi\right)d\phi$$

$$= 2\pi\left(-\frac{8}{3}\cos\phi + 2\sin^2\phi\right)\Big|_{\pi/4}^{\pi/2}$$

$$= 2\pi\left(0 + 2 + \frac{8}{3\sqrt{2}} - 1\right) = \frac{2\pi(3 + 4\sqrt{2})}{3} .$$

29. $m = 8\int_0^{\pi/2}\int_0^2\int_0^r zr\,dz\,dr\,d\theta$

$= 8\int_0^{\pi/2}\int_0^2 \left.\frac{rz^2}{2}\right|_0^r dr\,d\theta$

$= 4\int_0^{\pi/2}\int_0^2 r^3 dr\,d\theta$

$= 4\int_0^{\pi/2} \left.\frac{r^4}{4}\right|_0^2 d\theta$

$= 4\int_0^{\pi/2} 4\,d\theta$

$= 16 \cdot \frac{\pi}{2} = 8\pi$

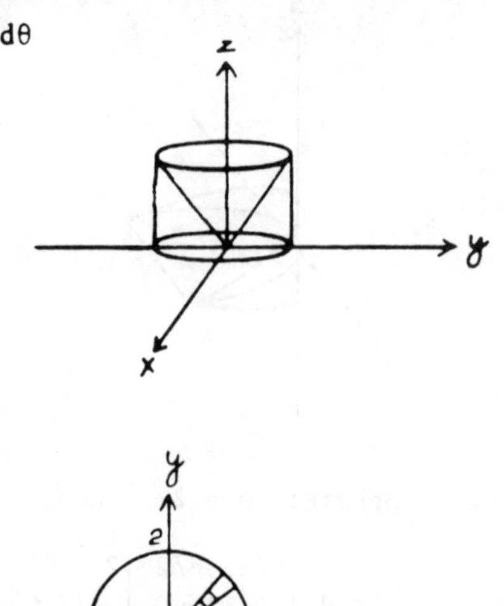

33. $\bar{y} = a,\ A = \pi a^2$

$A\bar{y} = M_x$

$\pi a^3 = \int_{-a}^{a}\int_{a-\sqrt{a^2-x^2}}^{a+\sqrt{a^2-x^2}} y\,dy\,dx.$

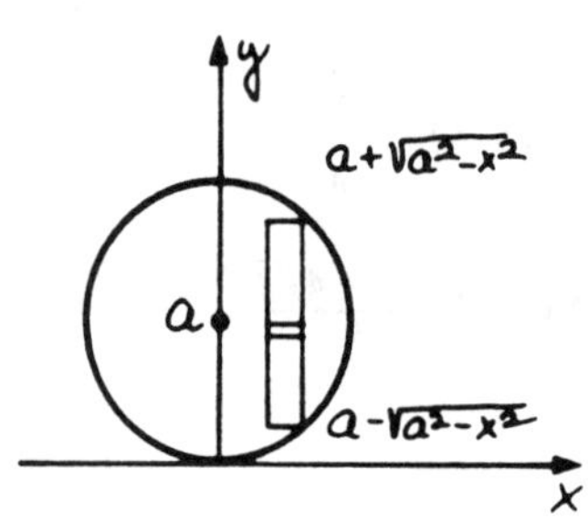

37. (a) $F(x,y) = y - f(x) = 0$

$F(x, \pm\sqrt{y^2 + z^2}) = 0$

$\pm\sqrt{y^2 + z^2} - f(x) = 0, \quad y^2 + z^2 = [f(x)]^2.$

(b) $2zz_x = 2ff', \qquad z_x = ff'/z$

$2y + 2zz_x = 0, \qquad z_y = -y/z$

$$z_x{}^2 + z_y{}^2 = \frac{f^2(f')^2}{z^2} + \frac{y^2}{z^2} = \frac{f^2(f')^2}{z^2} + \frac{f^2}{z^2}$$

$$= \frac{f^2}{z^2}[1 + (f')^2]$$

(c) $S = 4\iint_R \sqrt{z_x{}^2 + z_y{}^2 + 1}\, dA$ (See Chapter 20, Section 5.)

$$= 4\int_a^b \int_0^{f(x)} \frac{f\sqrt{1 + (f')^2}}{\sqrt{f^2 - y^2}}\, dy\, dx$$

(d) $y = f(x)\sin\theta, \; dy = f(x)\cos\theta\, d\theta$

$$= 4\int_a^b \int_0^{\pi/2} \frac{f\sqrt{1 + (f')^2}}{f\cos\theta} f\cos\theta\, d\theta$$

$$= 2\pi\int_a^b f(x)\sqrt{1 + [f'(x)]^2}\, dx.$$

Chapter 22
Line and Surface Integrals

CHAPTER 22, SECTION 1 (pp. 1087-1089)

1. $\int_0^2 (4t - 9t)2\,dt = -10\,\frac{t^2}{2}\Big|_0^2$

$= -20.$

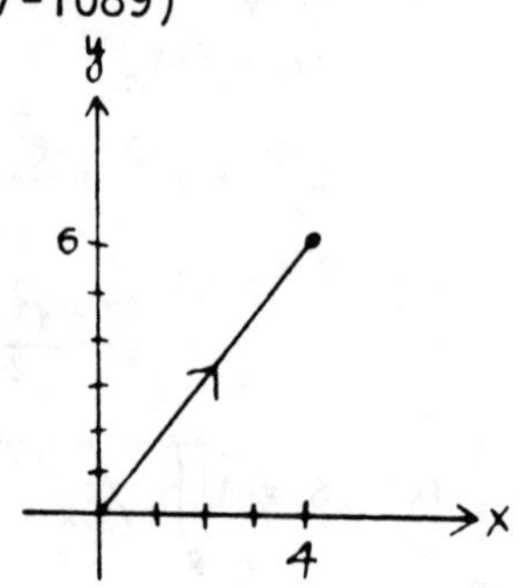

5. $\int_0^1 [2t^2(1 - t)(-1) - t^4 \cdot 2t]dt$

$= \int_0^1 (2t^3 - 2t^2 - 2t^5)dt$

$= \frac{1}{2}t^4 - \frac{2}{3}t^3 - \frac{1}{3}t^6\Big|_0^1$

$= \frac{1}{2} - \frac{2}{3} - \frac{1}{3} = -\frac{1}{2}.$

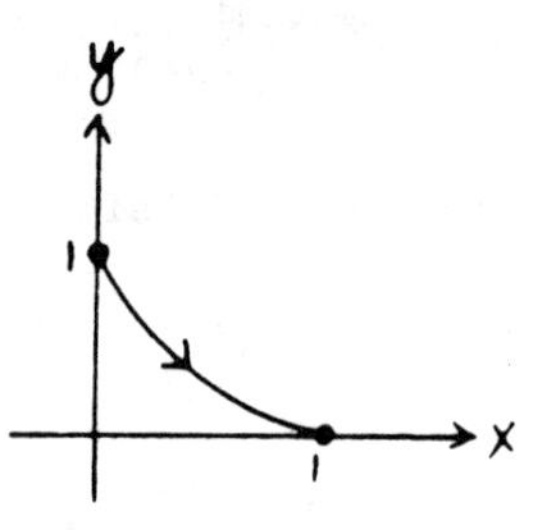

9. $\int_C \mathbf{f} \cdot d\mathbf{r} = \int_C 0dx + y(x + 1)dy$

$= \int_0^2 y(1 + 1)dy + \int_1^2 (x + 1)(x + 1)dx$

$= y^2\Big|_0^2 + \frac{(x + 1)^3}{3}\Big|_1^2$

$= 4 + 9 - \frac{8}{3} = \frac{31}{3}.$

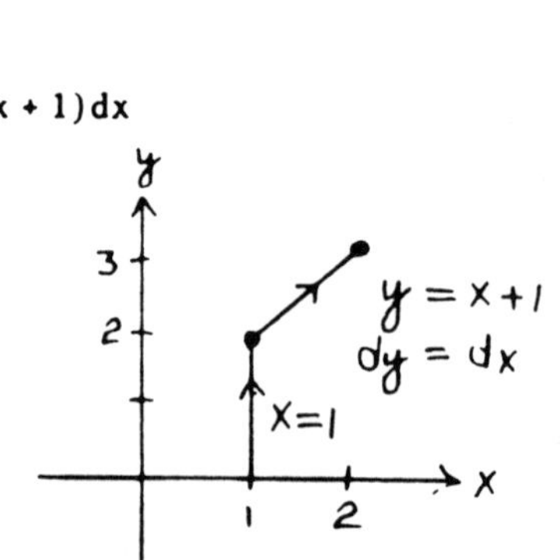

13. $\frac{dx}{dt} = e^t(\sin t + \cos t)$, $\frac{dy}{dt} = e^t(\cos t - \sin t)$

$$ds = \sqrt{e^{2t}(\sin t + \cos t)^2 + e^{2t}(\cos t - \sin t)^2}\, dt$$

$$= \sqrt{2}\, e^t\, dt$$

$$A = \int_0^{\pi/2} (e^{2t}\sin^2 t + e^{2t}\cos^2 t)\sqrt{2}\, e^t\, dt$$

$$= \frac{\sqrt{2}}{3} e^{3t}\Big|_0^{\pi/2} = \frac{\sqrt{2}}{3}(e^{3\pi/2} - 1).$$

17. $\frac{2}{3}x^{-1/3} + \frac{2}{3}y^{-1/3}\frac{dy}{dx} = 0$, $\frac{dy}{dx} = -\frac{y^{1/3}}{x^{1/3}}$

$$ds = \sqrt{1 + \frac{y^{2/3}}{x^{2/3}}}\, dx = x^{-1/3}\sqrt{x^{2/3} + y^{2/3}}\, dx$$

$$= x^{-1/3}\sqrt{a^{2/3}}\, dx$$

$$= a^{1/3}x^{-1/3}\, dx$$

$$A = \int_0^a (2 + x)a^{1/3}x^{-1/3}\, dx$$

$$= a^{1/3}\int_0^a (2x^{-1/3} + x^{2/3})dx$$

$$= a^{1/3}(3x^{2/3} + \frac{3}{5}x^{5/3})\Big|_0^a$$

$$= \frac{3}{5}a(5 + a).$$

21. $ds = \sqrt{(\frac{dx}{dt})^2 + (\frac{dy}{dt})^2}\, dt$

$$= \sqrt{(24t^2)^2 + (8t)^2}\, dt$$

$$\int_C ds = \int_0^1 8t\sqrt{9t^2 + 1}\, dt$$

$$= \frac{8}{18} \int_0^1 18t \sqrt{9t^2 + 1}\, dt$$

$$= \frac{4}{9} \cdot \frac{2}{3} (9t^2 + 1)^{3/2} \Big|_0^1$$

$$= \frac{8}{27} (10\sqrt{10} - 1) .$$

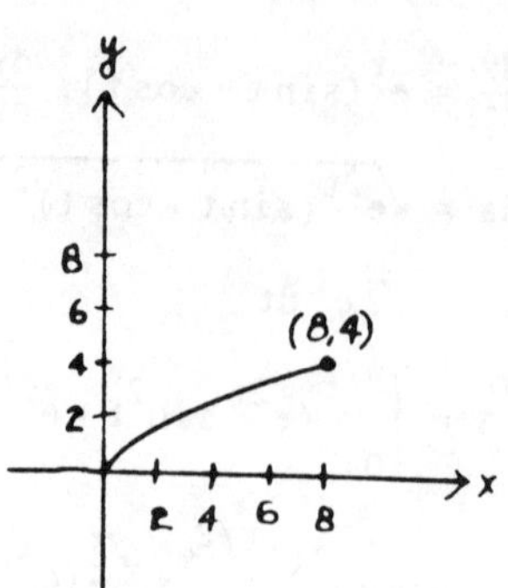

25. $\displaystyle\int_C \frac{ydx - xdy}{x^2 + y^2}$

$$= \int_0^{\pi/2} \frac{\sin^3 t(-3 \cos^2 t \sin t) - \cos^3 t(3 \sin^2 t \cos t)}{\cos^6 t + \sin^6 t} dt$$

$$= - 3 \int_0^{\pi/2} \frac{(\sin^2 t + \cos^2 t)\sin^2 t \cos^2 t}{\cos^6 t + \sin^6 t} dt$$

$$= -3 \int_0^{\pi/2} \frac{\dfrac{\sin^2 t}{\cos^4 t}}{1 + \left(\dfrac{\sin t}{\cos t}\right)^6} dt$$

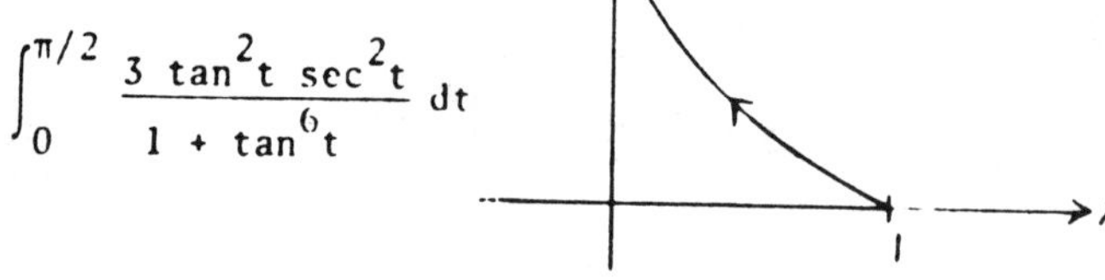

$$= - \int_0^{\pi/2} \frac{3 \tan^2 t \sec^2 t}{1 + \tan^6 t} dt$$

$$\begin{pmatrix} u = \tan^3 t, \ du = 3 \tan^2 t \sec^2 t\, dt \\ t = 0, \ u = 0; \ t = \frac{\pi}{2}, \ u = +\infty \end{pmatrix}$$

$$= - \int_0^{+\infty} \frac{du}{1 + u^2} = - \text{Arctan } u \Big|_0^{+\infty} = - \frac{\pi}{2} .$$

29. $T = \frac{r'(t)}{|r'(t)|} = \frac{1}{|r'(t)|}(x'(t)i + y'(t)j)$,

$$|r'(t)| = \sqrt{[x'(t)]^2 + [y'(t)]^2} = s'(t) .$$

$$\int_C f \cdot T ds$$

$$= \int_a^b \{\frac{P(x(t),y(t))x'(t) + Q(x(t),y(t))y'(t)}{|r'(t)|}\} s'(t)dt$$

$$= \int_a^b [P(x(t),y(t))x'(t) + Q(x(t),y(t))y'(t)]dt$$

$$= \int_C f \cdot dr .$$

CHAPTER 22, SECTION 2 (pp. 1096-1098)

1. $\int_C 2y\ dx - 3x\ dy = \iint_R (-3 - 2)dA = -5\pi(1)^2 = -5\pi$.

5. $\int_C (e^{x^2} + y^2)dx + (e^{y^2} + x^2)dy$

$$= \int_0^4 \int_0^{4-x} (2x - 2y)dy\ dx = \int_0^4 [2x(4 - x) - (4 - x)^2]dx$$

$$= \int_0^4 (16x - 3x^2 - 16)dx = 8x^2 - x^3 - 16x \Big|_0^4$$

$$= 128 - 64 - 64 = 0 .$$

9. $\int_{C_1-C_2} Pdx + Qdy$

$= \iint_R (Q_x - P_y)dA = 0,$

so $\int_{C_1} Pdx + Qdy = \int_{C_2} Pdx + Qdy$.

13. $A = -\int_C y\,dx$

$= -\int_{-1}^{2} x^2 dx + \int_{-1}^{2} (x+2)dx$

$= -\frac{x^3}{3}\Big|_{-1}^{2} + \left(\frac{x^2}{2} + 2x\Big|_{-1}^{2}\right)$

$= -\frac{8}{3} - \frac{1}{3} + 6 + \frac{3}{2} = \frac{9}{2}$.

17. $V = 2\pi \iint_R x^2\, dA$

$= 2\pi \int_C \frac{1}{3}x^3 dy$

$= \frac{2\pi}{3}\int_C x^3 dy.$

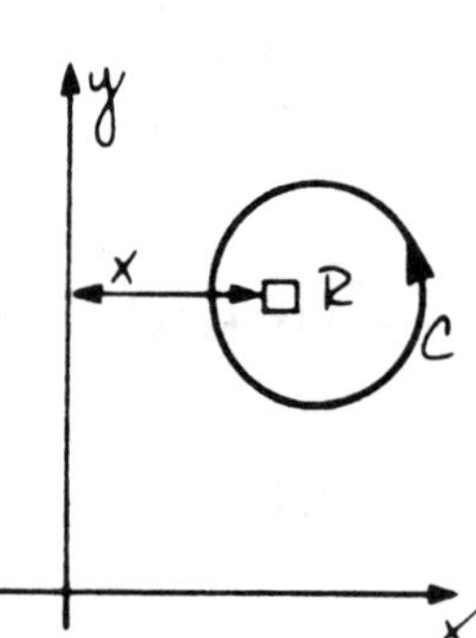

CHAPTER 22, SECTION 3 (pp. 1105-1107)

1. $\frac{\partial P}{\partial y} = 2y$, $\frac{\partial Q}{\partial x} = 2y$; exact.

$F = \int (y^2 + 3x^2)dx = xy^2 + x^3 + f(y)$

$\frac{\partial F}{\partial y} = 2xy + f'(y) = 2xy$, $f'(y) = 0$, $f(y) = C$

$F = xy^2 + x^3 + C$.

5. $\frac{\partial P}{\partial y} = 2x$, $\frac{\partial Q}{\partial x} = 2x$; exact.

$$F = \int \left(\frac{1}{x} + 2xy\right)dx = \ln|x| + x^2y + f(y)$$

$$\frac{\partial F}{\partial y} = x^2 + f'(y) = \frac{1}{y} + x^2 + 3y^2$$

$$f'(y) = \frac{1}{y} + 3y^2, \quad f(y) = \ln|y| + y^3 + C$$

$$F = \ln|x| + x^2y + \ln|y| + y^3 + C = \ln|xy| + x^2y + y^3 + C.$$

9. $\frac{\partial P}{\partial y} = 2xy\,e^{xy^2} + 2x^2y^3e^{xy^2} + 2xy\,e^{xy^2} + 2$

$$\frac{\partial Q}{\partial x} = 4xy\,e^{xy^2} + 2x^2y^3e^{xy^2} + 2; \text{ exact}$$

$$F = \int (xy^2e^{xy^2} + e^{xy^2} + 2y)dx = x\,e^{xy^2} + 2xy + C(y)$$

$$\frac{\partial F}{\partial y} = 2x^2y\,e^{xy^2} + 2x + C'(y) = 2x^2y\,e^{xy^2} + 2x - 12y^2$$

$$C'(y) = -12y^2, \; C(y) = -4y^3 + K$$

$$x\,e^{xy^2} + 2xy - 4y^3 + K = 0.$$

13. $\frac{\partial P}{\partial y} = e^{xy} + xye^{xy} + 4x$, $\frac{\partial Q}{\partial x} = xye^{xy} + e^{xy} + 4x$; exact.

$$F = \int (ye^{xy} + 4xy + 2x - 2)dx = e^{xy} + 2x^2y + x^2 - 2x + C(y)$$

$$\frac{\partial F}{\partial y} = xe^{xy} + 2x^2 + C'(y) = xe^{xy} + 2x^2 + 1$$

$$C'(y) = 1, \quad C(y) = y + K$$

$$e^{xy} + 2x^2y + x^2 - 2x + y + K = 0.$$

17. $\frac{1}{\cos y}$ is an integrating factor.

$$\tan y\,dx + x\sec^2 y\,dy = 0$$

$$F = \int \tan y\,dx = x\tan y + C(y)$$

$$\frac{\partial F}{\partial y} = x\sec^2 y + C'(y) = x\sec^2 y,$$

$C'(y) = 0, \quad C(y) = K$

$x \tan y + K = 0.$

21. $x^{m+1}y^{n+2}dx + (3x^{m+2}y^{n+1} - 8x^m y^{n+3})dy$

$\frac{\partial P}{\partial y} = (n+2)x^{m+1}y^{n+1}, \ \frac{\partial Q}{\partial x} = 3(m+2)x^{m+1}y^{n+1} - 8mx^{m-1}y^{n+3}$

$n + 2 = 3m + 6, \ 0 = -8m$

$m = 0, \ n = 4.$

25. $\frac{\partial P}{\partial y} = 3x^2, \ \frac{\partial Q}{\partial x} = 3x^2$; exact. The integral is independent the path. C is a path from $(0,0)$ to $(\pi^2,0)$. Choose the segment $[0,\pi^2]$ of the x axis as the path C' of integrati

Then $\int_{C'} (x^3 + 3x^2y)dx + (x^3e^y)dy$

$$= \int_0^{\pi^2} [(x^3 + 3x^2 \cdot 0) \cdot 1 + (x^3 + e^0) \cdot 0]dx$$

$$= \frac{x^4}{4}\bigg|_0^{\pi^2} = \frac{\pi^8}{4}.$$

CHAPTER 22, SECTION 4 (pp. 1113-1115)

1. $W = \int_C f \cdot dr = \int_0^2 [(4 - x) \cdot 1 + (x + 2(4 - x^2))(-2x)]dx$

$$= \int_0^2 (4x^3 - 2x^2 - 17x + 4)dx = x^4 - \frac{2}{3}x^3 - \frac{17}{2}x^2 + 4x\bigg|_0^2$$

$$= -\frac{46}{3}.$$

5. $W = \int_C f \cdot dr = \int_0^2 [x^2 + (e^{-x} + e^x)e^x]dx$

$$= \int_0^2 (x^2 + 1 + e^{2x})dx = \frac{1}{3}x^3 + x + \frac{1}{2}e^{2x}\bigg|_0^2$$

$$= \frac{8}{3} + 2 + \frac{1}{2}e^4 - \frac{1}{2} = \frac{25}{6} + \frac{1}{2}e^4.$$

9. $\int_C \mathbf{f} \cdot \mathbf{n} ds = \int_C P dy - Q dx = \int_C (2y - x^2) dy - (-x - y^2) dx$

$$\iint_R (P_x + Q_y) dA = \iint_R (-2x - 2y) dA = -2\int_{-2}^{2}\int_{-\sqrt{4-y^2}}^{\sqrt{4-y^2}} (x + y) dx\, dy$$

$$= -2 \int_{-2}^{2} (\tfrac{1}{2} x^2 + xy \Big|_{-\sqrt{4-y^2}}^{\sqrt{4-y^2}}) dy$$

$$= -2 \int_{-2}^{2} [\tfrac{1}{2} (4 - y^2 - 4 + y^2) + y(\sqrt{4 - y^2} + \sqrt{4 - y^2})] dy$$

$$= -2 \int_{-2}^{2} (2y\sqrt{4 - y^2}) dy = \frac{4}{3} (4 - y^2)^{3/2} \Big|_{-2}^{2} = 0 .$$

13. C: $x = 2 \cos \theta$, $y = 2 \sin \theta$, $\frac{\pi}{6} \le \theta \le \frac{\pi}{4}$.

$$\int_C \mathbf{f} \cdot d\mathbf{r} = \int_{\pi/6}^{\pi/4} (\frac{-2 \sin \theta}{2 \cos \theta} + \frac{2 \cos \theta}{2 \sin \theta}) d\theta$$

$$= \int_{\pi/6}^{\pi/4} (- \tan \theta + \cot \theta) d\theta = \ln \cos \theta + \ln \sin \theta \Big|_{\pi/6}^{\pi/4}$$

$$= 2 \ln \frac{1}{\sqrt{2}} - \ln \frac{\sqrt{3}}{2} - \ln \frac{1}{2}$$

$$= - \ln \frac{\sqrt{3}}{2} .$$

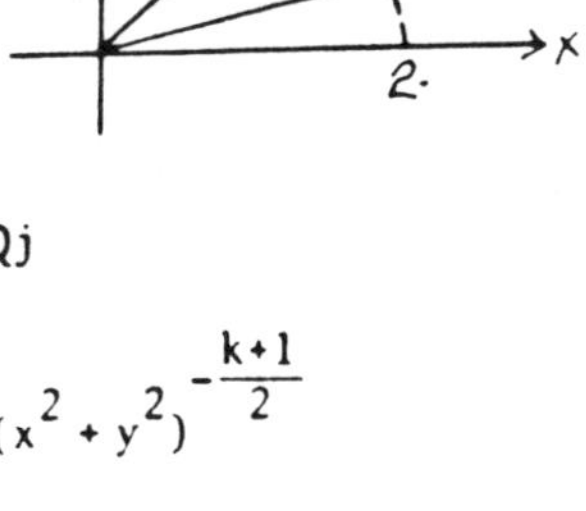

17. $\mathbf{F} = - \frac{cx}{r^{k+1}} \mathbf{i} - \frac{cy}{r^{k+1}} \mathbf{j} = P\mathbf{i} + Q\mathbf{j}$

$$P = -cx(x^2 + y^2)^{-\frac{k+1}{2}} , \quad Q = -cy(x^2 + y^2)^{-\frac{k+1}{2}}$$

$$\frac{\partial P}{\partial y} = \frac{k + 1}{2} \cdot cx \cdot 2y(x^2 + y^2)^{-\frac{k+1}{2} - 1}$$

$$\frac{\partial Q}{\partial x} = \frac{k+1}{2} \cdot cy \cdot 2x(x^2 + y^2)^{-\frac{k+1}{2} - 1}$$

$\frac{\partial P}{\partial y} = \frac{\partial Q}{\partial x}$ so the field is conservative.

From problem 16, if $F(r) = \frac{c}{k-2} r^{-k+2}$, then

$\nabla F = -\frac{c}{r^k}(xi + yj)$, $k \neq 2$.

If $k = 2$, let $F(r) = -\ln r = -\frac{1}{2}\ln(x^2 + y^2)$. Then

$\nabla \ln r = -\frac{1}{2} \cdot \frac{2x}{r^2} i - \frac{1}{2}\frac{2y}{r^2} j = -\frac{1}{r^2}(xi + yj)$.

21. $W(\lambda) = \int_C F \cdot dr$

$$= \int_0^1 [6 \cdot \lambda t(t-1) + 60t \cdot \lambda t(t-1)\lambda(2t-1)]dt$$

$$= \lambda[2t^3 - 3t^2 + 60\lambda(\tfrac{2}{5}t^5 - \tfrac{3}{4}t^4 + \tfrac{1}{3}t^3)]\Big|_0^1$$

$$= -\lambda - \lambda^2$$

$W(\lambda) = 0 : \lambda = 0, -1$

$W'(\lambda) = -1 - 2\lambda = 0, \lambda = -\frac{1}{2}$

$W''(\lambda) = -2 < 0$, W_{maximum} for $\lambda = -\frac{1}{2}$.

CHAPTER 22, SECTION 5 (pp. 1120-1121)

1. $x = t, \quad y = 2 - t, \quad z = t^2, \quad 0 \le t \le 1$

$$\int_C x^2 dx + xz\, dy + yz\, dz$$

$$= \int_0^1 [t^2 \cdot 1 + t \cdot t^2(-1) + (2 - t)t^2 \cdot 2t]dt$$

$$= \int_0^1 (t^2 + 3t^3 - 2t^4)dt = \frac{1}{3} t^3 + \frac{3}{4} t^4 - \frac{2}{5} t^5 \Big|_0^1 = \frac{41}{60} .$$

5. $P = 2xy + z^2, \quad Q = x^2 - 2, \quad R = 2xz + 1$

$$\frac{\partial P}{\partial y} = 2x , \quad \frac{\partial P}{\partial z} = 2z , \quad \frac{\partial Q}{\partial x} = 2x , \quad \frac{\partial Q}{\partial z} = 0 ,$$

$$\frac{\partial R}{\partial x} = 2z , \quad \frac{\partial R}{\partial y} = 0 , \quad \frac{\partial P}{\partial y} = \frac{\partial Q}{\partial x}, \quad \frac{\partial Q}{\partial z} = \frac{\partial R}{\partial y} ,$$

$\frac{\partial R}{\partial x} = \frac{\partial P}{\partial z}$; exact.

$$F = \int P dx = x^2 y + xz^2 + g(y,z)$$

$$\frac{\partial F}{\partial y} = x^2 + \frac{\partial g}{\partial y} = x^2 - 2 .$$

$$\frac{\partial g}{\partial y} = -2 , \quad g = -2y + h(z)$$

$$F = x^2 y + xz^2 - 2y + h(z)$$

$$\frac{\partial F}{\partial z} = 2xz + h'(z) = 2xz + 1 , \quad h'(z) = 1 , \quad h(z) = z + K$$

$$F = x^2 y + xz^2 - 2y + z + K .$$

9. $z = 4 - x - y$

$$S = \int_0^4 \int_0^{4-x} \sqrt{(-1)^2 + (-1)^2 + 1}\, dy\, dx$$

$$= \sqrt{3} \int_0^4 (4 - x)dx$$

$$= \sqrt{3}\, (4x - \frac{1}{2} x^2 \Big|_0^4) = 8\sqrt{3} .$$

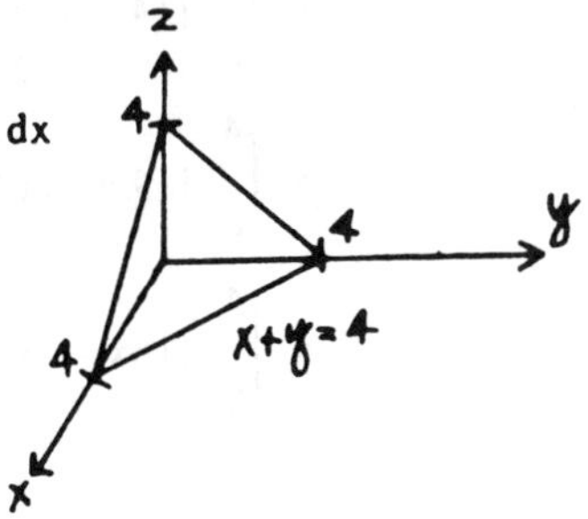

z
4
4
y
4
x+y=4
x

13. $S = \int_0^a \int_0^{2z} \sqrt{1 + \frac{z^2}{a^2 - z^2}}\, dy\, dz$

$= a \int_0^a \int_0^{2z} \frac{1}{\sqrt{a^2 - z^2}}\, dy\, dz$

$= a \int_0^a \frac{2z}{\sqrt{a^2 - z^2}}\, dz$

$= -2a \sqrt{a^2 - z^2}\Big|_0^a = 2a^2 .$

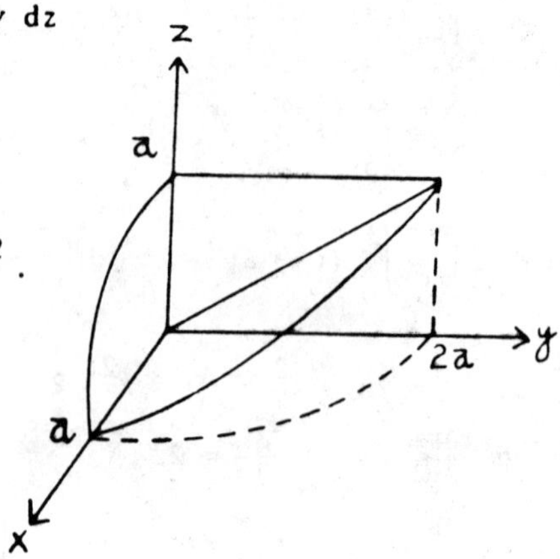

17. $x + y + z = 4 , \quad n = \frac{1}{\sqrt{3}} (i + j + k) .$

$\text{Flux} = \iint_S f \cdot n\, dS$

$= \int_0^4 \int_0^{4-x} \frac{1}{\sqrt{3}} [2x - 3y + (4 - x - y)]\sqrt{3}\, dy\, dx$

$= \int_0^4 \int_0^{4-x} (4 + x - 4y)dy\, dx$

$= \int_0^4 [(4 + x)y - 2y^2\Big|_0^{4-x}]dx$

$= \int_0^4 [(4 + x)(4 - x) - 2(4 - x)^2]dx$

$= \int_0^4 (16x - 3x^2 - 16)dx = 8x^2 - x^3 - 16x\Big|_0^4 = 0 .$

21. C_1 = segment from $(0,0,2)$ to $(2,0,0)$:

$$x = t\,, \quad y = 0\,, \quad z = 2 - t\,, \quad 0 \le t \le 2$$

C_2 = segment from $(2,0,0)$ to $(1,1,1)$:

$$x = 2 - t,\ y = t\,,\ z = t\,,\ 0 \le t \le 1$$

C_3 = segment from $(1,1,1)$ to $(0,0,2)$:

$$x = 1 - t\,, \quad y = 1 - t\,, \quad z = 1 + t\,, \quad 0 \le t \le 1$$

$$\int_{C_1} z^2dx - 2xz\ dy + x^2dz = \int_0^2 [(2 - t)^2 + t^2(-1)]dt$$

$$= \int_0^2 (4 - 4t)dt = 4t - 2t^2\Big|_0^2 = 0\,,$$

$$\int_{C_2} z^2dx - 2xz\ dy + x^2dz = \int_0^1 [t^2(-1) - 2(2-t)t + (2-t)^2]dt$$

$$= \int_0^1 (2t^2 - 8t + 4)dt = \frac{2}{3}t^3 - 4t^2 + 4t\Big|_0^1 = \frac{2}{3}\,,$$

$$\int_{C_3} z^2dx - 2xz\ dy + x^2dz$$

$$= \int_0^1 [(1 + t)^2(-1) - 2(1 - t)(1 + t)(-1) + (1 - t)^2]dt$$

$$= \int_0^1 (-2t^2 - 4t + 2)dt = -\frac{2}{3}t^3 - 2t^2 + 2t\Big|_0^1 = -\frac{2}{3}\,,$$

$$\int_C z^2dx - 2xz\ dy + x^2dz = 0 + \frac{2}{3} - \frac{2}{3} = 0\,.$$

Notice that the differential expression $z^2dx - 2xz\ dy + x^2dz$ is not exact.

CHAPTER 22, SECTION 6 (pp. 1129-1131)

1. $\nabla \cdot f = 0 + 2xz + 2z = 2z(x + 1)\,.$

5. $$\nabla \times f = \begin{vmatrix} i & j & k \\ \partial/\partial x & \partial/\partial y & \partial/\partial z \\ xy^2 & x^2y & z^2 \end{vmatrix}$$

$$= (0 - 0)i - (0 - 0)j + (2xy - 2xy)k = 0i + 0j + 0k .$$

9. $$\int_C f \cdot dr = \int_C - y\, dx + x\, dy + z\, dz$$

$$= a^2 \int_0^{2\pi} [(-\sin\theta)(-\sin\theta) + \cos\theta \cdot \cos\theta + 0]d\theta$$

$$= 2\pi a^2$$

$n = k$, $\nabla \times f \cdot n = 2k \cdot k = 2$

$$\iint_R \nabla \times f \cdot n\, dS = 2\pi a^2 .$$

13. div $f = 0 - 1 + 1 = 0$, $\iiint_V \nabla \cdot f\, dV = 0$.

In the xy plane, $z = 0$, $n = -k$, $f \cdot n = -xy$,

$$\int_0^2 \int_0^3 (-xy)dy\, dx = \int_0^2 (-x \left.\frac{y^2}{2}\right|_0^3)dx$$

$$= -\frac{9}{2}\int_0^2 x\, dx = -\frac{9}{2} \cdot \left.\frac{x^2}{2}\right|_0^2 = -9 .$$

In the xz plane, $y = 0$, $n = -j$, $f \cdot n = -z$,

$$\int_0^2 \int_0^1 (-z)dz\, dx = \int_0^2 (- \left.\frac{z^2}{2}\right|_0^1)dx = -\frac{1}{2} \cdot 2 = -1 .$$

In the yz plane, $x = 0$, $n = -i$, $f \cdot n = -2y$,

$$\int_0^1 \int_0^3 (-2y)dy\, dz = \int_0^1 (-y^2\Big|_0^3)dz = -9 .$$

In the plane $z = 1$, $n = k$, $f \cdot n = xy + 1$,

$$\int_0^2 \int_0^3 (xy + 1)dy\ dx = \int_0^2 \left(x \frac{y^2}{2} + y\Big|_0^3\right)dx = \int_0^2 \left(\frac{9}{2} x + 3\right)dx = 15.$$

In the plane $y = 3$, $n = j$, $f \cdot n = z - 3$,

$$\int_0^2 \int_0^1 (z - 3)dz\ dx = \int_0^2 \left(\frac{z^2}{2} - 3z\Big|_0^1\right)dx = -5\ .$$

In the plane $x = 2$, $n = i$, $f \cdot n = 2y$,

$$\int_0^1 \int_0^3 2y\ dy\ dz = 9\ .$$

$$\iint_S f \cdot n\ dA = -9 - 1 - 9 + 15 - 5 + 9 = 0.$$

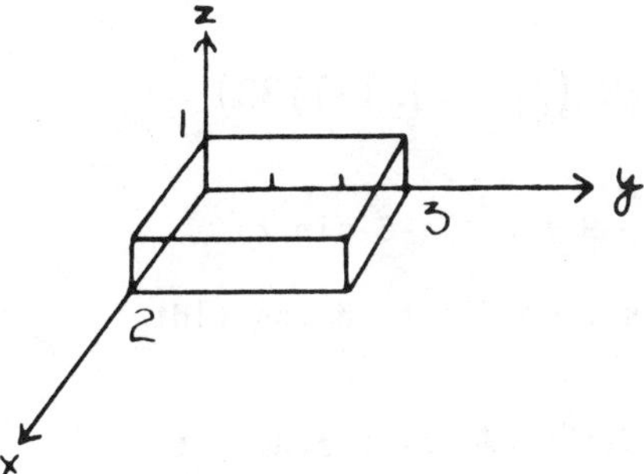

17. $f = Pi + Qj + Rk$, P, Q and R have continuous second partial derivatives.

$$\nabla \cdot \nabla \times f = \nabla \cdot \left[\left(\frac{\partial R}{\partial y} - \frac{\partial Q}{\partial z}\right)i + \left(\frac{\partial P}{\partial z} - \frac{\partial R}{\partial x}\right)j + \left(\frac{\partial Q}{\partial x} - \frac{\partial P}{\partial y}\right)k\right]$$

$$= \frac{\partial}{\partial x}\left(\frac{\partial R}{\partial y} - \frac{\partial Q}{\partial z}\right) + \frac{\partial}{\partial y}\left(\frac{\partial P}{\partial z} - \frac{\partial R}{\partial x}\right) + \frac{\partial}{\partial z}\left(\frac{\partial Q}{\partial x} - \frac{\partial P}{\partial y}\right)$$

$$= \frac{\partial^2 R}{\partial x \partial y} - \frac{\partial^2 Q}{\partial x \partial z} + \frac{\partial^2 P}{\partial y \partial z} - \frac{\partial^2 R}{\partial y \partial x} + \frac{\partial^2 Q}{\partial z \partial x} - \frac{\partial^2 P}{\partial z \partial y}$$

= 0, since the mixed second partial derivatives of P, Q, and R are equal.

21. $\nabla \times (x\mathbf{i} + y\mathbf{j} + z\mathbf{k}) = 0\mathbf{i} + 0\mathbf{j} + 0\mathbf{k}$.

25. $$\iint_S \mathbf{n} \cdot (u \nabla v)dS = \iiint_V \nabla \cdot (u \nabla v)dV$$

$$\iint_S \mathbf{n} \cdot (v \nabla u)dS = \iiint_V \nabla \cdot (v \nabla u)dV .$$

Subtracting,

$$\iint_S \mathbf{n} \cdot (u \nabla v - v \nabla u)dS = \iiint_V (\nabla u \cdot \nabla v + u\nabla^2 v - \nabla v \cdot \nabla u - v\nabla^2 u)dV$$

$$= \iiint_V (u\nabla^2 v - v\nabla^2 u)dV .$$

CHAPTER 22, REVIEW (pp. 1131-1133)

1. $$\int_0^{\pi/2} [(9 \cos^2 t - 9 \sin^2 t)(-3 \sin t) - 2 \cdot 3 \cos t \cdot 3 \sin t \cdot 3 \cos t]dt$$

$$= 27 \int_0^{\pi/2} (\sin^3 t - 3 \cos^2 t \sin t)dt$$

$$= 27 \int_0^{\pi/2} (\sin t - 4 \cos^2 t \sin t)dt$$

$$= 27(\frac{4}{3} \cos^3 t - \cos t)\Big|_0^{\pi/2} = 27(0 - \frac{4}{3} + 1) = -9 .$$

5. $$\int_C y^2 dx - x^2 dy$$

$$= \int_0^1 \int_0^{3x^2} (-2x - 2y)dy\, dx$$

$$= \int_0^1 (-2xy - y^2 \Big|_0^{3x^2})dx$$

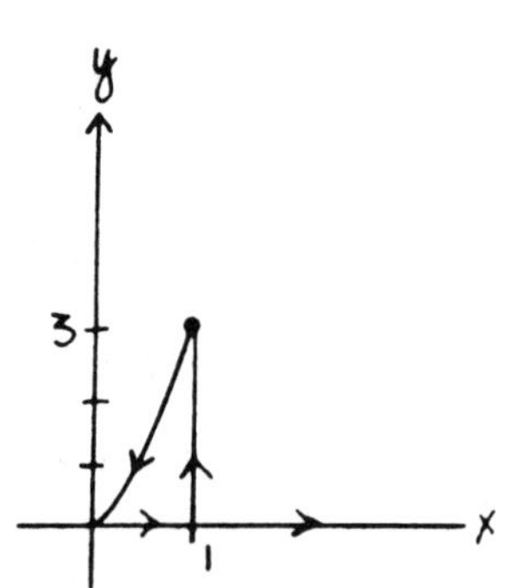

$$= - \int_0^1 (6x^3 + 9x^4)\,dx$$

$$= - \left(\frac{3}{2} x^4 + \frac{9}{5} x^5 \Big|_0^1\right) = - \frac{3}{2} - \frac{9}{5} = - \frac{33}{10} .$$

9. $\frac{\partial P}{\partial y} = 2xy, \quad \frac{\partial Q}{\partial x} = 2xy;$ exact.

$$F = \int (x^2 + xy^2)\,dx = \frac{x^3}{3} + \frac{x^2y^2}{2} + f(y)$$

$$\frac{\partial F}{\partial y} = x^2y + f'(y) = x^2y + y^2, \quad f'(y) = y^2 , \quad f(y) = \frac{y^3}{3} + C ,$$

$$F = \frac{x^3}{3} + \frac{y^3}{3} + \frac{x^2y^2}{2} + C .$$

13. $W = \int_C f \cdot dr = \int_C (1 + y)\,dx + (x - y)\,dy$

$$= \int_0^2 \{[1 + (3x + 2)] + [x - (3x + 2)] \cdot 3\}\,dx$$

$$= \int_0^2 (-3 - 3x)\,dx = -3x - \frac{3}{2} x^2 \Big|_0^2 = -6 - 6 = -12 .$$

17. $\frac{\partial R}{\partial y} = 1 = \frac{\partial Q}{\partial z}, \quad \frac{\partial P}{\partial z} = 2x = \frac{\partial R}{\partial x}, \quad \frac{\partial Q}{\partial x} = 0 = \frac{\partial P}{\partial y};$ exact.

$$F = \int (x^2 + 2xz)\,dx = \frac{1}{3} x^3 + x^2z + g(y,z)$$

$$\frac{\partial F}{\partial y} = \frac{\partial g}{\partial y} = z - y^2 , \quad g(y,z) = yz - \frac{1}{3} y^3 + h(z)$$

$$F = \frac{1}{3} x^3 + x^2z + yz - \frac{1}{3} y^3 + h(z)$$

$$\frac{\partial F}{\partial z} = x^2 + y + h'(z) = x^2 + y + z^2$$

$$h'(z) = z^2 , \quad h(z) = \frac{1}{3} z^3 + K$$

$$F = \frac{1}{3} x^3 + x^2z + yz - \frac{1}{3} y^3 + \frac{1}{3} z^3 + K .$$

21. $\iint_S f \cdot n \, dS = \iiint_V \nabla \cdot f \, dV = \iiint_V (2x + 2y - 2) dV$

$= 2 \int_0^{2\pi} \int_0^2 \int_0^5 (r \cos \theta + r \sin \theta - 1) r \, dz \, dr \, d\theta$

$= 10 \int_0^{2\pi} \int_0^2 (r^2 \cos \theta + r^2 \sin \theta - r) dr \, d\theta$

$= 10 \int_0^{2\pi} (\frac{r^3}{3} \cos \theta + \frac{r^3}{3} \sin \theta - \frac{r^2}{2} \Big|_0^2) d\theta$

$= 10 \int_0^{2\pi} (\frac{8}{3} \cos \theta + \frac{8}{3} \sin \theta - 2) d\theta$

$= \frac{80}{3} \sin \theta - \frac{80}{3} \cos \theta - 20\theta \Big|_0^{2\pi} = -40\pi$.

25. On C_1, $ds = \sqrt{e^{2t}(\sin t + \cos t)^2 + e^{2t}(\cos t - \sin t)^2}$

$= \sqrt{2}\, e^t$

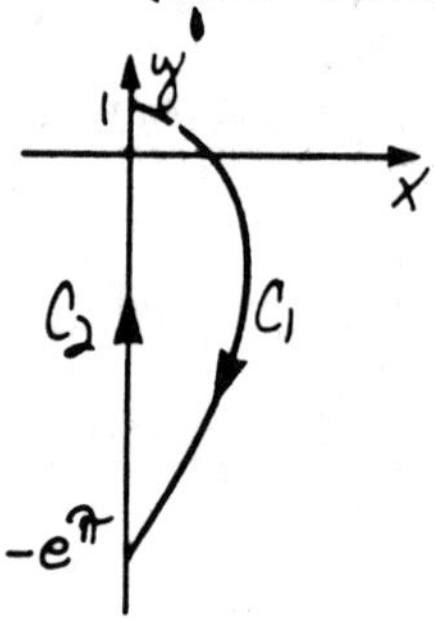

$\int_{C_1} (2x - y) ds$

$= \int_0^{\pi} (2e^t \sin t - e^t \cos t) \sqrt{2}\, e^t dt$

$= \sqrt{2} \int_0^{\pi} (2e^{2t} \sin t - e^{2t} \cos t) dt$

(Formulas 119, 120)

$= \sqrt{2} [\frac{2e^{2t}}{5} (2 \sin t - \cos t) - \frac{e^{2t}}{5} (2 \cos t - \sin t] \Big|_0^{\pi}$

$= \frac{\sqrt{2}}{5} (4e^{2\pi} + 4)$.

On C_2, $ds = dy$ and $x = 0$

$$\int_{C_2} (2x - y)ds$$

$$= \int_{-e^{\pi}}^{1} (0 - y)dy = -\frac{y^2}{2}\Big|_{-e^{\pi}}^{1}$$

$$= \frac{e^{2\pi}}{2} - \frac{1}{2}$$

$$\int_C (2x - y)ds = \frac{e^{2\pi}}{10}(8v_2 + 5) + \frac{4\sqrt{2}}{5} - \frac{1}{2}.$$

29. $$d(\arctan \frac{y}{x}) = \frac{\frac{x\,dy - y\,dx}{x^2}}{1 + \frac{y^2}{x^2}} = \frac{x\,dy - y\,dx}{x^2 + y^2}$$

Let (x_0,y_0) be the initial and terminal point of the simple closed curve C. Then

$$\int_C \frac{x\,dy - y\,dx}{x^2 + y^2} = \int_{(x_0,y_0)}^{(x_0,y_0)} \frac{x\,dy - y\,dx}{x^2 + y^2}$$

$$= \arctan \frac{y_0}{x_0}\Big|_{(x_0,y_0)}^{(x_0,y_0)} = 2n\pi, \text{ since}$$

the values of arctan $\frac{y_0}{x_0}$ differ by integer multiples of 2π.

If C is a circle with center at the origin, then

$\int_C \frac{x\,dy - y\,dx}{x^2 + y^2} = \pm1$, +1 if the circle is traversed in the counterclockwise sense, -1 if the circle is traversed in the clockwise sense.

Absolute Values and Inequalities A

APPENDIX A, SECTION 1 (p. 1140)

1. $2x + 5 < 3$

   $2x < -2$

   $x < -1$

5. $2x + 2 \le x - 4$

   $x \le -6$

9. $3x + 1 \ge 2x + 2$

   $x \ge 1$

13. $1 \le 2x + 1 \le 4$

   $0 \le 2x \le 3$

   $0 \le x \le \frac{3}{2}$

17. $2x - 1 \le x + 4$

   $x \le 5$

   $x + 4 \le 3x + 1$

   $-2x \le -3$

   $x \ge \frac{3}{2}$

   $\frac{3}{2} \le x \le 5$

21. Case I: $x - 4 < 0$

   $2x + 1 \ge x - 4$

   $x \ge -5$ and $x < 4$

   $-5 \le x < 4$

   Case II: $x - 4 > 0$

   $2x + 1 \le x - 4$

   $x \le -5$ and $x > 4$

   No solution in this case.

25. $\frac{a}{ab} < \frac{b}{ab}$, $\quad \frac{1}{b} < \frac{1}{a}$, $\quad \cdot \; \frac{1}{a} > \frac{1}{b}$

APPENDIX A, SECTION 2 (p. 1142)

1. $|x + 2| = 5$

   $x + 2 = \pm 5$

   $x = -2 \pm 5 = 3, -7$

5. $|x - 2| = 0$

   $x - 2 = 0$

   $x = 2$

9. $\left|\frac{x - 2}{x + 1}\right| = 3$

   $\frac{x - 2}{x + 1} = \pm 3$

   $x - 2 = 3x + 3$ or $x - 2 = -3x - 3$

   $-2x = 5$; $4x = -1$

   $x = -\frac{5}{2}$; $x = -\frac{1}{4}$

13. $|x + 3| < 1$

    $-1 < x + 3 < 1$

    $-4 < x < -2$

17. $|2x - 5| \le 4$

    $-4 \le 2x - 5 \le 4$

    $1 \le 2x \le 9$

    $\frac{1}{2} \le x \le \frac{9}{2}$

21. $|3x + 1| \ge 4$

    $3x + 1 \le -4$ or $3x + 1 \ge 4$

    $3x \le -5$; $3x \ge 3$

    $x \le -\frac{5}{3}$; $x \ge 1$

25. $|f(x) - 3| = |x + 1 - 3| = |x - 2| < 1$.

APPENDIX A, SECTION 3 (pp. 1145-1146)

1. $|x^2 + 1| \le |x|^2 + 1 < 4 + 1 = 5$

5. $|x^4 + 4| \le |x|^4 + 4 < 5^4 + 4 = 629$

9. If $-5 < x < 1$, then $|x| < 5$.

$|x^2 - 3| \le |x|^2 + 3 < 5^2 + 3 = 28$

13. If $|x - 1| < \delta$ and $\delta \le 1$, then $|x - 1| < 1$.

$-1 < x - 1 < 1$, $2 < x + 2 < 4$, $|x + 2| < 4$

$|x^2 + x - 2| = |x + 2| \cdot |x - 1| < 4\delta$

17. If $|x - 1| < \delta$ and $\delta \le \frac{3}{4}$, then $|x - 1| < \frac{3}{4}$.

$-\frac{3}{4} < x - 1 < \frac{3}{4}$, $\frac{1}{4} < x < \frac{7}{4}$

$|x| > \frac{1}{4}$ or $\frac{1}{|x|} < 4$

$\left|\frac{1}{x} - 1\right| = \left|\frac{1 - x}{x}\right| = \frac{1}{|x|} \cdot |x - 1| < 4\delta$

Trigonometry Review C

APPENDIX C (pp. 1159-1161)

1. $45° = \frac{\pi}{4}$, $-210° = -\frac{7\pi}{6}$, $270° = \frac{3\pi}{2}$, $30° = \frac{\pi}{6}$, $-180° = -\pi$, $-60° = -\frac{\pi}{3}$, $135° = \frac{3\pi}{4}$, $150° = \frac{5\pi}{6}$.

5. 1.

9. $$\begin{aligned}(\sin u + \cos u)^2 &= \sin^2 u + 2 \sin u \cos u + \cos^2 u \\ &= (\sin^2 u + \cos^2 u) + 2 \sin u \cos u \\ &= 1 + \sin 2u.\end{aligned}$$

13. $$\begin{aligned}\sec^3\theta \tan^3\theta &= \sec \theta(\tan^2\theta + 1)\tan \theta \cdot \tan^2\theta \\ &= \sec \theta \tan \theta[\sec^2\theta(\sec^2\theta - 1)] \\ &= \sec \theta \tan \theta(\sec^4\theta - \sec^2\theta).\end{aligned}$$

17. $$\sec 2\theta = \frac{1}{\cos 2\theta} = \frac{1}{2\cos^2\theta - 1} = \frac{1}{\frac{2}{\sec^2\theta} - 1} = \frac{\sec^2\theta}{2 - \sec^2\theta}.$$

21. $$\begin{aligned}\sin^2 x \cos^2 x &= (1 - \cos^2 x)\cos^2 x = \cos^2 x - \cos^4 x \\ &= (\tfrac{1}{2} + \tfrac{1}{2}\cos 2x) - \tfrac{3}{8} - \tfrac{1}{2}\cos 2x - \tfrac{1}{8}\cos 4x \\ &= \tfrac{1}{8} - \tfrac{1}{8}\cos 4x.\end{aligned}$$

25.

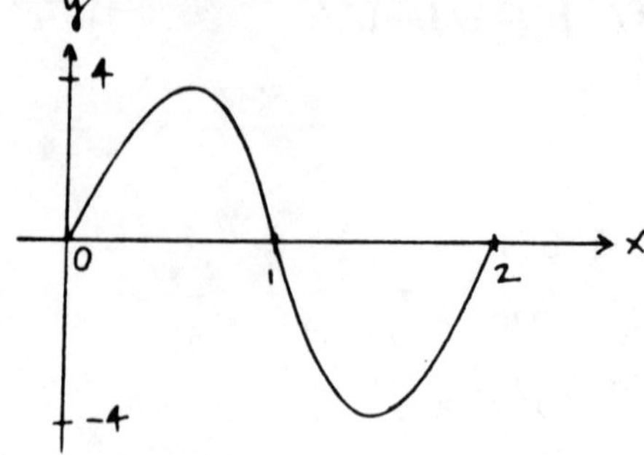

29.

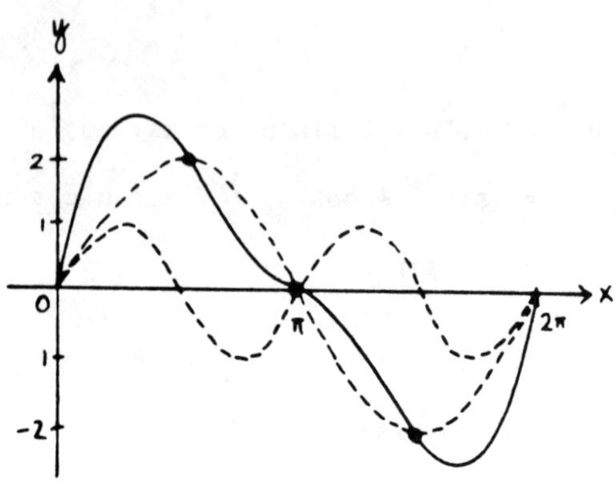

33.

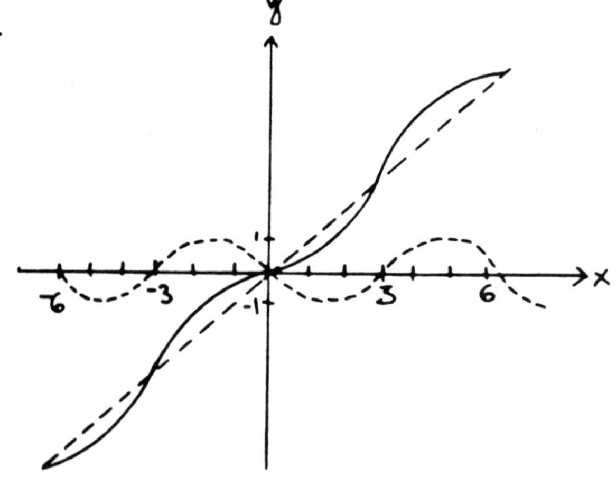

37. $\tan(\theta + \phi) = \dfrac{\sin(\theta + \phi)}{\cos(\theta + \phi)} = \dfrac{\sin\theta\cos\phi + \cos\theta\sin\phi}{\cos\theta\cos\phi - \sin\theta\sin\phi}$

$$\frac{\dfrac{\sin\theta\cos\phi}{\cos\theta\cos\phi} + \dfrac{\cos\theta\sin\phi}{\cos\theta\cos\phi}}{\dfrac{\cos\theta\cos\phi}{\cos\theta\cos\phi} - \dfrac{\sin\theta\sin\phi}{\cos\theta\cos\phi}} = \frac{\tan\theta + \tan\phi}{1 - \tan\theta\tan\phi}.$$